Modeling and Simulation of Metallurgical Processes in Ironmaking and Steelmaking

Modeling and Simulation of Metallurgical Processes in Ironmaking and Steelmaking

Editors

Thomas Echterhof
Ko-Ichiro Ohno
Ville-Valtteri Visuri

MDPI • Basel • Beijing • Wuhan • Barcelona • Belgrade • Manchester • Tokyo • Cluj • Tianjin

Editors
Thomas Echterhof
RWTH Aachen University
Germany

Ko-Ichiro Ohno
Kyushu University
Japan

Ville-Valtteri Visuri
University of Oulu
Finland

Editorial Office
MDPI
St. Alban-Anlage 66
4052 Basel, Switzerland

This is a reprint of articles from the Special Issue published online in the open access journal *Metals* (ISSN 2075-4701) (available at: https://www.mdpi.com/journal/metals/special_issues/modeling_simulation_ironmaking_steelmaking).

For citation purposes, cite each article independently as indicated on the article page online and as indicated below:

LastName, A.A.; LastName, B.B.; LastName, C.C. Article Title. *Journal Name* **Year**, *Volume Number*, Page Range.

ISBN 978-3-0365-5153-1 (Hbk)
ISBN 978-3-0365-5154-8 (PDF)

Contents

About the Editors

Thomas Echterhof

Thomas Echterhof is an academic director and deputy head of the Department for Industrial Furnaces and Heat Engineering at RWTH Aachen University, Germany. He received his Ph.D. (Dr.-Ing.) in engineering science in 2010. He is heading the research group on mass and energy balances with a special emphasis on the sustainability of electric arc furnace steelmaking including circular economy aspects. His main research interests include the optimization of energy and resource efficiency and the environmental impact of energy and resource-intensive processes such as EAF steelmaking. He is also involved in teaching and is giving a lecture on electric arc furnace technology at RWTH Aachen University. He has authored or co-authored more than 70 scientific papers in journals and conference proceedings.

Ko-Ichiro Ohno

Ko-ichiro Ohno is a professor in the Department of Materials in the Faculty of Engineering at Kyushu University. Before his current position, he worked at Kyushu University, as an assistant professor from 2006 to 2012. At the same time, stayed at RWTH Aachen University, Germany, as a postdoctoral researcher from 2010 to 2011. He worked at the Kyushu university as an associate professor from 2013 to 2021. Ohno´s field of specialization is metallurgical process engineering based on kinetics, transport phenomenology, and thermodynamics. His research fields are kinetics analysis and thermodynamics of the ironmaking process by "in situ" observation technique. The future target is to develop a novel technology in the iron-making process for the reduction of CO_2 emissions through our study and education. He has authored or co-authored over 80 peer-reviewed journal and conference papers. Ohno holds a doctoral degree in engineering as well as master's and bachelor's degrees from Tohoku University.

Ville-Valtteri Visuri

Ville-Valtteri Visuri is an associate professor and deputy head of the Process Metallurgy Research Unit at the University of Oulu. Before his current position, he worked as an R&D manager at Outokumpu Stainless Oy and as a senior research fellow and postdoctoral researcher at the University of Oulu. He has also had two research visits to RWTH Aachen University. Visuri's research is focused on the chemical and physical fundamentals of steelmaking processes, particularly the modeling of unit processes. He has authored or co-authored 59 peer-reviewed journal and conference papers. Alongside his main occupation, he serves as the chairman of the board in the Division of Metallurgical Engineers at the Finnish Association of Mining and Metallurgical Engineers. Visuri holds the title of Docent in Process Metallurgy, a doctoral degree in process engineering as well as master's and bachelor's degrees in Industrial Engineering and Management, all from the University of Oulu.

Preface to "Modeling and Simulation of Metallurgical Processes in Ironmaking and Steelmaking"

This Special Issue focuses on the modeling and simulation of ironmaking and steelmaking processes, providing a selection of peer-reviewed papers on topical research questions. The idea for co-editing a Special Issue was borne out of the bilateral collaborations between RWTH Aachen University, Kyushu University and the University of Oulu, which have included both experimental and modeling aspects.

The manuscript submission was opened amidst the COVID-19 pandemic in August 2020 and closed in March 2022 with still no end to the pandemic in sight. For this reason, all aspects of the Special Issue had to be organized without meeting in person and face-to-face marketing of the Special Issue was limited. Nevertheless, we managed to receive good visibility for our Special Issue and were thus able to attract a fair number of manuscript submissions for peer review. Eventually, a total of 14 manuscripts were accepted for publication based on positive recommendations from external subject matter experts. These papers and an accompanying editorial were published from 2021 to 2022 in volumes 11 and 12 of *Metals* and are reprinted in this Special Issue.

We are grateful to both the peer reviewers and authors, without whom this Special Issue could not have materialized. Mr. Toliver Guo from MDPI is acknowledged for taking care of all the practicalities related to the publication of the Special Issue both in digital and printed form. Finally, we hope that the Special Issue will make interesting reading and give our readers new ideas and tools for driving scientific and practical progress in the iron and steel industry.

Thomas Echterhof, Ko-Ichiro Ohno, and Ville-Valtteri Visuri
Editors

metals

Editorial

Modeling and Simulation of Metallurgical Processes in Ironmaking and Steelmaking

Thomas Echterhof [1,*], Ko-Ichiro Ohno [2,*] and Ville-Valtteri Visuri [3,*]

1 Department for Industrial Furnaces and Heat Engineering, RWTH Aachen University, Kopernikustr. 10, 52074 Aachen, Germany
2 Department of Materials Science and Engineering, Faculty of Engineering, Kyushu University, 744 Motooka, Nishiku, Fukuoka 819-0395, Japan
3 Process Metallurgy Research Unit, University of Oulu, P.O. Box 4300, 90014 Oulu, Finland
* Correspondence: echterhof@iob.rwth-aachen.de (T.E.); ohno.ko-ichiro.084@m.kyushu-u.ac.jp (K.-I.O.); ville-valtteri.visuri@oulu.fi (V.-V.V.)

Citation: Echterhof, T.; Ohno, K.-I.; Visuri, V.-V. Modeling and Simulation of Metallurgical Processes in Ironmaking and Steelmaking. *Metals* **2022**, *12*, 1185. https://doi.org/10.3390/met12071185

Received: 6 July 2022
Accepted: 6 July 2022
Published: 12 July 2022

1. Introduction and Scope

The UN's 2030 Sustainable Development Goals, the Paris Agreement, and the European Green Deal, among other goals, all aim to improve the sustainability of industrial production and reduce CO_2 emissions. The European Union, for example, aims to reach carbon neutrality and a circular economy by 2050. This goal cannot be achieved without a significant reduction in the CO_2 emissions created by the steel industry. To reach this goal, further process optimizations regarding energy and resource efficiency, as well as the development of new processes or process routes, are needed.

The parameters necessary for the analysis and optimization of the existing and new metallurgical processes in ironmaking and steelmaking often cannot be measured directly because of the harsh conditions inside the furnaces and metallurgical vessels. Typically, the direct information sources in ironmaking and steelmaking are off-gas analysis and spot measurements, which suffer from the delay associated with the analysis of the sample. Due to the harsh environment, opportunities to determine the flow conditions in the vessels by measurements are even more limited. While new methods for the direct and continuous measurement of some of these parameters are under development, for many processes, they are currently unavailable. Furthermore, plant trials that would be necessary to evaluate the impact of different optimization strategies may be impossible because of the prohibitive cost or safety concerns.

Modelling and simulation have established themselves as an invaluable source of information regarding otherwise unknown process parameters, and as an alternative to plant trials with a lower associated cost, risk, and duration. Models are also applicable for model-based control of metallurgical processes.

This Special Issue aimed to cover recent advances in the modelling and simulation of unit processes in ironmaking and steelmaking. To this end, fourteen articles have been published in the present Special Issue of *Metals*. The articles give voice to a total of 67 authors, representing affiliations from ten countries (Figure 1a) with the majority of contributions originating from universities (Figure 1b). The subjects include reviews on the fundamentals of modelling and simulation of metallurgical processes, as well as the fields of iron reduction/ironmaking, steelmaking via the primary and secondary route, and continuous casting.

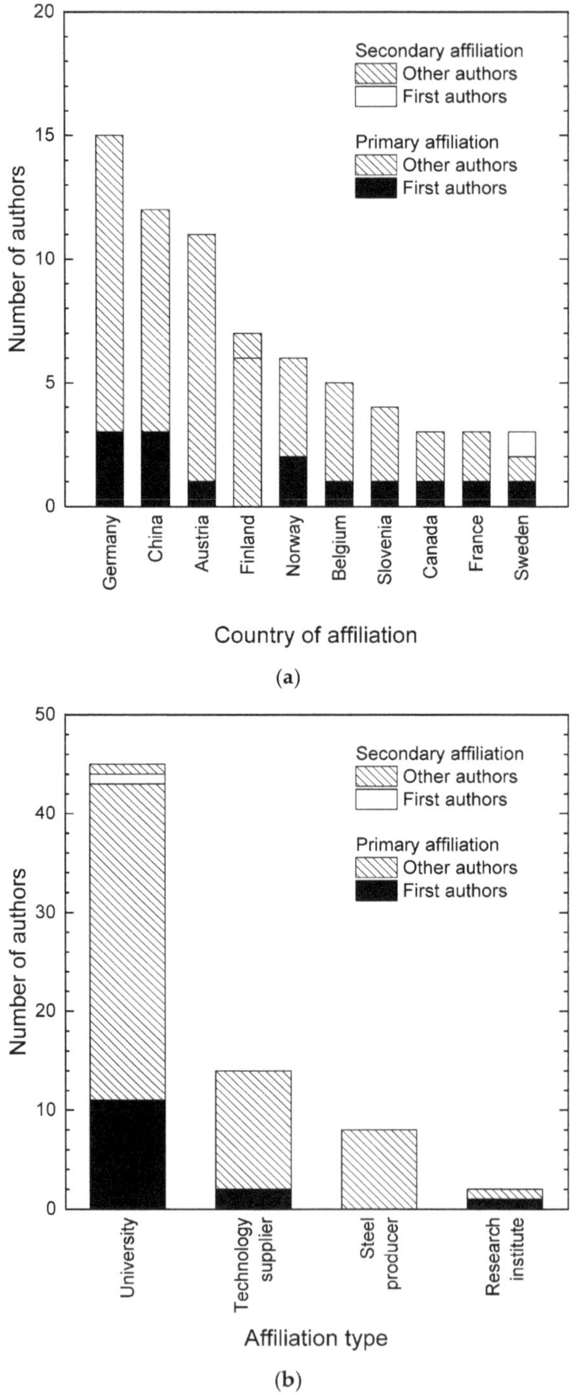

Figure 1. Authors of the contributions of the Special Issue: (**a**) by country of affiliation; (**b**) by affiliation type.

2. Contributions

The fundamentals of modelling and simulation are covered by several articles. Wang et al. [1] review the application of turbulence modelling in metallurgical applications. The article focuses on the validation of models by experiments and the applicability of models to industrial cases. Bubble-induced turbulence, supersonic jet transport, and electromagnetic suppression of turbulence are three specific flow problems discussed in the article. Haas et al. [2] present "A Review of Bubble Dynamics in Liquid Metals". The article presents a detailed discussion of the state of knowledge of bubble dynamics in liquid metals, measurement methods, and bubble formation mechanism at nozzles and purging plugs, and discusses modelling uncertainties using the steel ladle as an example. It is concluded that, while a lot of valuable information has been extracted from both aqueous and liquid metal systems, there are still plenty of uncertainties regarding the behaviour of bubbles in liquid steel. The uncertainties highlighted in the paper include the effect of impurities present in steel, the interaction of gas bubbles in bubble swarms, and the deformation of bubbles due to injectors or walls.

Three articles explore the reduction processes in ironmaking. Quatravaux et al. [3] investigated the Midrex NGTM process and adapted a blast furnace operating diagram to describe the direct reduction of iron oxide in the Midrex shaft furnace. The authors employed the developed graphical tool for a sensitivity study of the reduction in the shaft furnace. Wang et al. [4] present a combination of numerical and experimental investigations of the charging of carbon composite briquettes in a blast furnace. The authors studied the replacement of ore by the carbon composite briquettes and report on carbon conversion, coke rate, blast furnace operation, and productivity, etc. Liu et al. [5] studied the reduction process within a pre-reduction rotary kiln employing numerical simulation methods. The numerical model was validated based on measurement data and then used to investigate process details such as temperature gradients or the progress of the reduction process within the rotary kiln.

The field of steelmaking is covered by eight articles. The first two are more related to the primary steelmaking route, including the blast furnace and basic oxygen furnace while the following five concern steelmaking in the electric arc furnace (EAF). The final article addresses the overarching topic of steelmaking slags. Linnestad et al. [6] report on the development of a process model describing the Composition Adjustment by Sealed argon bubbling with Oxygen Blowing (CAS-OB) process. In this study, a simplified mathematical model for the heat-up stage of the CAS-OB process was coupled with adaptive estimation of state variables using a Kalman filter. The authors present the model's capabilities, as well as model results from the prediction of a data series with more than 1000 heats based on the model implementation at two CAS-OB stations at SSAB Europe Oy in Raahe, Finland. The predicted temperature profiles are displayed for the operator to assist in their decision-making, thus helping to reduce the number of rejected heats. Gao et al. [7] investigated the melting of scrap in hot metal and determined mass transfer coefficients during scrap melting based on laboratory-scale experiments. Subsequently, the authors determined the influence of the variables molten pool stirring rate, bath temperature, and scrap type on the mass transfer coefficient. Tomažič et al. [8] studied the optimization of the energy consumption in an EAF employing data-driven modelling approaches. The models were developed and validated based on industrial data from an EAF steel plant and were used to determine the optimal duration of the transformer profile during melting to reduce energy consumption, thereby increasing EAF efficiency. Jawahery et al. [9] present a first-principles-based process model of the EAF steelmaking designed for online real-time optimization of the process. The model focuses on the integration of auxiliary process data to predict energy efficiency and heat transfer limitations in the EAF. The model was validated using steel temperature and weight measurements achieving reasonable agreement. Al Nasser et al. [10] developed a simplified arc impingement model to study the direct-current electric arc in computational fluid dynamics simulations. The model was used to investigate the influence of the factors arc gap, the density of the gas and total electric current on the behaviour of the arc, and the overall process, including arc impingement depth, velocity magnitude, and arc stability. Reimann et al. [11] investigated statistical modelling approaches based on operational data from five industrial-size EAFs for the prediction of the electric energy demand. Results of the modelling approaches applied to the different industrial EAFs are presented, compared and differences are discussed. Schubert et al. [12] describe the development of a fast modelling approach that was subsequently used to simulate and predict the scrap preheating of a continuously charged EAF. Modelling applications, assumptions, possible enhancements, and limitations, as well as initial simulation results, are presented and discussed. Finally, Safavi Nick et al. [13] studied the modelling and simulation of slag heat recovery to design an optimized heat recovery system. The simulation focused on the heat exchanger and fluid medium used to transfer the heat. This study is a good example of how modern computational fluid dynamics (CFD)-based simulation tools can be used as a design tool for heat recovery processes.

The final field – covered by one article – is continuous casting. Guthrie and Isac [14] present a historical review of casting methods used for sheet steel production. The review shows the development from conventional continuous casting to thin-slab casting as well as different concepts of direct strip casting, and discusses the pros and cons of current methods.

3. Conclusions and Outlook

A variety of connected topics have been compiled in the present Special Issue of *Metals*, providing an overview of recent developments in different aspects of modelling and simulation of metallurgical processes in ironmaking and steelmaking. Despite being a long-term research interest, modelling and simulation could become even more important with the upcoming transformation to carbon-neutral ironmaking and steelmaking processes.

As Guest Editors of this Special Issue, we are very pleased with the contributions received and hope that the presented articles will be useful to researchers and operators of metallurgical plants working to optimize plant operation. We would like to thank all the authors for their contributions, as well as all reviewers for their efforts in maintaining the high quality of published papers. We would also like to give special thanks to all staff at the *Metals* Editorial Office, especially to Toliver Guo, Assistant Editor, who managed and facilitated the publication process of this Special Issue.

Conflicts of Interest: The authors declare no conflict of interest.

References

1. Wang, Y.; Cao, L.; Cheng, Z.; Blanpain, B.; Guo, M. Mathematical Methodology and Metallurgical Application of Turbulence Modelling: A Review. *Metals* **2021**, *11*, 1297. [CrossRef]
2. Haas, T.; Schubert, C.; Eickhoff, M.; Pfeifer, H. A Review of Bubble Dynamics in Liquid Metals. *Metals* **2021**, *11*, 664. [CrossRef]
3. Quatravaux, T.; Barros, J.; Gardin, P.; Lucena, G. Adaptation of the Rist Operating Diagram as a Graphical Tool for the Direct Reduction Shaft. *Metals* **2021**, *11*, 1953. [CrossRef]
4. Wang, Q.; Liu, Z.; Tang, H.; Guo, Z. Experimental and Numerical Investigations on Charging Carbon Composite Briquettes in a Blast Furnace. *Metals* **2021**, *11*, 1669. [CrossRef]
5. Liu, C.; Ding, X.; Liu, H.; Yan, X.; Dong, C.; Wang, J. Numerical Analysis on Characteristics of Reduction Process within a Pre-Reduction Rotary Kiln. *Metals* **2021**, *11*, 1180. [CrossRef]
6. Linnestad, K.; Ollila, S.; Wasbø, S.O.; Bogdanoff, A.; Rotevatn, T. Adaptive First Principles Model for the CAS-OB Process for Real-Time Applications. *Metals* **2021**, *11*, 1554. [CrossRef]
7. Gao, M.; Gao, J.-T.; Zhang, Y.-L. Evaluation of Mass Transfer Coefficient during Scrap Melting. *Metals* **2021**, *11*, 1368. [CrossRef]
8. Tomažič, S.; Andonovski, G.; Škrjanc, I.; Logar, V. Data-Driven Modelling and Optimization of Energy Consumption in EAF. *Metals* **2022**, *12*, 816. [CrossRef]
9. Jawahery, S.; Visuri, V.-V.; Wasbø, S.O.; Hammervold, A.; Hyttinen, N.; Schlautmann, M. Thermophysical Model for Online Optimization and Control of the Electric Arc Furnace. *Metals* **2021**, *11*, 1587. [CrossRef]
10. Al-Nasser, M.; Kharicha, A.; Barati, H.; Pichler, C.; Hackl, G.; Gruber, M.; Ishmurzin, A.; Redl, C.; Wu, M.; Ludwig, A. Toward a Simplified Arc Impingement Model in a Direct-Current Electric Arc Furnace. *Metals* **2021**, *11*, 1482. [CrossRef]
11. Reimann, A.; Hay, T.; Echterhof, T.; Kirschen, M.; Pfeifer, H. Application and Evaluation of Mathematical Models for Prediction of the Electric Energy Demand Using Plant Data of Five Industrial-Size EAFs. *Metals* **2021**, *11*, 1348. [CrossRef]

12. Schubert, C.; Büschgens, D.; Eickhoff, M.; Echterhof, T.; Pfeifer, H. Development of a Fast Modeling Approach for the Prediction of Scrap Preheating in Continuously Charged Metallurgical Recycling Processes. *Metals* **2021**, *11*, 1280. [CrossRef]
13. Safavi Nick, R.; Leinonen, V.; Mäyrä, J.; Björkvall, J. Towards Greener Industry: Modelling of Slag Heat Recovery. *Metals* **2021**, *11*, 1144. [CrossRef]
14. Guthrie, R.I.L.; Isac, M.M. Continuous Casting Practices for Steel: Past, Present and Future. *Metals* **2022**, *12*, 862. [CrossRef]

metals

MDPI

Review

Mathematical Methodology and Metallurgical Application of Turbulence Modelling: A Review

Yannan Wang, Lingling Cao *, Zhongfu Cheng, Bart Blanpain and Muxing Guo

Department of Materials Engineering, Katholieke Universiteit (KU) Leuven, 3000 Leuven, Belgium; wynheut@gmail.com (Y.W.); zhongfu.cheng@kuleuven.be (Z.C.); bart.blanpain@kuleuven.be (B.B.); muxing.guo@kuleuven.be (M.G.)

* Correspondence: meynell0809@gmail.com

Abstract: This paper focusses on three main numerical methods, i.e., the Reynolds-Averaged Navier-Stokes (RANS), Large Eddy Simulation (LES), and Direct Numerical Simulation (DNS) methods. The formulation and variation of different RANS methods are evaluated. The advantage and disadvantage of RANS models to characterize turbulent flows are discussed. The progress of LES with different subgrid scale models is presented. Special attention is paid to the inflow boundary condition for LES modelling. Application and limitation of the DNS model are described. Different experimental techniques for model validation are given. The consistency between physical experimentation/modelling and industrial cases is discussed. An emphasis is placed on the model validation through physical experimentation. Subsequently, the application of a turbulence model for three specific flow problems commonly encountered in metallurgical process, i.e., bubble-induced turbulence, supersonic jet transport, and electromagnetic suppression of turbulence, is discussed. Some future perspectives for the simulation of turbulent flow are formulated.

Keywords: turbulence modelling; RANS/LES/DNS; inflow condition; model validation; model application

Citation: Wang, Y.; Cao, L.; Cheng, Z.; Blanpain, B.; Guo, M. Mathematical Methodology and Metallurgical Application of Turbulence Modelling: A Review. *Metals* **2021**, *11*, 1297. https://doi.org/10.3390/met11081297

Academic Editors: Thomas Echterhof, Ko-Ichiro Ohno and Ville-Valtteri Visuri

Received: 10 June 2021
Accepted: 21 July 2021
Published: 17 August 2021

1. Introduction

Since turbulence plays a significant role in flow transport phenomena, considerable efforts have been devoted to understanding flow turbulence in both engineering and academic communities [1–4]. Nevertheless, turbulence is still not fully understood due to its complexities in nature, e.g., instantaneous and intermittency characteristics, strong nonlinearity, and a wide range of temporal and spatial scales [4,5].

Turbulent flow can be numerically resolved with different levels of accuracy. Many numerical approaches for solving turbulence have been proposed, such as the Reynolds-Averaged Navier–Stokes (RANS) [6–11], the Large Eddy Simulation (LES) [12–14], and Direct Numerical Simulation (DNS) approaches [15]. Among these numerical methods, the RANS approach, specifically the Eddy Viscosity Model (EVM), is widely used for calculating turbulent flows thanks to its relatively high accuracy in predicting the mean flow features and its more limited computational demands. However, this approach suffers from several weaknesses, e.g., compromised accuracy and uncertainties due to assumptions in the model construction and insufficient incorporation of the fluid physics. In the LES approach, the whole eddy range is separated into two parts, namely, the large-scale eddy and subgrid-scale (SGS) eddy. The former can be directly resolved, while the latter is computed using the SGS model. As the computing power rapidly increases, this approach is extensively used to study turbulence physics and to resolve low-to-medium Reynolds number flows. In order to obtain meaningful results, additional attention is paid to inflow boundary conditions and mesh density for LES modelling. Compared to the RANS and LES, the DNS approach is the most accurate numerical method because it directly resolves all the turbulent eddies without using any models. However, DNS requires extensive computational demands to solve turbulent engineering flows, which is difficult to be satisfied by the current computing power. In addition, the large amount of data generated by DNS should be carefully analyzed.

Due to the uncertainties of the RANS method, many model variants have been proposed to solve specific turbulent flows [6–10], which can easily confuse the users of Computational Fluids Dynamics (CFD) codes to choose an appropriate one for their own cases. A comprehensive and well-organized description of the model formulation and development is very helpful in judging the applicability of different models, although there are some works with respect to model description [1,3]. Apart from SGS models of the LES method, the inflow boundary condition is an indispensable part of the simulation. Considering a diversity of methods generating inflow information [16,17], an elaborate analysis and discussion on the methods is necessary and conducive for the LES model user. In order to clearly understand the potential of the DNS method, the capabilities and current limitations of DNS method need to be clarified. In addition to the numerical solutions for turbulent flows, model validation is needed to warrant the accuracy of the simulation. Depending on different studies, different physical experimentations assisted with measuring techniques are performed to validate the numerical model. There are a number of studies focusing on the fundamentals of different measuring techniques, data interpretation, and applications of the techniques [18,19]. However, the consistency between physical experimentation/modelling and industrial applications needs to be discussed. The limitations and development of the measuring techniques are critical for obtaining reliable data and should be reviewed.

In this work, we review the formulation and development of three main numerical approaches (i.e., the RANS, LES, and DNS) for turbulence modelling. The advantages and disadvantages of the approaches are systematically discussed. Different methods of the inflow boundary condition are described for LES modelling. For the purpose of validating numerical models, different physical experimentation methods are presented. The consistency between the physical modelling and industrial applications is discussed. Limitations and progress of the experimental validation techniques are shown. Three turbulence-related flow problems commonly encountered in metallurgical fields (i.e., bubble-induced turbulence (BIT), supersonic jet transport, and electromagnetic damping of turbulence) are discussed to demonstrate how to customize a conventional turbulence model for solving a specific flow problem. Finally, perspectives for modelling turbulent flows are proposed. With this review, we intend to help the current and potential CFD users to understand the modelling techniques for turbulence flows better and to expand the insight into the physics of turbulence.

2. Turbulence Simulation Approaches

2.1. RANS

In the RANS approach, instantaneous solution variables in the governing equations are decomposed into the mean and fluctuating components, as expressed in Equation (1).

$$f = \overline{f} + f' \tag{1}$$

Substituting this variable expression into the instantaneous continuity and Navier-Stokes (N-S) equations yields the ensemble- or time-averaged forms for single-phase Newtonian flow, as shown in Equations (2) and (3). Henceforth, repeated-suffix summation convention is used in the formulae.

$$\frac{\partial \rho}{\partial t} + \frac{\partial}{\partial x_i}(\rho \overline{u}_i) = 0 \tag{2}$$

$$\frac{\partial}{\partial t}(\rho \overline{u}_i) + \frac{\partial}{\partial x_j}(\rho \overline{u}_i \overline{u}_j) = -\frac{\partial \overline{p}}{\partial x_i} + \frac{\partial}{\partial x_j}\left(\mu \frac{\partial \overline{u}_i}{\partial x_j} - \rho \overline{u_i' u_j'}\right) \tag{3}$$

The fluctuating quantities are included in the Reynolds stress tensor ($-\rho\overline{u_i'u_j'}$) with six components. In order to close the equation set, the Reynolds stress tensor needs to be appropriately solved. One of the solutions for this closure problem employs the Boussinesq assumption [20], which relates the Reynolds stresses to the mean velocity gradients. The advantage of the Boussinesq assumption is the relatively low computational cost due to its simplicity. This works well for the engineering flows, which are dominated by only one turbulent shear stress such as the jet flow, wall boundary layer flow, and mixing layers flow. However, this approach is insensitive to the streamline curvature, rotation and body forces, and it exhibits a poor performance in the flows with a strong anisotropy or stress transport effect [21,22]. It also has difficulty in predicting transitional flows.

It is worth noting that the turbulent viscosity used in solving the Reynolds stress terms is a function of the space and flow features, rather than a physical parameter such as the fluid viscosity, which is dependent on the molecular structure of the fluid. Obviously, the turbulent viscosity needs to be solved before computing the Reynolds stress terms. In this paper, two-equation models, which include two additional transport equations, are reviewed. Usually the turbulence kinetic energy (k) is adopted as one equation and the turbulent kinetic energy dissipation rate (ε) or the specific dissipation rate (ω) as another one. The modified versions of the ε/ω model will also be presented here. Due to length restrictions, the zero- and one-equation models are not included in the article, but they may be found elsewhere [23–25].

2.1.1. The k-ε Model

The standard k-ε (SKE) model was originally proposed by Launder and Spalding [6]. The model has been widely applied for resolving turbulent flows without a severe pressure gradient or strong swirling effect (e.g., plane jet, mixing layer, and boundary layer flows) because of its relatively high robustness, low computational cost, and reasonable accuracy. Equations (4) and (5) show the general form of k and ε. By solving these two transport equations, the turbulent viscosity can be calculated as expressed in Equation (6).

$$\frac{\partial}{\partial t}(\rho k)+\frac{\partial}{\partial x_i}(\rho k\overline{u}_i)=\frac{\partial}{\partial x_j}\left[\left(\mu+\frac{\mu_t}{\sigma_k}\right)\frac{\partial k}{\partial x_j}\right]-\rho\overline{u_i'u_j'}\frac{\partial\overline{u}_i}{\partial x_j}-\rho\varepsilon+S_k \tag{4}$$

$$\frac{\partial}{\partial t}(\rho\varepsilon)+\frac{\partial}{\partial x_i}(\rho\varepsilon\overline{u}_i)=\frac{\partial}{\partial x_j}\left[\left(\mu+\frac{\mu_t}{\sigma_\varepsilon}\right)\frac{\partial\varepsilon}{\partial x_j}\right]-C_{1\varepsilon}\frac{\varepsilon}{k}\rho\overline{u_i'u_j'}\frac{\partial\overline{u}_i}{\partial x_j}-C_{2\varepsilon}\rho\frac{\varepsilon^2}{k}+S_\varepsilon \tag{5}$$

$$\mu_t=C_\mu\rho\frac{k^2}{\varepsilon} \tag{6}$$

The terms from left to right in Equations (6) and (7) are the respective local time derivative, convection, diffusion, production, sink, and source terms. The SKE model is derived from a fully turbulent flow with high Reynolds numbers. The viscous effect is ignored in the model. However, this cannot be applied in the vicinity of the wall, where the viscous force dominates the flow characteristics. In order to deal with this problem, either a wall function is adopted with the SKE model, or a low Reynolds number model is used. The former confuses the users' judgement whether the weakness of this method lies in the basic SKE model itself or in the wall function. The latter requires additional functions to modify the standard transport equations. With respect to the low Reynolds number models [26–32], an example proposed by Lam and Bremhorst (LB model) [31] is presented in Equations (7) and (8).

$$\frac{\partial}{\partial t}(\rho k)+\frac{\partial}{\partial x_i}(\rho k\overline{u}_i)=\frac{\partial}{\partial x_j}\left[\left(\mu+\frac{\mu_t}{\sigma_k}\right)\frac{\partial k}{\partial x_j}\right]+\mu_t\left(\frac{\partial\overline{u}_i}{\partial x_j}+\frac{\partial\overline{u}_j}{\partial x_i}\right)\frac{\partial\overline{u}_i}{\partial x_j}-\rho\varepsilon \tag{7}$$

$$\frac{\partial}{\partial t}(\rho\varepsilon)+\frac{\partial}{\partial x_i}(\rho\varepsilon\overline{u}_i)=\frac{\partial}{\partial x_j}\left[\left(\mu+\frac{\mu_t}{\sigma_\varepsilon}\right)\frac{\partial\varepsilon}{\partial x_j}\right]+C_{1\varepsilon}f_1\frac{\varepsilon}{k}\mu_t\left(\frac{\partial\overline{u}_i}{\partial x_j}+\frac{\partial\overline{u}_j}{\partial x_i}\right)\frac{\partial\overline{u}_i}{\partial x_j}-C_{2\varepsilon}f_2\rho\frac{\varepsilon^2}{k} \tag{8}$$

where:

$$\mu_t = C_\mu f_\mu \rho \frac{k^2}{\varepsilon} \tag{9}$$

Compared with the SKE model, different formulations of functions f_1, f_2 and f_μ are developed in the LB model to describe the near-wall behavior better. It has been confirmed that the function f_μ has a predominant influence on the model performance, and functions f_1, f_2 play a secondary role in the performance [11].

There are also other modified versions of the SKE model, amongst which the widely used Renormalization Group (RNG) k-ε model [7] and the Realizable k-ε (RKE) model [8] are introduced in the following section. The main differences between the RNG k-ε model and the SKE model are the modifications of the turbulent viscosity and ε sink term. A differential equation is analytically derived for effective viscosity μ_{eff} to account for the low Reynolds number effect. This feature can improve the predictive ability of the RNG k-ε model for low Reynolds number flows or near-wall flows. Additionally, a new ε destruction term is used to account for the rapid strain by modifying the constant of this term. The RKE model is modified mainly with regard to the turbulent viscosity and the ε equation. By defining a variable C_μ [8,33,34] instead of a constant value in turbulent viscosity formulation, the RKE model satisfies the realizability constraints, i.e., positive values for the normal stresses and the Schwartz inequality for the shear stresses. In order to increase the robustness of the model, a new ε equation is employed based on a dynamic equation for fluctuating vorticity. The new ε equation describes turbulent vortex stretching and turbulent dissipation more appropriately compared to the ε equation in the SKE model. With the modified ε equation, the well-known round-jet anomaly that is a poor prediction of the spreading rate of round/axisymmetric jet may be solved [8].

2.1.2. The k-ω Model

The k-ω model is widely used for turbulence modelling [35]. Different versions of this model have been developed in the last decades [9,10,36–40]. In this paper, the most well-known k-ω model proposed by Wilcox [9] is reviewed. The transport equations of this model are presented in Equations (10) and (11), where the calculation of β_1 refers to the work of Wilcox [9].

$$\frac{\partial}{\partial t}(\rho k) + \frac{\partial}{\partial x_i}(\rho k \overline{u}_i) = \frac{\partial}{\partial x_j}\left[\left(\mu + \frac{\mu_t}{\sigma_k}\right)\frac{\partial k}{\partial x_j}\right] - \rho\overline{u_i' u_j'}\frac{\partial \overline{u}_i}{\partial x_j} - \rho\beta^* k\omega + S_k \tag{10}$$

$$\frac{\partial}{\partial t}(\rho \omega) + \frac{\partial}{\partial x_i}(\rho \omega \overline{u}_i) = \frac{\partial}{\partial x_j}\left[\left(\mu + \frac{\mu_t}{\sigma_\omega}\right)\frac{\partial \omega}{\partial x_j}\right] - \alpha\frac{\omega}{k}\rho\overline{u_i' u_j'}\frac{\partial \overline{u}_i}{\partial x_j} - \rho\beta_1\omega^2 + S_\omega \tag{11}$$

Unlike the k-ε model, the k-ω model can be integrated through the viscous sublayer without any damping function to account for the low Reynolds number effect with high numerical stability. Therefore, it is well applied in aerodynamic flows [10,35]. However, the k-ω model is highly sensitive to the empirical value of ω at the free edge of the turbulent shear layer, which can lead to a large prediction error. In order to solve this problem, a modified model was proposed with combination of the original k-ω model and the SKE model by adding a blending function [10]. This new model is termed the Baseline k-ω model, which applies the original k-ω model in the near-wall region and switches to the SKE model in the outer region. The Baseline k-ω model has a similar performance to the original k-ω model in boundary layer flows, but the former one avoids the strong freestream dependence. However, both k-ω models fail to predict the onset and amount of separation in adverse pressure gradient flows. Based on the Baseline k-ω model, further modification to eddy viscosity is proposed to account for the transport effects of the principal turbulent shear stress, leading to a significant improvement in predicting the adverse pressure gradient flows [10]. However, the introduced blending function depends on empiricism (e.g., the distance to wall), limiting its application to flows in complex geometries.

2.1.3. Advanced Eddy Viscosity Model

All the aforementioned methods belong to the category of the Eddy Viscosity Model (EVM). Some other advanced EVMs were developed [41–45]. In order to account for the strong anisotropy in the near-wall region, Durbin [41] adopted a new turbulent viscosity term defined in Equation (12), which is considered to be more appropriate than that defined in Equation (6) in a near-wall region. A separate transport equation for a wall-normal turbulent stress $\overline{v^2}$ was proposed and solved with the aid of the elliptic relaxation concept. This model is termed the v^2-f model, where f represents the elliptic relaxation function. Subsequently, several modified versions were proposed with respect to the velocity scale [44], the characteristic length [42], the function f [45], and the variable $\overline{v^2}$ [43]. This model category performs well for pressure-induced separating flow, buoyancy impairing turbulent flow, and backstep flow [44–46].

$$\mu_t = C_\mu \rho \overline{v^2} \tau \tag{12}$$

where:

$$\tau = \max\left(\frac{k}{\varepsilon}, C_\tau \left(\frac{\mu}{\rho\varepsilon}\right)^{0.5}\right) \tag{13}$$

In addition to the modified versions of the linear EVM, the idea of non-linear EVM has been substantially used [47–53]. Even though these modified models demonstrated certain improvements over linear EVMs in predicting flows with a strong streamline curvature or turbulent stresses in the near-wall sublayer, they are still inferior to the more advanced model, e.g., the RSM model as seen in Section 2.1.4.

2.1.4. Reynolds Stress Model

In order to overcome the limitations of the EVM, Second-Moment Closure (SMC) models abandoning the Boussinesq assumption have been developed. The SMC model directly solves the transport equation for each of the Reynolds stress terms. Since the SMC approach considers the effects of streamline curvature, rotation, and rapid change of strain rate in a more rigorous manner, it is long expected to replace the currently widely applied two-equation models. The SMC model class consists of the Algebraic Stress Model (ASM) and the Differential Stress Model (DSM). ASM is derived from differential stress transport equations by invoking the weak-equilibrium assumption [54–56]. It ignores the transport terms of the anisotropy by assuming that the transport of the Reynolds stress is proportional to that of turbulent kinetic energy [57,58]. In general, the ASM is considered an intermediate tool between the LEVM and the DSM. Due to the space limitation, only the DSM is presented in this paper. A symbolic representation of the stress transport equation is expressed in Equation (14). In addition, a scale-determining equation, i.e., the ε equation, is needed to complete the DSM.

$$L_{ij} + C_{ij} = D_{T,ij} + D_{L,ij} + P_{ij} + G_{ij} + \phi_{ij} + \varepsilon_{ij} + s_{ij} \tag{14}$$

The terms from left to right represent the local time derivative of Reynolds stress, convection, turbulent diffusion, molecular diffusion, stress production, buoyancy production, pressure strain, dissipation and source, respectively. The s_{ij} term is user-defined for a specific stress transport source. If there is no source, this term becomes zero. It is required to model $D_{T,ij}$, G_{ij}, ϕ_{ij} and ε_{ij} to close the equation, while it is not necessary to model L_{ij}, C_{ij}, $D_{L,ij}$ and P_{ij}.

The $D_{T,ij}$ term includes the velocity transport and the pressure transport. The velocity triple moments can be measured, whereas the pressure transport is intractable. Usually, the pressure transport is considered to be negligible [59]. Therefore, the model is mainly designed for the velocity triple moments. The most popular model is the generalized gradient-diffusion model proposed by Daly and Harlow (DH) [60]. The DH model has a symmetry problem in the indices, leading to dependence on the coordinate frame. Subsequently, some variants of this model were developed by Hanjalic and Launder [61], Shir [62], Mellor and Herring [63]. More complex models were also put forward by Nagano and Tagawa [64] and Magnaudet [65]. Due to uncertainties in the modelling equations, the complex models may not necessarily outperform the simplified models. Given more computing resources consumed by the complex models and a poor convergence, application of these models in engineering has been doubted [3,66].

Compared to the ε_{ij} and G_{ij} terms, it is necessary to pay extra attention to model ϕ_{ij}. Usually, the ϕ_{ij} term is decomposed into three parts, namely, the slow pressure-strain term $\phi_{ij,1}$, rapid pressure-strain term $\phi_{ij,2}$, and wall-reflection term $\phi_{ij,w}$. Not all of the models introduced below include the third term. Rotta [67] proposed a linear model for $\phi_{ij,1}$, which considers that $\phi_{ij,1}$ is proportional to the stress anisotropy tensor. However, this linear model is unable to satisfy the realizability constraints. A general quadratic model [68] is proposed to solve this problem. Linear [61,69–71] and nonlinear models [72–74] were proposed to model the $\phi_{ij,2}$ term. Even though the nonlinear model is considered to be theoretically advanced, the complexity of the formulation prohibits its application for engineering computation. The turbulence anisotropy is enhanced due to the damping effect of the normal stress by the wall. This damping affects both the pressure-strain terms. In order to account for the damping effect, a commonly used model [62,75] is presented to model the wall-reflection term $\phi_{ij,w}$. However, this model involves a variable, i.e., the normal distance to the wall. This is believed to be a major weakness. For purpose of overcoming this weakness, the elliptic relaxation concept and elliptic-blending method were proposed to account for the near-wall inhomogeneity, which is described in Section 2.1.3, and more information on that can be found in [76,77].

The DSM is the most elaborate model in the RANS approach, which has an indisputable superiority over the rudimentary two-equation models in predicting complex flows, e.g., highly swirling and rotating flow, separating flow, and secondary flow. However, its application is limited by (1) a high degree of uncertainty in modelling the high-order correlation terms (e.g., pressure-strain and dissipative correlation) due to an insufficient knowledge of physics; (2) a high demand for computational resource (approximately 50–150% more computing time than a two-equation EVM [21]). Fortunately, due to the use of more advanced models (e.g., the DNS), the accuracy and robustness of the DSM have been improved. A rapid development in computer science (e.g., parallel processing and improved performance) satisfies the high computational need for the use of DSM model. The DSM has received more attention recently because of the failure of the EVM in predicting complex turbulent flows.

2.2. LES Approach

2.2.1. Formulation and Subgrid-Scale Model

Turbulent flow features a wide range of eddy scales from the Kolmogorov length scale to the size comparable to the characteristic length of the mean flow. The large eddies contain most of the turbulent energy and are mainly responsible for the momentum and energy transfer. They are strongly affected by boundary conditions. The small eddies tend to be more isotropic and homogeneous, and their dissipation process is linked to fluid viscosity. For this reason, it is very difficult for the RANS approach to model all the eddies in a single model. The LES approach separates the large eddies from the small ones by employing a spatial filtering method [78] for the instantaneous governing equations. After that, the large eddies are directly resolved by the filtered equation, and the small ones (i.e., the Subgrid-Scale (SGS) eddies) are modelled by the SGS model. The filtered variable (donated by an overbar) is defined by Equation (15). The resulting continuity and momentum equations are expressed in Equations (16) and (17), showing similar forms but a different physical meaning for those in the RANS approach.

$$\overline{f}(x) = \int_D f(x')G(x, x'; \overline{\Delta})dx' \quad (15)$$

$$\frac{\partial \rho}{\partial t} + \frac{\partial}{\partial x_j}(\rho \overline{u}_i) = 0 \tag{16}$$

$$\frac{\partial}{\partial t}(\rho \overline{u}_i) + \frac{\partial}{\partial x_j}(\rho \overline{u}_i \overline{u}_j) = -\frac{\partial \overline{p}}{\partial x_i} + \frac{\partial \sigma_{ij}}{\partial x_j} - \frac{\partial \tau_{ij}}{\partial x_j} \tag{17}$$

where σ_{ij} refers to the stress tensor due to molecular viscosity, and τ_{ij} represents the SGS stress term. Most of current SGS models adopt the Boussinesq assumption (termed the Eddy-Viscosity model), which relates the SGS stress to the large-scale strain-rate tensor, as shown in Equation (18). Based on the definition of the eddy viscosity, various SGS models [12–14,79,80] have been proposed. Smagorinsky [12] developed the first SGS model by assuming a local energy equilibrium between the large scale and the subgrid scale. The eddy viscosity in this model is defined in Equation (20). This model becomes very popular to date due to its simplicity, numerical robustness, and stability. However, it has several drawbacks: (1) the model constant varies with different flows; (2) the model cannot predict the inverse energy transfer (i.e., backscatter) due to its purely dissipative nature; (3) the model has difficulty in reproducing the correct mean quantities (e.g., SGS dissipation) as the grid scale approaches the integral scale; (4) the model does not yield a zero eddy viscosity in near-wall regions.

$$\tau_{ij} - \frac{1}{3}\tau_{kk}\delta_{ij} = -2\mu_t \overline{S}_{ij} \tag{18}$$

where:

$$\overline{S}_{ij} = \frac{1}{2}\left(\frac{\partial \overline{u}_i}{\partial x_j} + \frac{\partial \overline{u}_j}{\partial x_i}\right) \tag{19}$$

$$\mu_t = \rho (C_s \Delta)^2 |\overline{S}| \tag{20}$$

In Equation (20), where:

$$|\overline{S}| = \sqrt{2\overline{S}_{ij}\overline{S}_{ij}} \tag{21}$$

$$\Delta = (\Delta x \Delta y \Delta z)^{1/3} \tag{22}$$

where Δ represents the local grid scale. In order to solve the model constant problem, a Dynamic Smagorinsky–Lilly (DSL) model was proposed [13,14], where the model constant is dynamically calculated by using the resolved eddies with the scale size between the grid filter and test filter. The main advantage of this model is that it is not necessary to prescribe and/or tune the model constant. However, the DSL model is subjected to a numerical instability and a variable model constant. Germano [13] proposed an averaging method to overcome this weakness. A good performance was achieved in a channel flow simulation [81]. Another variant of Smagorinsky–Lilly model is the Dynamic Kinetic Energy (DKE) model [82–84]. Unlike the algebraic form in Smagorinsky–Lilly and DSL models, the DKE model solves an additional transport equation for the SGS turbulent kinetic energy instead of adopting the local equilibrium assumption. This model can better account for the energy transfer from the large-scale eddy at the cost of computational expenses. Some other variants of the Smagorinsky–Lilly model were formulated to solve the low Reynolds number effect in the near-wall region, one of which was based on the square of the velocity gradient tensor named the Wall-Adapting Local Eddy-viscosity (WALE) model [79]. Compared to the original Smagorinsky–Lilly model, the WALE model can produce a zero eddy viscosity in the vicinity of the wall or in a pure shear flow. Hence, this model does not need a damping function. In addition to the WALE model, a hybrid model [85] was proposed by combining the Smagorinsky–Lilly model with a damping function [86] to improve the predictive capability for wall-bounded flows. This hybrid model demonstrated a good performance in plane channel flow with different Reynolds numbers. However, this model involves a variable, i.e., the wall normal distance. The determination of this wall normal distance requires an empirical approach for specific flow.

In the framework of the eddy-viscosity SGS model, there are several alternatives to the Smagorinsky-type SGS model, such as Vreman's model [87], the QR model [88,89], the σ-model [90], and the S3PQR model [91]. Compared to the Smagorinsky-type SGS model, Vreman's model can predict zero eddy viscosity in near-wall regions or in transitional flows without explicit filtering, averaging or clipping procedures. However, it was found that the model coefficient in Vreman's model is far from universal. To solve this problem, two procedures were proposed to dynamically determine the model coefficient, i.e., the one based on the global equilibrium between the subgrid-scale dissipation and the viscous dissipation [92,93] and the other one based on the Germano identity [94]. It was reported that the latter is better suited for transient flows [94]. The QR model, which is a minimum-dissipation eddy-viscosity model, gives the minimum eddy dissipation required to dissipate the energy of sub-filter scales. The advantages of this model lie in appropriately switching off for laminar and transitional flows, the low computational complexity, and consistency with the exact sub-filter tensor on isotropic grids. The disadvantage of this model is the insufficient eddy dissipation, which can be corrected by increasing the model constant. Moreover, the QR model requires an approximation of the filter width to be consistent with the exact sub-filter tensor on anisotropic grids. It was noted that the accuracy of the model result for anisotropic grids is highly dependent on the used filter width approximation. By modifying the Poincaré inequality used in the QR model, the dependence can be removed, leading to the construction of an anisotropic minimum-dissipation model that generalizes the desirable properties of the QR model to anisotropic grids [95]. For the purpose of meeting a set of properties based on the practical/physical considerations, the σ-model based on the singular values of the velocity gradient tensor was developed [90]. Owing to its unique properties, ease of implementation, and low computational cost, the σ-model is considered to be suitable for complex flow configurations. Subsequently, through comparison between static and dynamic σ-models, it was found that the local dynamic procedure is not suited for the σ-model, and a global dynamic procedure is suggested [96]. Trias et al. [91] built a general framework for LES eddy-viscosity models, which is based on the 5D phase space of invariants. By imposing appropriate restrictions in this space, a new eddy-viscosity model, i.e., the S3PQR model, was developed. In addition to meeting a set of desirable properties such as positiveness, locality, Galilean invariance, proper near-wall behavior, and automatic switch-off for laminar, 2D and axisymmetric flows, this new model is well-conditioned and has a low computational cost, with no intrinsic limitations for statistically inhomogeneous flows. Despite of all the merits of this model, special attention should be given to the calculation of the characteristic length scale and the determination of the model constant before engineering applications.

Alternatives to the eddy-viscosity SGS model are the similarity model [97–100], the velocity estimation model [101,102], the Approximate Deconvolution model (ADM) [103–105], and the regularization model [106–108]. The similarity model class adopts the idea that an accurate approximation for a SGS model can be reconstructed from the information contained in the resolved field. Therefore, the similarity models [97–100] approximate the SGS stress tensor by a stress tensor calculated from the resolved scales. Due to this nature, the similarity model can naturally account for the inverse energy transfer (i.e., backscatter). This is different from the eddy viscosity model, which only considers the global SGS dissipation (i.e., the net energy flux from the resolved scales to the subgrid scales). It is worth mentioning that, due to the importance of accurate prediction of the inverse energy transfer, a dynamic two-component SGS model was proposed to include the non-local and local interactions between the resolved scales and subgrid scales [100]. The model correctly predicted the breakdown of the net transfer into forward and inverse contributions in a priori tests. In some cases, however, the similarity model underestimated the SGS dissipation. An extra dissipative term was added to solve this issue. This new model formulated is also referred to as the mixed model. Furthermore, the similarity model and the mixed model need additional computational resources due to the implementation of the second filtering. Special attention should be paid to choose an appropriate filter type and size. Following the same idea, Domaradzki et al. [101] improved the SGS stress approximation by replacing the unknown unfiltered variables by approximately deconvolved filtered variables. Subsequently, this SGS model based on the estimation of the unfiltered velocity, which was originally formulated in spectral space, was extended to the physical space [102]. It was found that both versions of this velocity estimation model perform better than or are comparable to classical eddy viscosity models for most physical quantities. This model can account for backscatter without any adverse effects on the numerical stability. Several questions for improving the model need to be addressed, such as the modelling of nonequilibrium and high Reynolds number turbulence in three Cartesian directions. Stolz et al. [103–105] proposed a formulation of the ADM for LES, in which an approximation of the unfiltered solution is obtained from the filtered solution by a series expansion involving repeated filtering. Given a good approximation of the unfiltered solution at a time instant, the nonlinear flux terms of the filtered N-S equations can be computed directly, avoiding the explicit computation of the SGS closures. The effect of the non–represented scales is modelled by a relaxation regularization involving a second filtering and a dynamically estimated relaxation parameter. The ADM is evaluated for incompressible wall-bounded flow [104] and compressible flows [103,105]. The results showed that the ADM can have a significant improvement over the standard and dynamic Smagorinsky models, while at less computational cost compared with that of the dynamic models or the velocity estimation model. The high-Reynolds-number supersonic flow [109] and transitional flow [110] were investigated by the ADM. Agreement was observed between the ADM and experiments or DNS. For the former flow, a rescaling and recycling method was used to have a better control on the desired inflow data. Recently, the ADM was extended for a two-phase flow simulation [111]. By comparing the macroscopic flow characteristics, the ADM showed a better performance than the conventional LES model. However, further investigations should be performed on the relaxation term model for a two-phase simulation and microscopic characteristics of the dispersed phase in a 3D simulation. Another important class of the SGS model is the regularization model, which combines a regularization principle with an explicit filter and its inversion. The regularization model includes many versions, such as the Leray model [106], the Leray-α model [107,112], the Clark-α model [108], the Navier–Stokes-α (NS-α) model [113], etc. For the last one, several variants are proposed, including the NS-α deconvolution model [114–117] and the reduced order NS-α model [118,119]. It was found that the NS-α deconvolution model can significantly improve the prediction accuracy by carefully choosing the filtering radius and by correctly selecting the approximate deconvolution order [117]. Given the difficulties in efficient and stable simulation of the NS-α model for incompressible flows on coarse grids, the reduced order NS-α model is introduced by using deconvolution as an approximation to the filter inverse, reducing the fourth-order NS-α formulation to a second-order model. In spite of the success of the reduced order NS-α model, future work needs to be conducted on locally and dynamically choosing α and numerical testing on different benchmark flows, to name but a few. In addition, comparative studies have been performed between different regularization models [120–122], in which the capability of the regularization model has been demonstrated for a specific flow.

The LES is considered a compromise between the RANS and the DNS. It is more accurate than the RANS and it needs less computational resources than the DNS. However, the LES model has not reached the maturity stage as a numerical tool for the design or the parametric study of complex engineering flows, due to not only a high computational requirement, but also many unresolved issues such as ill-defined boundary conditions, wall-resolved flow, turbulent flow with chemical reactions, and compressible flow. Nevertheless, the LES model has been successfully applied in transitional flow [123–126], separated flow [127,128], and bubbly flow [129,130]. Figure 1 shows the calculation results in a separated boundary layer transition on a flat plate with a semi-circular leading edge of radius of 5 mm under elevated free-stream turbulence. A periodic boundary condition was adopted in the spanwise direction. Free-slip and no-slip conditions were used at lateral boundaries and the plate surface, respectively. The simulation agrees well with experimental data on mean and fluctuating streamwise velocities for an Enhanced-Turbulence-Level (ETL) case, demonstrating a good performance of the LES model for the transitional flow. Figure 2 shows a comparison between a modified k-ε model and the LES model in a bubbly flow. Compared with the experimental data, the LES model is superior in predicting the turbulence.

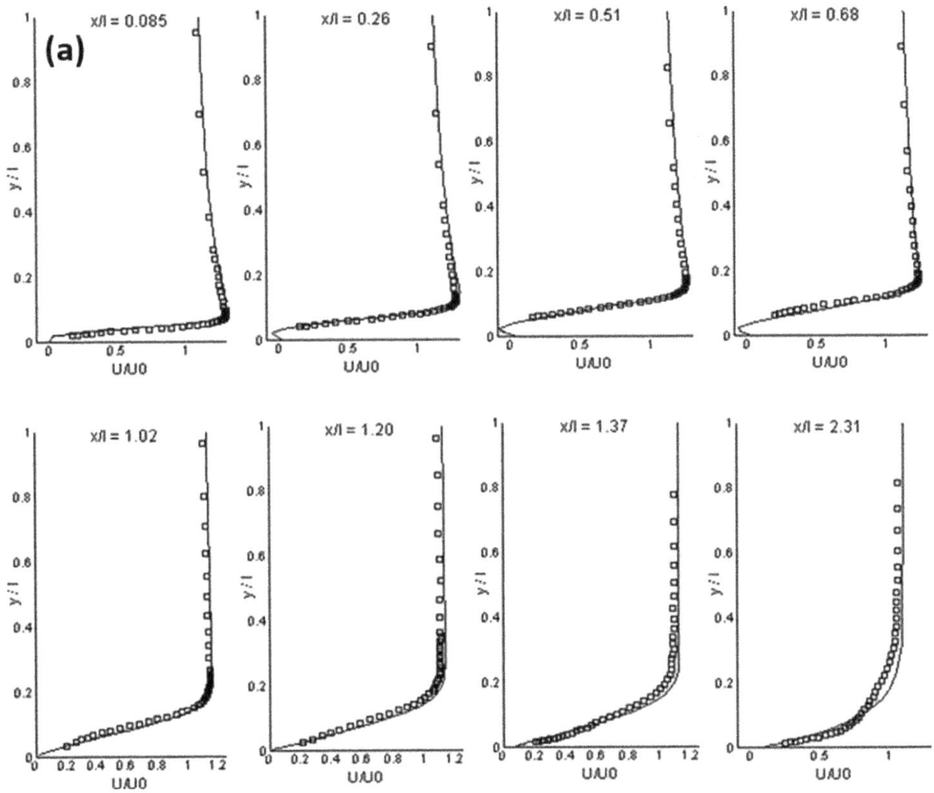

Figure 1. *Cont.*

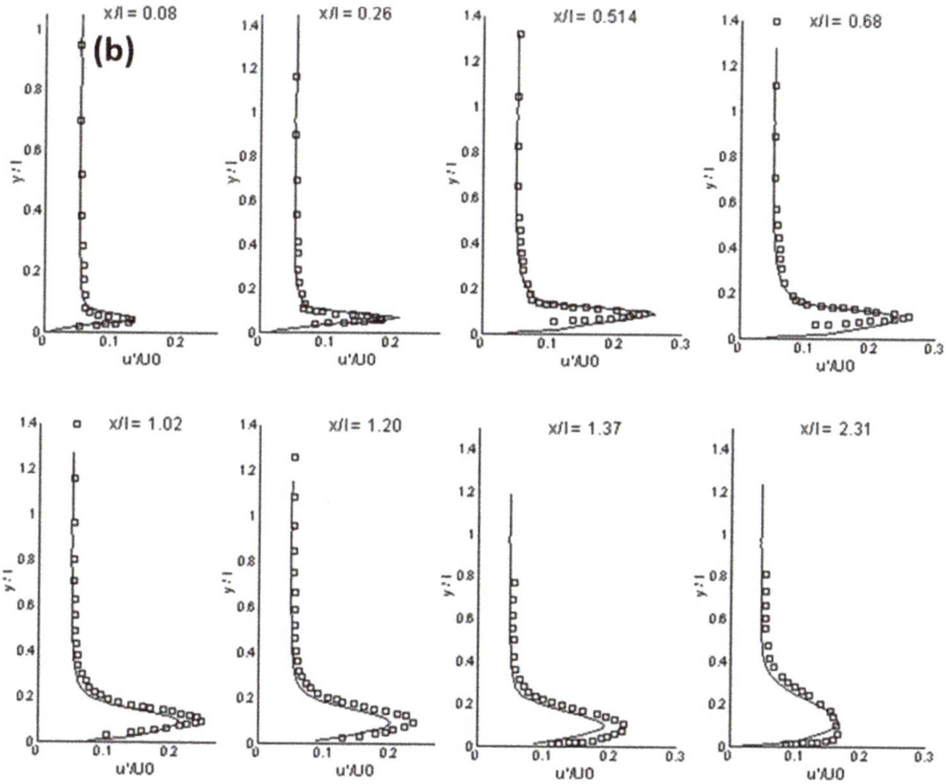

Figure 1. (**a**) Mean streamwise velocity, (**b**) rms streamwise velocity at different streamwise stations (ETL-case) (LES: solid lines, experimental data: symbols, x/l: normalized streamwise, y/l: wall-normal direction, U/U_0: normalized mean streamwise velocity, u'/U_0: normalized rms streamwise velocity) [126].

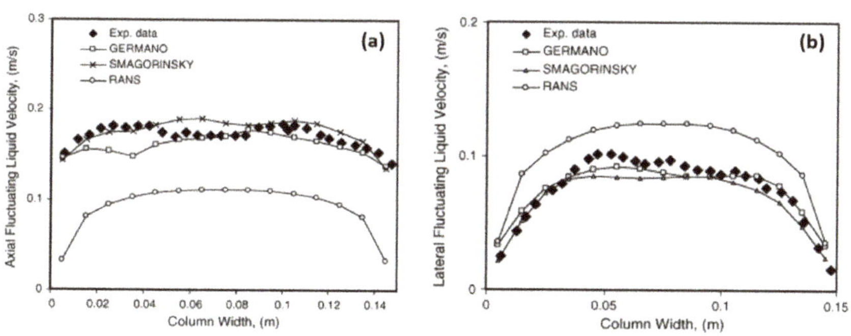

Figure 2. Comparison of model predictions with experimental data for (**a**) axial and (**b**) lateral liquid velocity fluctuations at a height of 0.25 m in a square cross-sectioned bubble column with a superficial gas velocity of 4.9×10^{-3} m/s [129].

2.2.2. Inlet Boundary Condition

The fluid behavior in the domain is largely determined by the inflow condition [131]. The treatment of the inflow condition is of significant importance for LES modelling. Currently, there are two main categories for generating the inflow data, namely, precursor simulation and a synthetic method. The former involves a separate simulation, where the periodic boundary condition or the recycling method can be used. The flow data are stored at each time step in this simulation and then introduced to the inlet boundary for modelling the flow of interest. The main advantage of this method is to obtain more realistic inflow conditions, which represent the required flow characteristics (e.g., velocity profile, turbulence intensity, shear stress, power spectrum, and turbulent structures). The inlet boundary, however, needs to be placed in an equilibrium region for scaling arguments in the precursor simulation, which may even not exist in some flows. This method may lead to a spurious periodicity for the time series [132]. In addition, running a separate simulation requires high computational costs especially for a high-Reynolds number flow. This restrains its application to complex engineering flows. The limitation can be reduced by an internal mapping method. This method integrates the precursor simulation into the main domain, mapping the data downstream of the inlet back to the inlet boundary [133,134].

The synthetic method as an alternative to the precursor simulation is expected to construct the inflow conditions for practical flows. The simplest way is to impose a white-noise random component on the inlet velocity. The magnitude of this random component is determined by the turbulent intensity. Since the turbulence-like component is rapidly dissipated due to the lack of spatial and temporal correlation, this white-noise method is inappropriate to generate the inflow data [135]. In order to impose realistic inflow data on the inlet boundary, other advanced synthetic methods have been developed. These advanced methods consist of the Fourier technique [136,137], principal orthogonal decomposition (POD) method [138,139], digital filtering technique [140,141], and synthetic eddy method (SEM) [142–144]. Several comparative studies have been performed on different synthetic methods [143–146]. Jarrin et al. [143] used the SEM in the hybrid RANS/LES simulations for turbulent flows from simple channel and square duct flows to the flow over an airfoil trailing edge. Compared to other synthetic methods (i.e., Batten's method (Fourier method) [136] and random method), the SEM can substantially reduce the inlet section, leading to a large decrease in the CPU time. Figure 3 shows a better performance of the SEM in the inlet velocity vector compared to that of the others. The SEM realistically reproduced the magnitude and length scale of the fluctuations. The fluctuations in the Batten's model decorrelated in space in the near-wall region due to the decomposition in Fourier's mode. Recently, Skillen et al. [144] improved the SEM of Jarrin et al. (Original SEM) [142] with respect to the normalization algorithm and the eddy placement. The former leads to an improvement over the original SEM model. The latter saves a cost of around 1–2 orders of magnitude. Figures 4 and 5 show the turbulent shear stresses and skin-friction coefficients from the original SEM [142], the improved SEM of Skillen et al. [144], and the precursor LES of Kaltenbach et al. [147]. In comparison with the original SEM, the improved SEM shows a better agreement with the precursor LES results both for the turbulent shear stress and the skin-friction coefficient. Although there is a rapid development of synthetic methods, the available synthetic methods are limited to construct the inflow condition with all required turbulence characteristics mentioned above. Further development is needed. It is unnecessary to claim which method is the best; however, the most appropriate method can be selected by considering the accuracy and the computational cost.

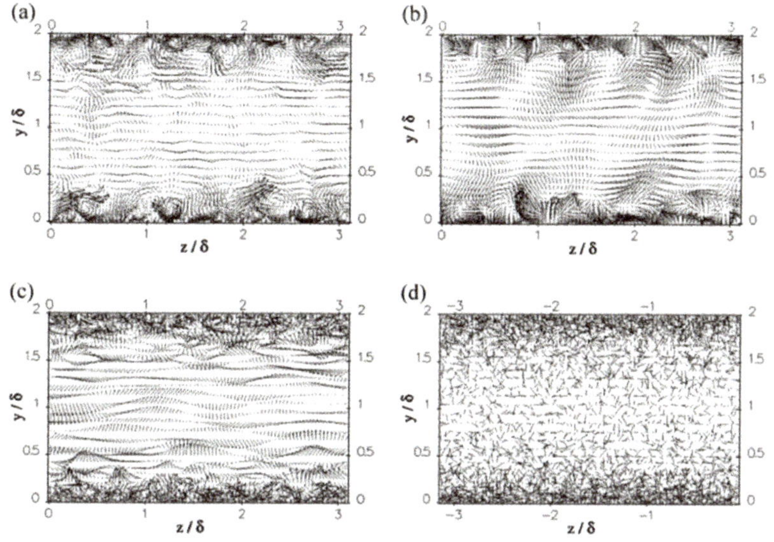

Figure 3. Velocity vectors of the LES inlet condition for hybrid simulation of channel flow: (**a**) precursor LES, (**b**) SEM, (**c**) Batten's method, (**d**) random method (y/δ: normalized y distance, z/δ: normalized z distance, δ: initial boundary layer thickness) [143].

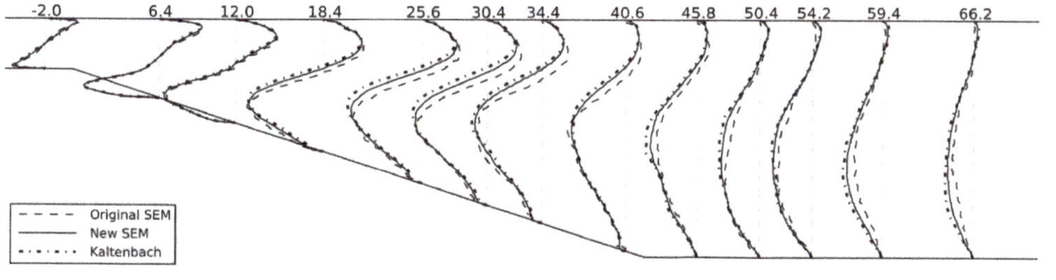

Figure 4. Profiles of turbulent shear stress (the number at the top of the figure indicates profile locations (the distance from the start of the incline, normalized by the initial boundary layer thickness)) [144].

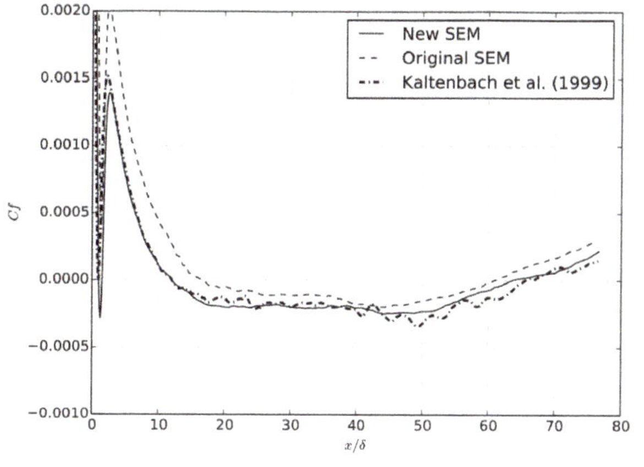

Figure 5. Wall shear stress for the inclined wall (Cf: wall shear stress, x/δ: normalized distance from the start of the incline) [144].

2.3. DNS Approach

The DNS model numerically solves the three-dimensional, time-dependent N-S equations without using a turbulence model. The DNS method captures all the turbulence scales present in the given flow by directly using a very fine mesh and very small time step. The application of the DNS approach is hindered by the requirement of extremely high computational resource. It has been estimated that the number of grid points is proportional to $Re^{9/4}$ in a DNS case [148]. Since the eddy scale in the near-wall region is much smaller than that in the outer domain, a refinement of the mesh is needed in the near-wall region to fully resolve the turbulence, which further increases the number of grid points. Given the large computational domain, complex geometry, and high Reynolds number in practical engineering, the application of DNS approach is currently impossible for most practical flows. However, with the continuous development of parallel computing technique [149], hybrid CPU + GPU computing architecture [150,151], and advanced numerical algorithm [152], a remarkably high computing performance has been reached for DNS. Meanwhile, new challenges have arisen regarding the high-performance computing. Discussion on this point is excluded due to the limited space in this work; however, readers can be directed to relevant work for more information [153,154].

To demonstrate how DNS resolves flow turbulence, an example case is presented for bubble-induced turbulence (BIT), which is one of the important research topics in gas–liquid flow. In the study by Feng et al. [155], a precursor DNS simulation was performed on a homogeneous single-phase turbulent flow. The results of this precursor simulation were used as the inflow condition to calculate the turbulent field around a fully resolved bubble. Figure 6a,b shows the mesh used in the BIT study and the turbulent eddy generated on the highly deformable bubble surface. The results showed that the bubble created new vortices in the wake region, leading to turbulence enhancement. The magnitude of the turbulence enhancement increased with the liquid turbulent intensity and the relative velocity [155]. Apart from the direct study of turbulence, DNS plays an indispensable role in evaluating and developing turbulence models [156–161] and in providing complementary information for experimental study [162–164].

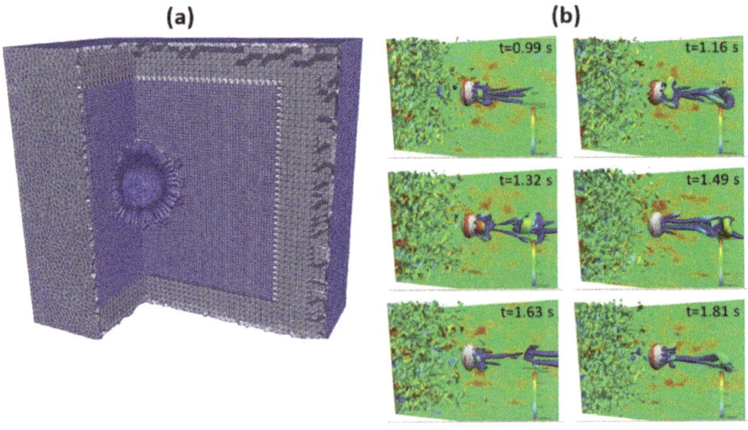

Figure 6. (**a**) Mesh used in the BIT study, (**b**) turbulent eddy generated on the highly deformable bubble surface (Q-criterion with a value of 100) [155].

3. Model Validation Technique

In order to guarantee the fidelity of the established model, model validation is a must before practical simulation. This can be done numerically or experimentally. In the numerical way, an advanced approach such as the DNS can be used to verify a less advanced model such as the RANS. The merit of this validation technique is to reproduce the real conditions of the practical flow, leading to more comprehensive and realistic results. Since the DNS is not fully ready for the simulation of complex engineering flows due to the computing power problem, this technique is less popular compared to the experimental technique. Table 1 lists several commonly used experimental techniques. Depending on the purpose, different experimental methods are employed for the validation, such as an optical fiber probe to measure solid phase holdup and solid velocity in a gas–solid system [165], LDA (Laser-Doppler Anemometry) and PIV (Particle Imaging Velocimetry) to measure phase velocity in a single- or two-phase system [166,167], an electroresistivity probe to monitor mixing behavior in metallurgical processes (converter and ladle) [168,169], and video recording to capture cavity shape and dimension (converter) [170].

Table 1. Experimental validation techniques.

System	Study Object	Measuring Technique and Refs.
Gas-solid	Solid velocity and holdup Gas velocity and holdup, bubble size Granular temperature Mass flux and heat flux	Optical fiber probe [165,166,171–176] LDA [165,166,177] PIV + DIA [164,167,178–180] PIV + DIA + Infrared [181–183]
Gas-liquid	Phase velocity Turbulence quantities	LDA [169,184–188] PIV [189–193]
Gas-liquid Gas-liquid-liquid	Mixing time	Electroresistivity probe [169,170,194–196]
Gas-liquid Gas-liquid-liquid	Cavity dimension	Video recording [170,197–201]

Note: DIA: Digital Image Analysis.

Experimental validation is very useful for understanding model development and turbulence physics, while it also suffers from many limitations. Improvement is needed to represent more realistic situations. In the metallurgical field, it is very difficult to measure flow velocity and flow pattern due to the aggressive and complex operating environment (e.g., high temperature, opaque vessel, and multiphase coexistence). Physical modelling experimentation is adopted based on the similarity principle to obtain flow information. This validation method, however, has several shortcomings: (1) A full-scale physical model is prohibitive to set up in laboratory study due to the difficulty in building a large industry-scale vessel, and the difficulty in mimicking the industrial operational conditions. In general, a scaled-down low temperature physical model is usually employed. This can only reproduce part of the flow dynamics since not all of necessary flow dimensionless numbers can be simultaneously satisfied [202]; (2) Most physical modelling experiments are conducted at room temperature, at which it is impossible to study the heat transfer and melt solidification behavior. In addition, the effect of temperature on the gas phase is not considered in such experiments; (3) The reliability of physical modelling depends on a selection of materials used in the simulation of the real system, resulting in an additional experimental error. For instance, water is frequently used to mimic liquid steel because the kinematic viscosity of water is very close to that of the liquid steel at 1600 °C. Other properties of water, however, such as the density and surface tension, are very different from liquid steel, making the similarity criteria difficult to be entirely fulfilled. Compared to water, low melting point alloys, such as Bi-Sn and Ga-In-Sn alloys, have a closer resemblance to the physical properties of liquid steel [203,204]. The Bi-Sn alloy system can be operated in a temperature range of 200–400 °C, compared to the GaInSn alloy system at room temperature. The effect of temperature on liquid viscosity can be investigated. Due to similar electrical conductivity between the low-melting point alloy and liquid steel, the alloy can be used to model the electromagnetic stirring or breaking in the continuous casting process. With the aid of measuring techniques such as Ultrasound Doppler Velocimetry (UDV) [205] and Contactless Inductive Flow Tomography

(CIFT) [206,207], a significant progress has been achieved with this physical modelling for simulating the real flow behavior in the continuous casting process.

There are other errors induced by the measurement techniques. For PIV and LDA measurement, a correct phase discrimination can reduce errors, improving the measurement quality. Different phase discrimination methods have been proposed. Kulick et al. [208] exploited the large difference in the amplitude of the Doppler burst pedestals obtained from the solid particle and the tracer and then took the ratio of the Doppler signal amplitude to the pedestal amplitude as the discriminator. Regarding the PIV technique, Kiger and Pan employed a two-dimensional median filter to correctly identify and separate the dispersed particles from the two-phase image [209]. Khalitov and Longmire [210] adopted a two-parameter (size and brightness) algorithm to separate the tracer from the solid phase. This algorithm has been proved to be applicable in gas–solid and liquid–liquid systems [210,211]. In addition, the Turbulence Kinetic Energy (TKE) dissipation rate is very difficult to be measured by the PIV or LDA technique since it strongly depends on the spatial resolution [212]. Tanaka and Eaton [213] performed High-Resolution PIV (HR-PIV) measurements with a spatial resolution smaller than the Kolmogorov scale η_k. A modified method was used for the phase identification algorithm to eliminate the common error in the HR-PIV measurement. It was concluded that the measurement error of the TKE dissipation rate can be reduced to a few percent if a proper spatial resolution is employed (in the range of $\eta_k/10$ to $\eta_k/2$). The HR-PIV is also used to measure the carrier-phase velocity and turbulence structures near the particle surface [214,215]. The attenuations of the TKE and its dissipation rate were experimentally obtained, leading to a reasonable prediction of the macroscopic turbulence modification. Even though the HR-PIV technique works well for diluted flow, further development of the HR-PIV technique is needed for investigating the dense flow or the flow region with a high volume fraction of the dispersed phase. By using the HR-PIV technique, Wang et al. [130,216] found a large deviation of the velocity magnitude in the bubbly zone of a top-submerged gas injection flow. This deviation was partly attributed to the less reliable HR-PIV experimental data caused by the refraction of gas bubbles. For brevity's sake, other techniques, for instance, the measurement of the pressure drop and fluctuation and holdup of solid phase are excluded from this work. Interested readers can find relevant studies elsewhere [217,218].

4. Applications

Most of the flows involved in industrial production are turbulent. Therefore, the choice of a proper turbulence model used to accurately represent the flow characteristics is essential to CFD applications. In order to draw attention to this point, the applications of turbulence models in solving three commonly encountered turbulence-related flow issues are discussed. It is necessary to mention that this work aims to demonstrate how to customize a conventional turbulence model for a specific flow problem instead of comprehensively reviewing the single phase or multiphase phase flow characteristics.

4.1. Bubble-Induced Turbulence (BIT)

In an engineering bubbly flow such as the flow in a bubble column or the flow in a metallurgical ladle, the BIT has to be taken into consideration to fully reproduce the flow characteristics. There are two common methods to account for the BIT: (1) the Effective Viscosity Modified Method (EVMM), where BIT is added in the effective viscosity term [191,219]; (2) the Transport Equation Modified Method (TEMM), where BIT is modelled with a source term in the turbulent transport equation [220–225]. Both methods can be used with the RANS-based approach (k-ε/ω/SST (Shear-Stress Transport) [224] and RSM, [226–228]) termed RANS-BIT, and LES approach, termed LES-BIT [129,191,229]. Many RANS-BIT comparative studies were performed on the EVMM and TEMM [224,225]. The results showed that the EVMM is not suitable for the turbulence prediction since it depends on an algebraic model (e.g., the model of Sato et al. [219]). LES-BIT studies were performed with the EVMM, and it was observed that including the BIT slightly changed the numerical results [129,191]. Since different experimental validations suggest different formulations of the source term, no universal formulation of the source term has been found. Compared to the k-ε/ω/SST-based BIT modelling, the RSM-based BIT modelling [225,228] considers the effect of the anisotropic property of turbulence. This improves the overall model performance in simulating both the void fraction distribution and turbulent kinetic energy. This model is more suitable for modelling the complex multiphase flow with prominent anisotropy. Nevertheless, it introduces more unknown parameters, making this model less readily applicable. Reliable experiments with individual Reynolds stress measurement are needed for the model validation. Niceno et al. [229] compared the EVMM and TEMM of the LES-based approach and concluded that both methods predict the turbulent kinetic energy well qualitatively, but the TEMM is superior for the quantitative prediction.

Recently, the DNS study became available for bubbly flows; several DNS studies [161,230,231] on disperse bubbly channel flows were performed, and the obtained TKE budgets were used to account for the BIT. With the aid of the DNS data, Ma et al. developed a model for the BIT and incorporated it as a source/sink term in the k-ω SST model transport equation. This model was adopted by Liao et al. to evaluate its performance on bubbly flows in containers and vertical pipes [232]. The results showed improvements regarding the radial gas volume fraction and velocity profile in high-volume fraction cases were achieved. Later, Ma et al. [231] extended the BIT model to a second moment level. During the development of this full SMC for BIT in the Euler-Euler framework, particular attention was given to the treatment of the pressure-strain term for bubbly flows and the form of the interfacial term to account for BIT. For the latter, an effective BIT source term was proposed, which largely simplified the modelling work. To understand the anisotropic behavior of the bubbly flow, an anisotropy-invariant analysis was conducted, based on which the BIT closure was improved. This new SMC with the proposed BIT model was compared with the experimental data of Akbar et al. [233]. A good agreement was achieved in predicting the gas void fraction, phase streamwise velocity, and liquid phase Reynolds stresses (see Figure 7). As the computational power increases rapidly, the DNS study will play a more important role in uncovering the turbulence physics of bubbly flows. Therefore, more DNS data are expected.

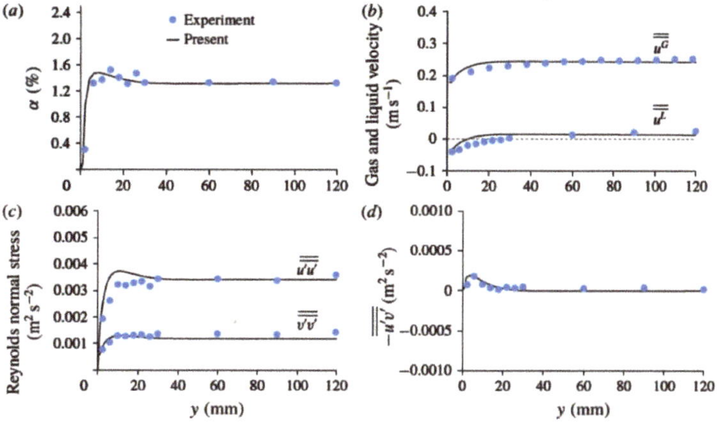

Figure 7. Comparison between the model of Ma et al. [231] and the experimental data of Akbar et al. [233]: (**a**) gas void fraction, (**b**) streamwise velocities of the liquid and gas phases, (**c**) Reynolds normal stresses of the liquid phase, (**d**) Reynolds shear stress of the liquid phase.

4.2. Supersonic Jet Transport

Supersonic gas jets are applied in Basic Oxygen Furnace (BOF) steelmaking processes [234,235]. Due to its vital role in the refining efficiency [236] and the service lifetime of the furnace lining [237], it is imperative to understand the behavior of the supersonic jet flow. During supersonic jet transport, it interacts with the surrounding medium to produce a turbulent mixing region, as shown in Figure 8. An accurate prediction of the growth rate of the mixing region is a challenge in modelling the jet transport. The standard k-ε turbulence model suffers from a typical weakness, i.e., an overestimated growth rate of the turbulent mixing region around the supersonic gas jet [238]. This leads to a large error in predicting the volume of cavity formed by the gas jet impinging on the liquid surface [239]. Several modifications have been proposed to solve this problem [240–242]. Sarkar et al. [240] considered that the compressible dissipation term should be included to account for the effect of the compressibility on the supersonic jet flow. Based on the asymptotic analysis and DNS data, an algebraic model was proposed for the compressible dissipation by using the turbulent Mach number. Subsequently, Sarkar [241] found that the reduced turbulence generation is the main cause of preventing the turbulent mixing layer from the growth rather than the effect of the dilatation term (pressure dilatation and compressible dissipation) on the growth. A gradient Mach number was induced to describe the inhibiting effect of compressibility on the turbulence growth rate. Heinz [242] used the gradient Mach number to evaluate the effect of compressibility on the turbulence distribution. An expression of the model constant C_μ as a function of the gradient Mach number was obtained. However, the k-ε turbulence model with the compressibility correction failed to predict the supersonic jet flow with a temperature gradient, such as the potential flow core length and the cavity shape and dimension [243]. This is because the temperature gradient between the gas jet and the ambient is ignored by the original turbulence model. To sensitize the turbulence model with temperature fluctuation, efforts [244–246] have been made to modify the turbulent viscosity term or turbulent heat flux term. Abdol-Hamid et al. [247] corrected the turbulent viscosity term with the temperature gradient for the case of a hot gas jet entering into low temperature ambient. The results of modelling agreed well with experimental data for subsonic and supersonic jet flows. However, this model did not give reasonable results for the BOF supersonic jet flow [243], under which a room temperature gas jet enters into a high temperature ambient (see Figure 9a). Alam et al. [243] modified Abdol-Hamid's model for the gas jet of BOF. The predicted distributions of axial velocity and dynamic pressure along the central axis of the jet closely agreed with the experimental data (see Figure 9b,c). It is worth mentioning that the data compared in Figure 9 were extracted along the central axis of the jet (see the dashed line in Figure 8). Wang et al. [248] adopted the model of Alam et al. to study the multiple supersonic oxygen jets in the BOF process, where a better prediction of the shape of cavity caused by the gas jet impinging on the liquid surface was achieved. To fit the empirical data available for a cold jet to hot environment [249], Lebon et al. [250] derived another expression of C_μ as a function of the enthalpy ratio of ambient gas to gas jet. It was found that the modified model is adequate to model a compressible jet to hot environment. In addition to the two-equation turbulence model, the LES model with the compressibility and temperature corrections is also widely used to study the jet noises [251]. Wang et al. [252] and Bodony et al. [253] critically reviewed the roles of the SGS model and inflow boundary condition in predicting jet noise. The open issues and future directions were also included in the papers. Therefore, the LES model for jet noise prediction is excluded from this article to avoid repetition.

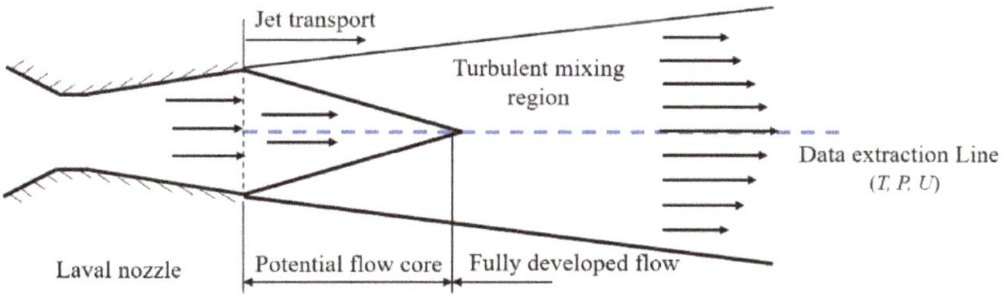

Figure 8. Illustration of the supersonic jet transport.

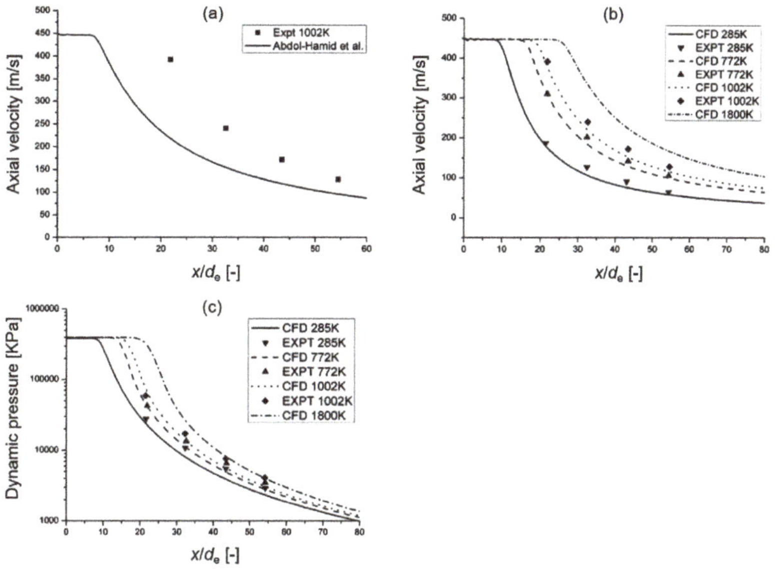

Figure 9. (**a**) and (**b**) Axial velocity distribution, (**c**) dynamic pressure distribution along the central axis of the jet (*x*: distance along the central axis, *de*: nozzle diameter) [243].

4.3. Electromagnetic Suppression of Turbulence

A static magnetic field has been commonly used to stabilize the turbulent flow in the continuous casting process, aiming to improve the product quality. The application of the magnetic field not only suppresses the mean flow, but also dampens the flow turbulence. For the former, including the electromagnetic force/Lorentz force in the momentum equations, as reported by many studies [254–258], solves the suppression of the mean flow. The latter, also called Joule dissipation, can be resolved by adding the electromagnetic damping effect in a conventional turbulence model. In order to tailor the turbulence model towards magnetohydrodynamic (MHD) flows, many fundamental investigations have been performed [259–263]. Ji and Gardner [259] modified a standard low Re k-ε model [27] to account for the Joule dissipation. Additional source/sink terms were added in the k and ε equations, as well as a damping factor for turbulent viscosity. Both the new terms and the damping factor contain an exponential expression e^{-CN}, where C is an empirical constant determined from experimental data, and N is the interaction parameter defined as the ratio of the time scale of large eddies (L/U) to the characteristic magnetic braking time ($\rho/\sigma B^2$). This interaction parameter represents the strength of turbulence damping due to the magnetic field. However, as reported by Kenjereš and Hanjalić [260], the use of the time scale of large eddies makes the model dependent on the bulk flow properties, restricting the model to homogeneous magnetic fields and simple geometries. Replacing the bulk time scale with the local turbulent time scale (k/ε) can overcome the above-mentioned deficiency. As noted in the work of Ji and Gardner [259], the modified turbulence model failed to accurately predict the turbulent kinetic energy when the strength of the magnetic field increased. This is because the model is formulated based on the assumption of isotropic turbulence, while MHD flows exhibit strong anisotropy caused by the damping of turbulent fluctuations nonparallel to the magnetic field. The anisotropy of the Joule dissipation has to be considered in the turbulence model. Widlund et al. [261] proposed a scalar dimensionality anisotropy parameter carrying the information with regard to the magnitude and anisotropy of the Joule dissipation tensor. A scalar transport equation for the anisotropy parameter was also proposed. With this scalar transport equation, it is easy to extend a conventional turbulence model towards MHD engineering applications. It should be noted that the scalar transport equation is based on phenomenological reasoning, and special attention should be paid to the hydrodynamic part of the proposed equation. Miao et al. [264] coupled the scalar transport equation with the RANS SST k-ω turbulence model [10] to solve the MHD flow in a continuous casting mold. The Joule dissipation was taken into account by adding source/sink terms in the k and ω equations, and the anisotropic behavior of the Joule dissipation was represented by the anisotropy parameter contained in the source/sink terms. The computational domain of the continuous casting mold is shown in Figure 10, and the measured region and data extraction line for liquid metal (low melting point alloy Ga68In20Sn12) velocity are depicted in the figure. The averaged velocities obtained from the three-equation turbulence model (i.e., k-ω-anisotropy parameter) qualitatively and quantitatively agreed with the experimental data (see Figures 11 and 12). The turbulence models described above are modified in the framework of the RANS approach. Due to the time-averaging treatment in the RANS approach, the change of turbulence structure and transient flow behavior are difficult to be captured by the RANS-based turbulence model. Given that engineering flows are usually very complex and have a high Re number, the RANS-based turbulence models are still considered practical candidates for solving MHD flows.

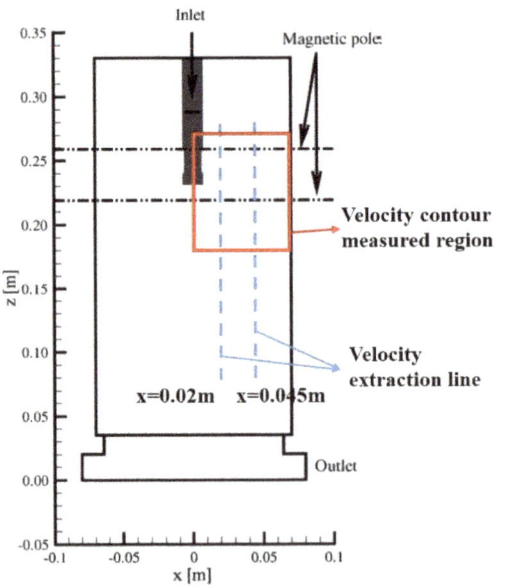

Figure 10. The computational domain of the continuous casting mold and the measured region and data extraction line (figure adapted from [264]).

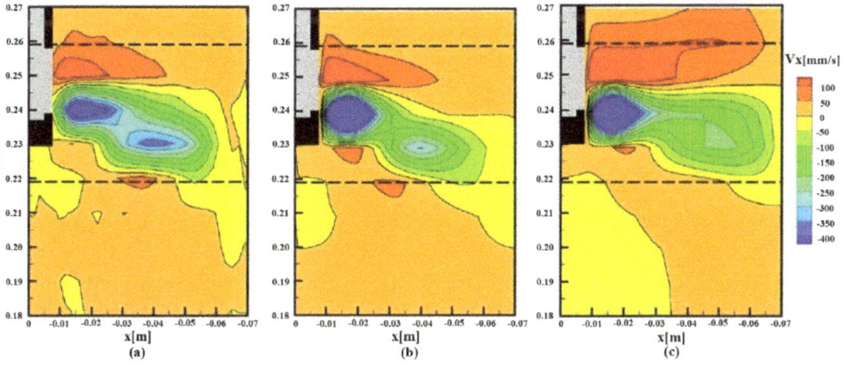

Figure 11. The time-averaged horizontal velocity in the midplane parallel to the mold wide face with electrically conducting mold walls, B0 = 310 mT (within the dotted lines): (**a**) experimental results, (**b**) numerical results from the SST k-ω model + Joule dissipation, (**c**) numerical results from the conventional SST k-ω model [264] (The measured region in Figure 10).

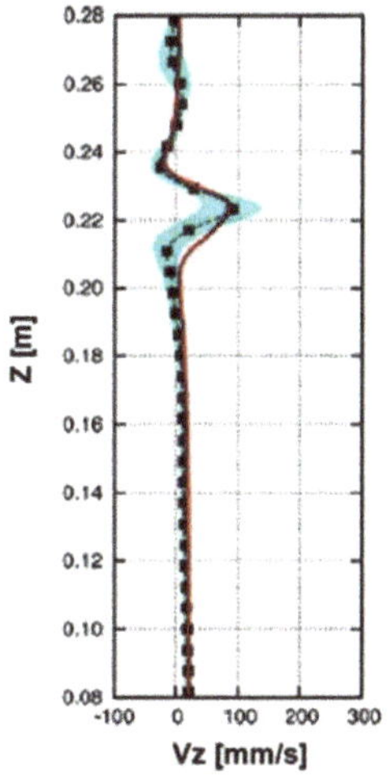

Figure 12. The time-averaged vertical velocity along a vertical line in the midplane at x = 0.045 m (B0 = 310 mT with electrically conducting mold) [264] (The data extraction line in Figure 10).

In addition to the RANS-based turbulence model, the LES model, which is able to capture turbulence structures, has been modified or developed to solve the MHD turbulent flow. Shimomura [262] incorporated the magnetic damping effect in the form of a locally determined damping factor for SGS eddy viscosity. Compared with the original Smagorinsky model [12], the new model showed better performance in both turbulent and laminar states. The new model also successfully predicted the anisotropic laminarization caused by the applied magnetic field, for which the RANS-based model is incompetent. Kobayashi [263] developed a new SGS model based on the coherent structure extracted by the second invariant of a velocity gradient tensor in grid scale flow field. Compared with the conventional Smagorinsky model [12], the new SGS model does not need an explicit wall-damping function or change the model parameter depending on the flow. Compared with the dynamic Smagorinsky model [13], the new SGS model is numerically stable due to the fact that the model parameter is always positive. Considering the advantages of the new SGS model, Singh et al. [265,266] used it for the MHD turbulent flow in a continuous caster. The numerical results closely matched the measured data. Although the LES models can capture transient flow behavior and turbulent structures, the computational cost required is much higher than that of the RANS-based turbulence model. With the fast increase in computing power, more LES simulations are expected for engineering flows in the near future. The DNS studies for MHD flows as presented by Chaudhary et al. [267,268] are also necessary since important information regarding turbulent quantities can be obtained for formulating the RANS or LES model. In addition, detailed experiment data, especially the turbulent quantities, are expected for validating numerical simulation of MHD flow.

5. Conclusions

Three main turbulence approaches (i.e., the RANS, LES, and DNS) have been reviewed in this paper. The formulations and variations of the RANS approach have been described, evaluated, and discussed. The eddy viscosity models still remain the most widely used methods for calculating simple engineering flows. The DSM overcomes the limitations of the eddy viscosity model and can be used to predict complex anisotropic flows. However, since a high degree of uncertainty is introduced in the DSM, special attention should be paid to its application. Although the LES is not completely ready for the calculation of high Re number engineering flow in the current stage, it can be used for the studies on fundamental turbulence physics and the low Re number flow in simple geometry. The LES is highly recommended to quantify the turbulent quantities and transient flow behavior. Since the inflow conditions are very important for LES modelling, a further improvement of the method is needed. With the improvement of the model formulation and accurate specification of inlet boundary conditions, the LES shows great potential to realistically solve complex turbulent flows. Due to the high computing power needed, the DNS is often adopted to understand the turbulence physics and to evaluate less advanced turbulence models, rather than being applied for a real case study. It contributes to the development of the turbulence model. Depending on the actual situation, the corresponding experimental validation should be adopted to ensure the fidelity of the used turbulence model before a formal simulation. To demonstrate the application of turbulence modelling, the conventional turbulence models have been customized to resolve three important turbulence-related flow issues, namely, BIT, supersonic jet transport, and electromagnetic damping. Success has been achieved by the customized turbulence model. The general outlook for turbulent flow simulation is listed as follows:

1. Due to insufficient knowledge of the turbulence physics, there is a high degree of uncertainty in modelling the higher-order correlations in a DSM simulation. It is necessary to perform fundamental studies on turbulence physics, and the DNS is an important method for those studies. Besides, advanced numerical schemes are expected to minimize the diffusive errors when solving the non-linear higher-order correlations.
2. An inflow condition method, which can generate the required realistic turbulence characteristics at a reasonable computational cost, is needed for an accurate LES simulation.
3. It is of significant importance to experimentally obtain accurate turbulence data such as the Reynolds stress term, turbulence kinetic energy, and its dissipation rate, especially for a densely dispersed flow. Advanced non-intrusive experimental techniques are needed to measure the flow field information and phase distribution. The obtained data can be used, for instance, to develop a reliable model considering the effect of the dispersed phase.
4. The BIT is important for the turbulent flow. Due to a small size of the bubble, the induced turbulence is characterized with small spatial and temporal scales. An approach (experimental technique and/or numerical simulation) is required to obtain insight into the induced turbulence.
5. Comprehensive and realistic modelling is indispensable for engineering applications, such as the inclusion of bubble coalescence and breakage, the consideration of temperature and pressure affecting the bubble behavior, the involvement of compressibility and the temperature gradient in supersonic jet flow, and the turbulence damping

caused by magnetic fields. The mechanisms of turbulence modulation with respect to these aspects remain open for fundamental investigation.

Author Contributions: Conceptualization, Y.W.; writing, Y.W., and L.C.; review, and editing, Z.C., B.B. and M.G.; supervision, M.G. and B.B. All authors have read and agreed to the published version of the manuscript.

Funding: This research was funded by Belgian IWT project, grant number 140514, and China Scholarship Council, grant number 201506460005.

Data Availability Statement: Not applicable.

Acknowledgments: The financial support from Belgian IWT project (No.140514) and China Scholarship Council (CSC) is highly acknowledged. The authors also would like to give their thanks to the anonymous reviewers whose comments helped to improve the manuscript.

Conflicts of Interest: The authors declare no conflict of interest.

Nomenclature

B	Magnetic field intensity, T.
$f,\ \overline{f},\ f'$	Instantaneous, ensemble/time-averaged, and fluctuating variable, respectively.
g	Gravitational acceleration, m/s^2.
k	Turbulent kinetic energy, m^2/s^2.
L	Characteristic length scale, m.
$p,\ p'$	Pressure, pressure fluctuation, Pa.
$Pr_t Re_k, Re_t S_{ij}$	Mean rate of strain tensor, s^{-1}.
$S_k,\ S_\varepsilon,\ S_\omega$	Source term in k, ε, and ω equation, respectively.
$u,\ \overline{u}, u'$	Instantaneous, mean (large scale), and fluctuating (subgrid scale) velocity, respectively, m/s.
U	Bulk velocity, m/s.
Greek Symbols	
δ_{ij}	Kronecker delta.
ε	Turbulent energy dissipation rate, m^2/s^3.
μ	Dynamic viscosity, Pa·s.
μ_{eff}	Effective viscosity, Pa·s.
μ_t	Turbulent viscosity, Pa·s.
ρ	Density, kg/m^3.
σ	Electrical conductivity, S/m.
ω	Specific dissipation rate
Subscripts	
i, j, k	Coordinate direction indices.

References

1. Markatos, N.C. The mathematical modelling of turbulent flows. *Appl. Math. Model.* **1986**, *10*, 190–220. [CrossRef]
2. Georgiadis, N.J.; Rizzetta, D.P.; Fureby, C. Large-eddy simulation: Current capabilities, recommended practices, and future research. *AIAA J.* **2010**, *48*, 1772–1784. [CrossRef]
3. Hanjalić, K. Advanced turbulence closure models: A view of current status and future prospects. *Int. J. Heat Fluid Flow* **1994**, *15*, 178–203. [CrossRef]
4. Argyropoulos, C.D.; Markatos, N.C. Recent advances on the numerical modelling of turbulent flows. *Appl. Math. Model.* **2015**, *39*, 693–732. [CrossRef]
5. Boris, J.P.; Grinstein, F.F.; Oran, E.S.; Kolbe, R.L. New insights into large eddy simulation. *Fluid Dyn. Res.* **1992**, *10*, 199–228. [CrossRef]
6. Launder, B.E.; Spalding, D.B. The numerical computation of turbulent flows. *Meth. Appl. Mech. Eng.* **1974**, *3*, 269–289. [CrossRef]
7. Yakhot, V.; Orszag, S.A. Renormalization group analysis of turbulence: Basic theory. *J. Sci. Comput.* **1986**, *1*, 3–51. [CrossRef]
8. Shih, T.H.; Liou, W.W.; Shabbir, A.; Yang, Z.; Zhu, J. A new k-ϵ eddy viscosity model for high Reynolds number turbulent flows. *Comput. Fluids* **1995**, *24*, 227–238. [CrossRef]
9. Wilcox, D.C. Reassessment of the scale-determining equation for advanced turbulence models. *AIAA J.* **1988**, *26*, 1299–1310. [CrossRef]
10. Menter, F.R. Two-equation eddy-viscosity turbulence models for engineering applications. *AIAA J.* **1994**, *32*, 1598–1605. [CrossRef]

11. Patel, V.C.; Rodi, W.; Scheuerer, G. Turbulence models for near-wall and low Reynolds number flows—A review. *AIAA J.* **1985**, *23*, 1308–1319. [CrossRef]
12. Smagorinsky, J. General circulation experiments with the primitive equations. *Month. Wea. Rev.* **1963**, *91*, 99–164. [CrossRef]
13. Germano, M.; Piomelli, U.; Moin, P.; Cabot, W.H. A dynamic subgrid-scale eddy viscosity model. *Phys. Fluids* **1991**, *3*, 1760–1765. [CrossRef]
14. Lilly, D.K. A proposed modification of the Germano subgrid-scale closure method. *Phys. Fluids* **1992**, *4*, 633–635. [CrossRef]
15. Moin, P.; Mahesh, K. Direct numerical simulation: A tool in turbulence Research. *Annu. Rev. Fluid Mech.* **1998**, *30*, 539–578. [CrossRef]
16. Montorfano, A.; Piscaglia, F.; Ferrari, G. Inlet boundary conditions for incompressible LES: A comparative study. *Math. Comput. Model.* **2013**, *57*, 1640–1647. [CrossRef]
17. Xie, B.; Gao, F.; Boudet, J.; Shao, L.; Lu, L. Improved vortex method for large-eddy simulation inflow generation. *Comput. Fluids* **2018**, *168*, 87–100. [CrossRef]
18. Tayebi, D.; Svendsen, H.F.; Jakobsen, H.A.; Grislingas, A. Measurement techniques and data interpretations for validating CFD multiphase reactor models. *Chem. Eng. Comm.* **2001**, *186*, 56–159. [CrossRef]
19. Van Ommen, J.R.; Mudde, R.F. Measuring the gas-solids distribution in fluidized beds-a review. *Int. J. Chem. React. Eng.* **2008**, *6*, 1–29. [CrossRef]
20. Hinze, J.O. *Turbulence*; McGraw-Hill Publishing Co.: New York, NY, USA, 1975.
21. Launder, B.E. Second moment closure and its use in modelling turbulent industrial flows. *Int. J. Numer. Meth. Fluids* **1989**, *9*, 963–985. [CrossRef]
22. Hanjalić, K. *Introduction to Turbulence Modelling*; Von Kaman Institute for Fluid Dynamics: Sint-Genesius-Rode, Belgium, 2004; pp. 1–75.
23. Prandtl, L. Bericht fiber untersuchungen zur ausgebildeten turbulent. *ZAMM* **1925**, *5*, 136–139. [CrossRef]
24. Spalart, P.R.; Allmaras, S.R. A one-equation turbulence model for aerodynamic flows. In Proceedings of the 30th Aerospace Sciences Meeting and Exhibit, Reno, NV, USA, 6–9 January 1992.
25. Nallasamy, M. Turbulence models and their applications to the prediction of internal flows: A review. *Comput. Fluids* **1987**, *15*, 151–194. [CrossRef]
26. Jones, W.P.; Launder, B.E. The prediction of laminarization with a two-equation model of turbulence. *Int. J. Heat Mass Trans.* **1972**, *15*, 301–314. [CrossRef]
27. Jones, W.P.; Launder, B.E. The calculation of low-Reynolds-number phenomena with a two-equation model of turbulence. *Int. J. Heat Mass Trans.* **1973**, *16*, 1119–1130. [CrossRef]
28. Launder, B.E.; Sharma, B.I. Application of the energy-dissipation model of turbulence to the calculation of flow near a spinning disc. *Lett. Heat Mass Trans.* **1974**, *1*, 131–138. [CrossRef]
29. Hoffmann, G.H. Improved form of the low Reynolds number $k-\varepsilon$ turbulence model. *Phys. Fluids* **1975**, *18*, 309–312. [CrossRef]
30. Hassid, S.; Poreh, M. A turbulent energy dissipation model for flows with drag reduction. *J. Fluids Eng.* **1978**, *100*, 107–112. [CrossRef]
31. Lam, C.K.G.; Bremhorst, K. A modified form of the k-ε model for predicting wall turbulence. *Tans. ASME* **1981**, *103*, 456–460. [CrossRef]
32. Chien, K.Y. Predictions of channel and boundary-layer flows with a low-Reynolds-number turbulence model. *AIAA J.* **1982**, *20*, 33–38. [CrossRef]
33. Reynolds, W.C. *Fundamentals of Turbulence for Turbulence Modeling and Simulation in Lecture Notes for Von Karman Institute*; AGARD Lecture Note Series: New York, NY, USA, 1987.
34. Shih, T.H.; Zhu, J.; Lumley, J.L. A new Reynolds stress algebraic model. *NASA TM* **1995**, *125*, 1–4.
35. Menter, F.R. Review of the shear-stress transport turbulence model experience from an industrial perspective. *Int. J. Comput. Fluid Dyn.* **2009**, *23*, 305–316. [CrossRef]
36. Wilcox, D.C. Comparison of two-equation turbulence models for boundary layers with pressure gradient. *AIAA J.* **1993**, *31*, 1414–1421. [CrossRef]
37. Wilcox, D.C. Simulation of transition with a two-equation turbulence model. *AIAA J.* **1994**, *32*, 247–255. [CrossRef]
38. Peng, S.H.; Davidson, L.; Holmberg, S. A modified low-Reynolds-number k-ω model for recirculating flows. *J. Fluid Eng.* **1997**, *119*, 867–875. [CrossRef]
39. Bredberg, J.; Peng, S.H.; Davidson, L. An improved k-ω turbulence model applied to recirculating flows. *Int. J. Heat Fluid Flow* **2002**, *23*, 731–743. [CrossRef]
40. Resende, P.R.; Pinho, F.T.; Younis, B.A.; Kim, K.; Sureshkumar, R. Development of a low-Reynolds-number k-ω model for FENE-P fluids. *Flow Turbul. Combust.* **2013**, *90*, 69–94. [CrossRef]
41. Durbin, P.A. Near-wall turbulence closure modeling without "damping functions". *Theor. Comput. Fluid Dyn.* **1991**, *3*, 1–13.
42. Durbin, P.A. On the k-ε stagnation point anomaly. *Int. J. Heat Fluid Flow* **1996**, *17*, 89–90. [CrossRef]
43. Laurence, D.R.; Uribe, J.C.; Utyuzhnikov, S.V. A robust formulation of the v2−f model. *Flow Turb. Combust.* **2004**, *73*, 169–185. [CrossRef]
44. Hanjalić, K.; Popovac, M.; Hadžiabdić, M. A robust near-wall elliptic-relaxation eddy-viscosity turbulence model for CFD. *Int. J. Heat Fluid Flow* **2004**, *25*, 1047–1051. [CrossRef]

45. Billard, F.; Laurence, D. A robust k-$\varepsilon\overline{v^2}k$ elliptic blending turbulence model applied to near-wall, separated and buoyant flows. *Int. J. Heat Fluid Flow* **2012**, *33*, 45–58. [CrossRef]
46. Billard, F.; Revell, A.; Craft, T. Application of recently developed elliptic blending based models to separated flows. *Int. J. Heat Fluid Flow* **2012**, *35*, 141–151. [CrossRef]
47. Speziale, C.G. On nonlinear kl and k-ε models of turbulence. *J. Fluid Mech.* **1987**, *178*, 459–475. [CrossRef]
48. Lien, F.S.; Leschziner, M.A. Assessment of turbulence-transport models including non-linear RNG eddy-viscosity formulation and second-moment closure for flow over a backward-facing step. *Comput. Fluids* **1994**, *23*, 983–1004. [CrossRef]
49. Craft, T.J.; Launder, B.E.; Suga, K. Development and application of a cubic eddy-viscosity model of turbulence. *Int. J. Heat Fluid Flow* **1996**, *17*, 08–115. [CrossRef]
50. Craft, T.J.; Launder, B.E.; Suga, K. Prediction of turbulent transitional phenomena with a nonlinear eddy-viscosity model. *Int. J. Heat Fluid Flow* **1997**, *18*, 15–28. [CrossRef]
51. Apsley, D.D.; Leschziner, M.A. A new low-Reynolds-number nonlinear two-equation turbulence model for complex flows. *Int. J. Heat Fluid Flow* **1998**, *19*, 209–222. [CrossRef]
52. Jongen, T.; Machiels, L.; Gatski, T.B. Predicting noninertial effects with linear and nonlinear eddy-viscosity, and algebraic stress models. *Flow Turb. Combust.* **1998**, *60*, 215–234. [CrossRef]
53. Hellsten, A.; Wallin, S. Explicit algebraic Reynolds stress and non-linear eddy-viscosity models. *Int. J. Comput. Fluid Dyn.* **2009**, *23*, 349–361. [CrossRef]
54. Gatski, T.B.; Speziale, C.G. On explicit algebraic stress models for complex turbulent flows. *J. Fluid Mech.* **1993**, *254*, 59–78. [CrossRef]
55. Gatski, T.B.; Jongen, T. Nonlinear eddy viscosity and algebraic stress models for solving complex turbulent flows. *Prog. Aerosp. Sci.* **2000**, *36*, 655–682. [CrossRef]
56. Gatski, T.B.; Wallin, S. Extending the weak-equilibrium condition for algebraic Reynolds stress models to rotating and curved flows. *J. Fluid Mech.* **2004**, *518*, 147–155. [CrossRef]
57. Rodi, W. The Prediction of Free Turbulent Shear Flows by Use of a 2-Equation Model of Turbulence. Ph.D. Thesis, University of London, London, UK, 1972.
58. Rodi, W. A new algebraic relation for calculating the Reynolds stresses. *ZAMM* **1976**, *56*, 219–221. [CrossRef]
59. Kim, J.; Moin, P.; Moser, R. Turbulence statistics in fully developed channel flow at low Reynolds number. *J. Fluid Mech.* **1987**, *177*, 133–166. [CrossRef]
60. Daly, B.J.; Harlow, F.H. Transport equations in turbulence. *Phys. Fluids* **1970**, *13*, 2634–2649. [CrossRef]
61. Hanjalić, K.; Launder, B.E. A Reynolds stress model of turbulence and its application to thin shear flows. *J. Fluid Mech.* **1972**, *52*, 609–638. [CrossRef]
62. Shir, C.C. A preliminary numerical study of atmospheric turbulent flows in the idealized planetary boundary layer. *J. Atmospher. Sci.* **1973**, *30*, 1327–1339. [CrossRef]
63. Mellor, G.L.; Herring, H.J. A survey of the mean turbulent field closure models. *AIAA J.* **1973**, *11*, 590–599. [CrossRef]
64. Nagano, Y.; Tagawa, M. Turbulence model for triple velocity and scalar correlations. In *Turbulent Shear Flows 7*; Springer: Berlin, Germany, 1991; Volume 7, pp. 47–62.
65. Magnaudet, J. The modelling of inhomogeneous turbulence in the absence of mean velocity gradients. In Proceedings of the 4th European Turbulence Conference, Delft, The Netherlands, 30 June–3 July 1992.
66. Launder, B.E. Phenomenological modelling: Present and future? In Proceedings of the Workshop Held at Cornell University, Ithaca, NY, USA, 22–24 March 1989.
67. Rotta, J.C. Statistische Theorie nichthomogener Turbulenz. *Z. Phys.* **1951**, *129*, 547–572. [CrossRef]
68. Lumley, J.L. Computational modeling of turbulent flows. *Adv. Appl. Mech.* **1979**, *18*, 123–176.
69. Naot, D. Interactions between components of the turbulent velocity correlation tensor. *Isr. J. Technol.* **1970**, *8*, 259–269.
70. Launder, B.E.; Reece, G.J.; Rodi, W. Progress in the development of a Reynolds-stress turbulence closure. *J. Fluid Mech.* **1975**, *68*, 537–566. [CrossRef]
71. Speziale, C.G.; Sarkar, S.; Gatski, T.B. Modelling the pressure-strain correlation of turbulence: An invariant dynamical systems approach. *J. Fluid Mech.* **1991**, *227*, 245–272. [CrossRef]
72. Shih, T.H.; Lumley, J.L. *Modeling of Pressure Correlation Terms in Reynolds Stress and Scalar Flux Equations*; Cornell University: Ithaca, NY, USA, 1985.
73. Craft, T.J.; Launder, B.E. Computation of impinging flows using second-moment closures. In Proceedings of the 8th Symposium on Turbulent Shear Flows, Munich, Germany, 9–11 September 1991; pp. 851–856.
74. Launder, B.E.; Tselepidakis, D.P. Progress and paradoxes in modelling near-wall turbulence. In Proceedings of the 8th Symposium on Turbulent Shear Flows, Munich, Germany, 9–11 September 1991; pp. 2911–2916.
75. Gibson, M.M.; Launder, B.E. Ground effects on pressure fluctuations in the atmospheric boundary layer. *J. Fluid Mech.* **1978**, *86*, 491–511. [CrossRef]
76. Durbin, P.A. A Reynolds stress model for near-wall turbulence. *J. Fluid Mech.* **1993**, *249*, 465–498. [CrossRef]
77. Manceau, R.; Hanjalić, K. Elliptic blending model: A new near-wall Reynolds-stress turbulence closure. *Phys. Fluids* **2002**, *14*, 744–754. [CrossRef]
78. Leonard, A. Energy cascade in large-eddy simulations of turbulent fluid flows. *Adv. Geophys.* **1975**, *18*, 237–248.

79. Nicoud, F.; Ducros, F. Subgrid-scale stress modelling based on the square of the velocity gradient tensor. *Flow Turb. Combust.* **1999**, *62*, 183–200. [CrossRef]
80. Schumann, U. Subgrid scale model for finite difference simulations of turbulent flows in plane channels and annuli. *J. Comput. Phys.* **1975**, *18*, 376–404. [CrossRef]
81. Piomelli, U. High Reynolds number calculations using the dynamic subgrid-scale stress model. *Phys. Fluids* **1993**, *5*, 1484–1490. [CrossRef]
82. Kim, W.W.; Menon, S. A new dynamic one-equation subgrid-scale model for large eddy simulations. In Proceedings of the AIAA 34th Aerospace Sciences Meeting and Exhibit, Reno, NV, USA, 15–18 January 1996.
83. Kajishima, T.; Nomachi, T. One-equation subgrid scale model using dynamic procedure for the energy production. *J. Appl. Mech.* **2006**, *73*, 368–373. [CrossRef]
84. Huang, S.; Li, Q.S. A new dynamic one equation subgrid scale model for large eddy simulations. *Int. J. Numer. Meth. Eng.* **2010**, *81*, 835–865. [CrossRef]
85. Shur, M.L.; Spalart, P.R.; Strelets, M.K.; Travin, A.K. A hybrid RANS-LES approach with delayed-DES and wall-modelled LES capabilities. *Int. J. Heat Fluid Flow* **2008**, *29*, 1638–1649. [CrossRef]
86. Piomelli, U.; Moin, P.; Ferziger, J.H. Model consistency in large eddy simulation of turbulent channel flows. *Phys. Fluids* **1988**, *31*, 884–1891. [CrossRef]
87. Vreman, A.W. An eddy-viscosity subgrid-scale model for turbulent shear flow: Algebraic theory and applications. *Phys. Fluids* **2004**, *16*, 3670–3681. [CrossRef]
88. Verstappen, R.W.C.P.; Bose, S.T.; Lee, J.; Choi, H.; Moin, P. A dynamic eddy-viscosity model based on the invariants of the rate-of-strain. In Proceedings of the Summer Program of the Center for Turbulence Research, Stanford, CA, USA, 2010; pp. 183–192.
89. Verstappen, R.W.C.P. When does eddy viscosity damp subfilter scales sufficiently? *J. Sci. Comput.* **2011**, *49*, 94–110. [CrossRef]
90. Nicoud, F.; Toda, H.B.; Cabrit, O.; Bose, S.; Lee, J. Using singular values to build a subgrid-scale model for large eddy simulations. *Phys. Fluids* **2011**, *23*, 085106. [CrossRef]
91. Trias, F.X.; Folch, D.; Gorobets, A.; Oliva, A. Building proper invariants for eddy-viscosity subgrid-scale models. *Phys. Fluids* **2015**, *27*, 065103. [CrossRef]
92. Park, N.; Lee, S.; Lee, J.; Choi, H. A dynamic subgrid-scale eddy viscosity model with a global model coefficient. *Phys. Fluids* **2006**, *18*, 125109. [CrossRef]
93. You, D.; Moin, P. A dynamic global-coefficient subgrid-scale eddy-viscosity model for large-eddy simulation in complex geometries. *Phys. Fluids* **2007**, *19*, 065110. [CrossRef]
94. Lee, J.; Choi, H.; Park, N. Dynamic global model for large eddy simulation of transient flow. *Phys. Fluids* **2010**, *22*, 075106. [CrossRef]
95. Rozema, W.; Bae, H.J.; Moin, P.; Verstappen, R. Minimum-dissipation models for large-eddy simulation. *Phys. Fluids* **2015**, *27*, 085107. [CrossRef]
96. Rieth, M.; Proch, F.; Stein, O.T.; Pettit, M.W.A.; Kempf, A.M. Comparison of the Sigma and Smagorinsky LES models for grid generated turbulence and a channel flow. *Comput. Fluids* **2014**, *99*, 172–181. [CrossRef]
97. Bardina, J.; Ferziger, J.H.; Reynolds, W.C. Improved subgrid-scale models for large-eddy simulation. In Proceedings of the AIAA 13th Fluid and Plasma Dynamics Coference, Colorado, CO, USA, 14–16 July 1980.
98. Liu, S.; Meneveau, C.; Katz, J. On the properties of similarity subgrid-scale models as deduced from measurements in a turbulent jet. *J. Fluid Mech.* **1994**, *275*, 83–119. [CrossRef]
99. Liu, S.; Meneveau, C.; Katz, J. Experimental study of similarity subgrid-scale models of turbulence in the far-field of a jet. *Appl. Sci. Res.* **1995**, *54*, 177–190. [CrossRef]
100. Akhavan, R.; Ansari, A.; Kang, S.; Mangiavacchi, N. Subgrid-scale interactions in a numerically simulated planar turbulent jet and implications for modelling. *J. Fluid Mech.* **2000**, *408*, 83–120. [CrossRef]
101. Domaradzki, J.A.; Saiki, E.M. A subgrid-scale model based on the estimation of unresolved scales of turbulence. *Phys. Fluids.* **1997**, *9*, 2148–2164. [CrossRef]
102. Domaradzki, J.A.; Loh, K.C. The subgrid-scale estimation model in the physical space representation. *Phys. Fluids* **1999**, *11*, 2330–2342. [CrossRef]
103. Stolz, S.; Adams, N.A. An approximate deconvolution procedure for large-eddy simulation. *Phys. Fluids* **1999**, *11*, 1699–1701. [CrossRef]
104. Stolz, S.; Adams, N.A.; Kleiser, L. An approximate deconvolution model for large-eddy simulation with application to incompressible wall-bounded flows. *Phys. Fluids* **2001**, *13*, 997–1015. [CrossRef]
105. Stolz, S.; Adams, N.A.; Kleiser, L. The approximate deconvolution model for large-eddy simulations of compressible flows and its application to shock-turbulent-boundary-layer interaction. *Phys. Fluids* **2001**, *13*, 2985–3001. [CrossRef]
106. Geurts, B.J.; Holm, D.D. Regularization modeling for large-eddy simulation. *Phys. Fluids* **2003**, *15*, 13–16. [CrossRef]
107. Cheskidov, A.; Holm, D.D.; Olson, E.; Titi, E.S. On a Leray-α model of turbulence. *Proc. R. Soc.* **2005**, *461*, 629–649. [CrossRef]
108. Cao, C.; Holm, D.D.; Titi, E.S. On the Clark-α model of turbulence: Global regularity and long-time dynamics. *J. Turb.* **2005**, *6*, 1–11. [CrossRef]

109. Stolz, S.; Adams, N.A. Large-eddy simulation of high-Reynolds-number supersonic boundary layers using the approximate deconvolution model and a rescaling and recycling technique. *Phys. Fluids* **2003**, *15*, 2398–2412. [CrossRef]
110. Schlatter, P.; Stolz, S.; Kleiser, L. LES of transitional flows using the approximate deconvolution model. *Int. J. Heat Fluid Flow* **2004**, *25*, 549–558. [CrossRef]
111. Saeedipour, M.; Vincent, S.; Pirker, S. Large eddy simulation of turbulent interfacial flows using approximate deconvolution model. *Int. J. Multiph. Flow* **2019**, *112*, 286–299. [CrossRef]
112. Ilyin, A.A.; Lunasin, E.M.; Titi, E.S. A modified-Leray-α subgrid scale model of turbulence. *Nonlinearity* **2006**, *19*, 879–897. [CrossRef]
113. Foias, C.; Holm, D.D.; Titi, E.S. The Navier-stokes-alpha model of fluid turbulence. *Phys. D* **2001**, *152–153*, 505–519. [CrossRef]
114. Rebholz, L.G. A family of new, high order NS-α models arising from helicity correction in Leray turbulence models. *J. Math. Anal. Appl.* **2008**, *342*, 246–254. [CrossRef]
115. Layton, W.; Rebholz, L.; Sussman, M. Energy and helicity dissipation rates of the NS-alpha and NS-alpha-deconvolution models. *J. Appl. Math.* **2010**, *75*, 56–74. [CrossRef]
116. Rebholz, L.G.; Sussman, M.M. On the high accuracy NS-alpha-deconvolution turbulence model. *Math. Models Methods Appl. Sci.* **2010**, *20*, 611–633. [CrossRef]
117. Breckling, S.; Neda, M. Numerical study of the Navier–Stokes-α deconvolution model with pointwise mass conservation. *Int. J. Comput. Math.* **2018**, *95*, 1727–1760. [CrossRef]
118. Cuff, V.M.; Dunca, A.A.; Monica, C.C.; Rebholz, L.G. The reduced order NS-α model for incompressible flow: Theory, numerical analysis and benchmark testing. *Math. Modell. Numer. Anal.* **2015**, *49*, 641–662. [CrossRef]
119. Bowers, A.L.; Rebholz, L.G. The reduced NS-α model for incompressible flow: A review of recent progress. *Fluids* **2017**, *38*, 38. [CrossRef]
120. Geurts, B.J.; Kuczaj, A.K.; Titi, E.S. Regularization modeling for large-eddy simulation of homogeneous isotropic decaying turbulence. *J. Phys. Math. Theor.* **2008**, *41*, 344008. [CrossRef]
121. Graham, J.P.; Holm, D.D.; Mininni, P.D.; Pouquet, A. Three regularization models of the navier-stokes equations. *Phys. Fluids* **2008**, *20*, 035107. [CrossRef]
122. Sheu, T.W.H.; Lin, Y.X.; Yu, C.H. Numerical study of two regularization models for simulating the turbulent flows. *Comput. Fluids* **2013**, *74*, 13–31. [CrossRef]
123. Huai, X.; Joslin, R.D.; Piomelli, U. Large-eddy simulation of transition to turbulence in boundary layers. *Theor. Comput. Fluid Dyn.* **1997**, *9*, 149–163. [CrossRef]
124. Yang, Z.; Voke, P.R. Large-eddy simulation of boundary-layer separation and transition at a change of surface curvature. *J. Fluid Mech.* **2001**, *439*, 305–333. [CrossRef]
125. Sayadi, T.; Moin, P. Large eddy simulation of controlled transition to turbulence. *Phys. Fluids* **2012**, *24*, 114103. [CrossRef]
126. Langari, M.; Yang, Z. Numerical study of the primary instability in a separated boundary layer transition under elevated free-stream turbulence. *Phys. Fluids* **2013**, *25*, 074106. [CrossRef]
127. Mittal, R.; Moin, P. Suitability of upwind-biased finite difference schemes for large-eddy simulation of turbulent flows. *AIAA J.* **1997**, *35*, 1415–1417. [CrossRef]
128. Kravchenko, A.G.; Moin, P. Numerical studies of flow over a circular cylinder at Re_D = 3900. *Phys. Fluids* **2000**, *12*, 403–417. [CrossRef]
129. Dhotre, M.T.; Niceno, B.; Smith, B.L. Large eddy simulation of a bubble column using dynamic sub-grid scale model. *Chem. Eng. J.* **2008**, *136*, 337–348. [CrossRef]
130. Wang, Y.; Vanierschot, M.; Cao, L.; Cheng, Z.; Blanpain, B.; Guo, M. Hydrodynamics study of bubbly flow in a top-submerged lance vessel. *Chem. Eng. Sci.* **2018**, *192*, 1091–1104. [CrossRef]
131. Nilsen, K.M.; Kong, B.; Fox, R.O.; Hill, J.C.; Olsen, M.G. Effect of inlet conditions on the accuracy of large eddy simulations of a turbulent rectangular wake. *Chem. Eng. J.* **2014**, *250*, 175–189. [CrossRef]
132. Spille-Kohoff, A.; Kaltenbach, H.J. Generation of turbulent inflow data with a prescribed shear-stress profile. In Proceedings of the 3rd AFOSR International conference on DNS/LES, Arlington, TX, USA, 5–9 August 2001.
133. Baba-Ahmadi, M.H.; Tabor, G.R. Inlet conditions for LES using mapping and feedback control. *Comput. Fluids* **2009**, *38*, 1299–1311. [CrossRef]
134. Baba-Ahmadi, M.H.; Tabor, G.R. Inlet conditions for large eddy simulation of gas-turbine swirl injectors. *AIAA J.* **2008**, *46*, 1782–1790. [CrossRef]
135. Aider, J.L.; Danet, A.; Lesieur, M. Large-eddy simulation applied to study the influence of upstream conditions on the time-dependant and averaged characteristics of a backward-facing step flow. *J. Turb.* **2007**, *8*, 1–30. [CrossRef]
136. Batten, P.; Goldberg, U.; Chakravarthy, S. Interfacing statistical turbulence closures with large-eddy simulation. *AIAA J.* **2004**, *42*, 485–492. [CrossRef]
137. Andersson, N.; Eriksson, L.E.; Davidson, L. LES prediction of flow and acoustic field of a coaxial jet. In Proceedings of the 11th AIAA/CES Aeroacoustics Conference, Monterey, CA, USA, 23–25 May 2005.
138. Druault, P.; Lardeau, S.; Bonnet, J.P.; Coiffet, F.; Delville, J.; Lamballais, E.; Largeau, J.F.; Perret, L. Generation of three-dimensional turbulent inlet conditions for large-eddy simulation. *AIAA J.* **2004**, *42*, 447–456. [CrossRef]

139. Perret, L.; Delville, J.; Manceau, R.; Bonnet, J.P. Turbulent inflow conditions for large-eddy simulation based on low-order empirical model. *Phys. Fluids* **2008**, *20*, 075107. [CrossRef]
140. Klein, M.; Sadiki, A.; Janicka, J. A digital filter based generation of inflow data for spatially developing direct numerical or large eddy simulations. *J. Comput. Phys.* **2003**, *186*, 652–665. [CrossRef]
141. Di Mare, L.; Klein, M.; Jones, W.P.; Janicka, J. Synthetic turbulence inflow conditions for large-eddy simulation. *Phys. Fluids* **2006**, *18*, 025107. [CrossRef]
142. Jarrin, N.; Benhamadouche, S.; Laurence, D.; Prosser, R. A synthetic-eddy-method for generating inflow conditions for large-eddy simulations. *Int. J. Heat Fluid Flow* **2006**, *27*, 585–593. [CrossRef]
143. Jarrin, N.; Prosser, R.; Uribe, J.C.; Benhamadouche, S.; Laurence, D. Reconstruction of turbulent fluctuations for hybrid RANS/LES simulations using a synthetic-eddy method. *Int. J. Heat Fluid Flow* **2009**, *30*, 435–442. [CrossRef]
144. Skillen, A.; Revell, A.; Craft, T. Accuracy and efficiency improvements in synthetic eddy methods. *Int. J. Heat Fluid Flow* **2016**, *62*, 386–394. [CrossRef]
145. Keating, A.; Piomelli, U.; Balaras, E.; Kaltenbach, H.J. A priori and a posteriori tests of inflow conditions for large-eddy simulation. *Phys. Fluids* **2004**, *16*, 4696–4712. [CrossRef]
146. Benhamadouche, S.; Jarrin, N.; Addad, Y.; Laurence, D. Synthetic turbulent inflow conditions based on a vortex method for large-eddy simulation. *Prog. Comput. Fluid Dyn.* **2006**, *6*, 50–57. [CrossRef]
147. Kaltenbach, H.J.; Fatica, M.; Mittal, R.; Lund, T.S.; Moin, P. Study of flow in a planar asymmetric diffuser using large-eddy simulation. *J. Fluid Mech.* **1999**, *390*, 151–185. [CrossRef]
148. Kasagi, N.; Shikazono, N. Contribution of direct numerical simulation to understanding and modelling turbulent transport. *Proc. R. Soc. Lond.* **1995**, *451*, 257–292.
149. Kim, J.W.; Sandberg, R.D. Efficient parallel computing with a compact finite difference scheme. *Comput. Fluids* **2012**, *58*, 70–87. [CrossRef]
150. Alfonsi, G.; Ciliberti, S.A.; Mancini, M.; Primavera, L. Performances of navier-stokes solver on a hybrid cpu/gpu computing system. In Proceedings of the International Conference on Parallel Computing Techonologies, Berlin/Heidelberg, Germany, 19–23 September 2011; pp. 404–416.
151. Alfonsi, G.; Ciliberti, S.A.; Mancini, M.; Primavera, L. Direct numerical simulation of turbulent channel flow on high-performance GPU computing system. *Computation* **2016**, *4*, 13. [CrossRef]
152. Alfonsi, G.; Ciliberti, S.A.; Mancini, M.; Primavera, L. GPGPU implementation of mixed spectral-finite difference computational code for the numerical integration of the three-dimensional time-dependent incompressible Navier–Stokes equations. *Comput. Fluids* **2014**, *102*, 237–249. [CrossRef]
153. Laizet, S.; Lamballais, E.; Vassilicos, J.C. A numerical strategy to combine high-order schemes, complex geometry and parallel computing for high resolution DNS of fractal generated turbulence. *Comput. Fluids* **2010**, *39*, 471–484. [CrossRef]
154. Alfonsi, G. On direct numerical simulation of turbulent flows. *Appl. Mech. Rev.* **2011**, *64*, 020802. [CrossRef]
155. Feng, J.; Bolotnov, I.A. Evaluation of bubble-induced turbulence using direct numerical simulation. *Int. J. Multiph. Flow* **2017**, *93*, 92–107. [CrossRef]
156. Abe, H.; Kawamura, H.; Matsuo, Y. Direct numerical simulation of a fully developed turbulent channel flow with respect to the Reynolds number dependence. *J. Fluids Eng.* **2001**, *123*, 382–393. [CrossRef]
157. Pirozzoli, S.; Grasso, F.; Gatski, T.B. Direct numerical simulation and analysis of a spatially evolving supersonic turbulent boundary layer at M = 2.25. *Phys. Fluids* **2004**, *16*, 530–545. [CrossRef]
158. Ma, M.; Lu, J.; Tryggvason, G. Using statistical learning to close two-fluid multiphase flow equations for a simple bubbly system. *Phys. Fluids* **2015**, *27*, 092101. [CrossRef]
159. Fang, J.; Cambareri, J.J.; Brown, C.S.; Feng, J.; Gouws, A.; Li, M.; Bolotnov, I.A. Direct numerical simulation of reactor two-phase flows enabled by high-performance computing. *Nucl. Eng. Des.* **2018**, *330*, 409–419. [CrossRef]
160. Singhal, A.; Cloete, S.; Radl, S.; Quinta-Ferreira, R.; Amini, S. Heat transfer to a gas from densely packed beds of monodisperse spherical particles. *Chem. Eng. J.* **2017**, *314*, 27–37. [CrossRef]
161. Feng, X.; Yang, C.; Mao, Z.S.; Lu, J.; Tryggvason, G. Bubble induced turbulence model improved by direct numerical simulation of bubbly flow. *Chem. Eng. J.* **2019**, *377*, 120001. [CrossRef]
162. Feng, J.; Bolotnov, I.A. Interfacial force study on a single bubble in laminar and turbulent flows. *Nucl. Eng. Des.* **2017**, *313*, 345–360. [CrossRef]
163. Suzuki, Y.; Kasagi, N. Evaluation of hot-wire measurements in wall shear turbulence using a direct numerical simulation database. *Exp. Therm. Fluid Sci.* **1992**, *5*, 69–77. [CrossRef]
164. Tang, Y.; Lau, Y.M.; Deen, N.G.; Peters, E.A.J.F.; Kuipers, J.A.M. Direct numerical simulations and experiments of a pseudo-2D gas-fluidized bed. *Chem. Eng. Sci.* **2016**, *143*, 166–180. [CrossRef]
165. Werther, J.; Hage, B.; Rudnick, C. A comparison of laser Doppler and single-fiber reflection probes for the measurement of the velocity of solids in a gas-solid circulating fluidized bed. *Chem. Eng. Process.* **1996**, *35*, 381–391. [CrossRef]
166. Zhang, M.; Qian, Z.; Yu, H.; Wei, F. The solid flow structure in a circulating fluidized bed riser/downer of 0.42-m diameter. *Powder Technol.* **2003**, *129*, 46–52. [CrossRef]
167. Link, J.; Zeilstra, C.; Deen, N.G.; Kuipers, J.A.M. Validation of a discrete particle model in a 2D spout-fluid bed using non intrusive optical measuring techniques. *Can. J. Chem. Eng.* **2004**, *82*, 30–36. [CrossRef]

168. Cao, L.; Wang, Y.; Liu, Q.; Feng, X. Physical and mathematical modeling of multiphase flows in a converter. *ISIJ Int.* **2018**, *58*, 573–584. [CrossRef]
169. Iguchi, M.; Nakamura, K.-I.; Tsujino, R. Mixing time and fluid flow phenomena in liquids of varying kinematic viscosities agitated by bottom gas injection. *Metall. Mater. Trans.* **1998**, *29*, 569–575. [CrossRef]
170. Cao, L.; Liu, Q.; Wang, Z.; Li, N. Interaction behaviour between top blown jet and molten steel during BOF steelmaking process. *Ironmak. Steelmak.* **2018**, *45*, 239–248. [CrossRef]
171. Zhang, M.; Qian, Z.; Yu, H.; Wei, F. The near wall dense ring in a large-scale down-flow circulating fluidized bed. *Chem. Eng. J.* **2003**, *92*, 161–167. [CrossRef]
172. Liu, J.; Grace, J.R.; Bi, X. Novel multifunctional optical-fiber probe: Development and validation. *AIChE J.* **2003**, *49*, 1405–1420. [CrossRef]
173. Liu, J.; Grace, J.R.; Bi, X. Novel multifunctional optical-fiber probe: High-density CFB measurements. *AIChE J.* **2003**, *49*, 1421–1432. [CrossRef]
174. Ellis, N.; Bi, H.T.; Lim, C.J.; Grace, J.R. Hydrodynamics of turbulent fluidized beds of different diameters. *Powder Technol.* **2004**, *141*, 124–136. [CrossRef]
175. Almuttahar, A.; Taghipour, F. Computational fluid dynamics of high density circulating fluidized bed riser: Study of modeling parameters. *Powder Technol.* **2008**, *185*, 11–23. [CrossRef]
176. Li, D.; Ray, M.B.; Ray, A.K.; Zhu, J. A comparative study on hydrodynamics of circulating fluidized bed riser and downer. *Powder Technol.* **2013**, *247*, 235–259. [CrossRef]
177. Tsuji, Y.; Morikawa, Y. LDV measurements of an air-solid two-phase flow in a horizontal pipe. *J. Fluid Mech.* **1982**, *120*, 385–409. [CrossRef]
178. Laverman, J.A.; Roghair, I.; Annaland, M.v.; Kuipers, J.A.M. Investigation into the hydrodynamics of gas-solid fluidized beds using particle image velocimetry coupled with digital image analysis. *Can. J. Chem. Eng.* **2008**, *86*, 523–535. [CrossRef]
179. Van Buijtenen, M.S.; Börner, M.; Deen, N.G.; Heinrich, S.; Antonyuk, S.; Kuipers, J.A.M. An experimental study of the effect of collision properties on spout fluidized bed dynamics. *Powder Technol.* **2011**, *206*, 139–148. [CrossRef]
180. Se Jong, J.F.; Odu, S.O.; Van Buijtenen, M.S.; Deen, N.G.; Annaland, M.V.; Kuipers, J.A.M. Development and validation of a novel digital image analysis method for fluidized bed particle image velocimetry. *Powder Technol.* **2012**, *230*, 193–202. [CrossRef]
181. Dang, T.Y.N.; Kolkman, T.; Gallucci, F.; Annaland, M.V. Development of a novel infrared technique for instantaneous, whole-field, non invasive gas concentration measurements in gas–solid fluidized beds. *Chem. Eng. J.* **2013**, *219*, 545–557. [CrossRef]
182. Patil, A.V.; Peters, E.A.J.F.; Sutkar, V.S.; Deen, N.G.; Kuipers, J.A.M. A study of heat transfer in fluidized beds using an integrated DIA/PIV/IR technique. *Chem. Eng. J.* **2015**, *259*, 90–106. [CrossRef]
183. Li, Z.; Janssen, T.C.E.; Buist, K.A.; Deen, N.G.; Annaland, M.V.; Kuipers, J.A.M. Experimental and simulation study of heat transfer in fluidized beds with heat production. *Chem. Eng. J.* **2017**, *317*, 242–257. [CrossRef]
184. Becker, S.; Sokolichin, A.; Eigenberger, G. Gas-liquid flow in bubble columns and loop reactors: Comparison of detailed experiments and flow simulations. *Chem. Eng. Sci.* **1994**, *49*, 5747–5762. [CrossRef]
185. Mudde, R.F.; Groen, J.S.; Van den Akker, H.E.A. Liquid velocity field in a bubble column: LDA experiments. *Chem. Eng. Sci.* **1997**, *52*, 4217–4224. [CrossRef]
186. Iguchi, M.; Kondoh, T.; Nakajima, K. Establishment time of liquid flow in a bath agitated by bottom gas injection. *Metall. Mater. Trans.* **1997**, *28B*, 605–612. [CrossRef]
187. Pfleger, D.; Gomes, S.; Gilbert, N.; Wagner, H.-G. Hydrodynamic simulations of laboratory scale bubble columns fundamental studies of the Eulerian-Eulerian modelling approach. *Chem. Eng. Sci.* **1999**, *54*, 5091–5099. [CrossRef]
188. Kulkarnia, A.A.; Joshia, J.B.; Kumarb, V.R.; Kulkarnib, B.D. Application of multiresolution analysis for simultaneous measurement of gas and liquid velocities and fractional gas hold-up in bubble column using LDA. *Chem. Eng. Sci.* **2001**, *56*, 5037–5048. [CrossRef]
189. Mudde, R.F.; Lee, D.J.; Reese, J.; Fan, L.-S. Role of coherent structures on Reynolds stresses in a 2-D bubble column. *AIChE J.* **1997**, *43*, 913–926. [CrossRef]
190. Delnoij, E.; Kuipers, J.A.M.; van Swaaij, W.P.M.; Westerweel, J. Measurement of gas-liquid two-phase flow in bubble columns using ensemble correlation PIV. *Chem. Eng. Sci.* **2000**, *55*, 3385–3395. [CrossRef]
191. Deen, N.G.; Solberg, T.; Hjertager, B.H. Large eddy simulation of the gas-liquid flow in a square cross-sectioned bubble column. *Chem. Eng. Sci.* **2001**, *56*, 6341–6349. [CrossRef]
192. Ayati, A.A.; Kolaas, J.; Jensen, A.; Johnson, G.W. A PIV investigation of stratified gas–liquid flow in a horizontal pipe. *Int. J. Multiph. Flow* **2014**, *61*, 129–143. [CrossRef]
193. Rzehaka, R.; Krauß, M.; Kováts, P.; Zähringer, K. Fluid dynamics in a bubble column: New experiments and simulations. *Int. J. Multiph. Flow* **2017**, *89*, 299–312. [CrossRef]
194. Singh, R.P.; Ghosh, D.N. Cold model study of mixing and mass transfer in LBE process of steel-making. *ISIJ Int.* **1990**, *30*, 955–960. [CrossRef]
195. Conejo, A.N.; Kitamura, S.; Maruoka, N.; Kim, S.-J. Effects of top layer, nozzle arrangement, and gas flow rate on mixing time in agitated ladles by bottom gas injection. *Metall. Mater. Trans.* **2013**, *44B*, 914–923. [CrossRef]
196. Zhou, X.; Ersson, M.; Zhong, L.; Jönsson, P. Optimization of combined blown converter process. *ISIJ Int.* **2014**, *54*, 2255–2262. [CrossRef]

197. Nordquist, A.; Kumbhat, N.; Jenssen, L.; Jönsson, P. The effect of nozzle diameter, lance height and flow rate on penetration depth in a top blown water model. *Steel Res. Int.* **2006**, *77*, 82–90. [CrossRef]
198. Hwang, H.; Irons, G.A. A water model study of impinging gas jets on liquid surfaces. *Metall. Mater. Trans.* **2012**, *43B*, 302–315. [CrossRef]
199. Zhou, X.; Ersson, M.; Zhong, L.; Yu, J.; Jönsson, P. Mathematical and physical simulation of a top blown converter. *Steel Res. Int.* **2014**, *85*, 273–281. [CrossRef]
200. Sabah, S.; Brooks, G. Splashing in oxygen steelmaking. *ISIJ Int.* **2014**, *54*, 836–844. [CrossRef]
201. Li, M.; Li, Q.; Kuang, S.; Zou, Z. Determination of cavity dimensions induced by impingement of gas jets onto a liquid bath. *Metall. Mater. Trans.* **2016**, *47B*, 116–126. [CrossRef]
202. Sahai, Y.; Emi, T. Criteria for water modeling of melt flow and inclusion removal in continuous casting tundishes. *ISIJ Int.* **1996**, *36*, 1166–1173. [CrossRef]
203. Timmel, K.; Eckert, S.; Gerbeth, G.; Stefani, F.; Wondrak, T. Experimental modeling of the continuous casting process of steel using low melting point metal alloys, the limmcast program. *ISIJ Int.* **2010**, *50*, 1134–1141. [CrossRef]
204. Lopez, P.E.R.; Jalali, P.N.; Björkvall, J.; Sjöström, U.; Nilsson, C. Recent developments of a numerical model for continuous casting of steel: Model theory, setup and comparison to physical modelling with liquid metal. *ISIJ Int.* **2014**, *54*, 342–350. [CrossRef]
205. Eckert, S.; Gerbeth, G.; Melnikov, V.I. Velocity measurements at high temperatures by ultrasound Doppler velocimetry using an acoustic wave guide. *Exp. Fluids* **2003**, *35*, 381–388. [CrossRef]
206. Stefani, F.; Gerbeth, G. A contactless method for velocity reconstruction in electrically conducting fluids. *Meas. Sci. Technol.* **2000**, *11*, 758–765. [CrossRef]
207. Stefani, F.; Gundrum, T.; Gerbeth, G. Contactless inductive flow tomography. *Phys. Rev.* **2004**, *70*, 056306. [CrossRef]
208. Kulick, J.D.; Fessler, J.R.; Eaton, J.K. Particle response and turbulence modification in fully developed channel flow. *J. Fluid Mech.* **1994**, *277*, 109–134. [CrossRef]
209. Kiger, K.T.; Pan, C. PIV technique for the simultaneous measurement of dilute two-phase flows. *J. Fluids Eng.* **2000**, *122*, 811–818. [CrossRef]
210. Khalitov, D.A.; Longmire, E.K. Simultaneous two-phase PIV by two-parameter phase discrimination. *Exp. Fluids* **2002**, *32*, 252–268. [CrossRef]
211. Longmire, E.K.; Norman, T.L.; Gefroh, D.L. Dynamics of pinch-off in liquid/liquid jets with surface tension. *Int. J. Multiph. Flow* **2001**, *27*, 1735–1752. [CrossRef]
212. Saarenrinne, P.; Piirto, M. Turbulent kinetic energy dissipation rate estimation from PIV velocity vector fields. *Exp. Fluids Suppl.* **2000**, *29*, S300–S307. [CrossRef]
213. Tanaka, T.; Eaton, J.K. A correction method for measuring turbulence kinetic energy dissipation rate by PIV. *Exp. Fluids* **2007**, *42*, 893–902. [CrossRef]
214. Tanaka, T.; Eaton, J.K. Sub-Kolmogorov resolution partical image velocimetry measurements of particle-laden forced turbulence. *J. Fluid Mech.* **2010**, *643*, 177–206. [CrossRef]
215. Hwang, W.; Eaton, J.K. Homogeneous and isotropic turbulence modulation by small heavy (St$\sim$50) particles. *J. Fluid Mech.* **2006**, *564*, 361–393. [CrossRef]
216. Wang, Y.; Cao, L.; Vanierschot, M.; Cheng, Z.; Blanpain, B.; Guo, M. Modelling of gas injection into a viscous liquid through a top-submerged lance. *Chem. Eng. Sci.* **2020**, *212*, 115359. [CrossRef]
217. Boyer, C.; Duquenne, A.-M.; Wild, G. Measuring techniques in gas–liquid and gas-liquid-solid reactors. *Chem. Eng. Sci.* **2002**, *57*, 3185–3215. [CrossRef]
218. Bai, B.; Gheorghiu, S.; van Ommen, J.R.; Nijenhuis, J.; Coppens, M.-O. Characterization of the void size distribution in fluidized beds using statistics of pressure fluctuations. *Powder Technol.* **2005**, *160*, 81–92. [CrossRef]
219. Sato, Y.; Sadatomi, M.; Sekoguchi, K. Momentum and heat transfer in two-phase bubble flow—Theory. *Int. J. Multiph. Flow* **1981**, *7*, 167–177. [CrossRef]
220. Liu, Z.; Li, B.; Jiang, M.; Tsukihashi, F. Euler-Euler-Lagrangian modeling for two-phase flow and particle transport in continuous casting mold. *ISIJ Int.* **2014**, *54*, 1314–1323. [CrossRef]
221. Pfleger, D.; Becker, S. Modelling and simulation of the dynamic flow behaviour in a bubble column. *Chem. Eng. Sci.* **2001**, *56*, 1737–1747. [CrossRef]
222. Troshko, A.A.; Hassan, Y.A. A two-equation turbulence model of turbulent bubbly flows. *Int. J. Multiph. Flow* **2001**, *27*, 1965–2000. [CrossRef]
223. Politano, M.S.; Carrica, P.M.; Converti, J. A model for turbulent polydisperse two-phase flow in vertical channels. *Int. J. Multiph. Flow* **2003**, *29*, 1153–1182. [CrossRef]
224. Rzehak, R.; Krepper, E. CFD modeling of bubble-induced turbulence. *Int. J. Multiph. Flow* **2013**, *55*, 138–155. [CrossRef]
225. Colombo, M.; Fairweather, M. Multiphase turbulence in bubbly flows: RANS simulations. *Int. J. Multiph. Flow* **2015**, *77*, 222–243. [CrossRef]
226. de Bertodano, M.L.; Lee, S.-J.; Lahey, R.T., Jr.; Drew, D.A. The prediction of two-phase turbulence and phase distribution phenomena using a Reynolds stress model. *J. Fluids Eng.* **1990**, *112*, 107–113. [CrossRef]
227. Chahed, J.; Roig, V.; Masbernat, L. Eulerian-Eulerian two-fluid model for turbulent gas-liquid bubbly flows. *Int. J. Multiph. Flow* **2003**, *29*, 23–49. [CrossRef]

228. Parekh, J.; Rzehak, R. Euler-Euler multiphase CFD-simulation with full Reynolds stress model and anisotropic bubble-induced turbulence. *Int. J. Multiph. Flow* **2018**, *99*, 231–245. [CrossRef]
229. Niceno, B.; Dhotre, M.T.; Deen, N.G. One-equation sub-grid scale (SGS) modelling for Euler-Euler large eddy simulation (EELES) of dispersed bubbly flow. *Chem. Eng. Sci.* **2008**, *63*, 3923–3931. [CrossRef]
230. Ma, T.; Santarelli, C.; Ziegenhein, T.; Lucas, D.; Fröhlich, J. Direct numerical simulation-based Reynolds-averaged closure for bubble-induced turbulence. *Phys. Rev. Fluids* **2017**, *2*, 034301. [CrossRef]
231. Ma, T.; Lucas, D.; Jakirlić, S.; Fröhlich, J. Progress in the second-moment closure for bubbly flow based on direct numerical simulation data. *J. Fluid Mech.* **2020**, *883*, A9. [CrossRef]
232. Liao, L.; Ma, T.; Krepper, E.; Lucas, D.; Fröhlich, J. Application of a novel model for bubble-induced turbulence to bubbly flows in containers and vertical pipes. *Chem. Eng. Sci.* **2019**, *202*, 55–69. [CrossRef]
233. Akbar, M.H.M.; Hayashi, K.; Hosokawa, S.; Tomiyama, A. Bubble tracking simulation of bubble-induced pseudoturbulence. *Multi. Sci. Techn.* **2012**, *24*, 197–222.
234. Cao, L.; Liu, Q.; Sun, J.; Lin, W.; Feng, X. Effect of slag layer on the multiphase interaction in a converter. *JOM* **2019**, *71*, 754–763. [CrossRef]
235. Cao, L.; Wang, Y.; Liu, Q.; Sun, L.; Liao, S.; Guo, W.; Ren, K.; Blanpain, B.; Guo, M. Assessment of gas-slag-metal interaction during a converter steelmaking process. In Proceedings of the 9th International Symposium on High-Temperature Metallurgical Processing, Phoenix, AZ, USA, 11–15 March 2018; Springer: New York, NY, USA, 2018; pp. 353–364.
236. Subagyo; Brooks, G.A.; Coley, K.S.; Irons, G.A. Generation of droplets in slag-metal emulsions through top gas blowing. *ISIJ Int.* **2003**, *43*, 983–989. [CrossRef]
237. Luomala, M.J.; Fabritius, T.M.J.; Virtanen, E.O.; Siivola, T.P.; Härkki, J.J. Splashing and spitting behaviour in the combined blown steelmaking converter. *ISIJ Int.* **2002**, *42*, 944–949. [CrossRef]
238. Pope, S.B. An explanation of the turbulent round-jet/plane-jet anomaly. *AIAA J.* **1978**, *16*, 279–281. [CrossRef]
239. Doh, Y.; Chapelle, P.; Jardy, A.; Djambazov, G.; Pericleous, K.; Ghazal, G.; Gardin, P. Toward a full simulation of the basic oxygen furnace: Deformation of the bath free surface and coupled transfer processes associated with the post-combustion in the gas region. *Metall. Mater. Trans.* **2013**, *44B*, 653–670. [CrossRef]
240. Sarkar, S.; Erlebacher, G.; Hussaini, M.Y.; Kreiss, H.O. The analysis and modelling of dilatational terms in compressible turbulence. *J. Fluid Mech.* **1991**, *227*, 473–493. [CrossRef]
241. Sarkar, S. The stabilizing effect of compressibility in turbulent shear flow. *J. Fluid Mech.* **1995**, *282*, 163–186. [CrossRef]
242. Heinz, S. A model for the reduction of the turbulent energy redistribution by compressibility. *Phys. Fluids* **2003**, *15*, 3580–3583. [CrossRef]
243. Alam, M.; Naser, J.; Brooks, G. Computational fluid dynamics simulation of supersonic oxygen jet behavior at steelmaking temperature. *Metall. Mater. Trans.* **2010**, *41B*, 636–645. [CrossRef]
244. Tam, C.K.W.; Ganesan, A. A modified k-ε model for calculating the mean flow and noise of hot jets. *AIAA* **2003**, *1064*, 2003.
245. Abe, K.; Kondoh, T.; Nagano, Y. A two-equation heat transfer model reflecting second-moment closures for wall and free turbulent flows. *Int. J. Heat Fluid Flow* **1996**, *17*, 228–237. [CrossRef]
246. Ronki, M.; Gatski, T.B. Predicting turbulent convective heat transfer in fully developed duct flows. *Int. J. Heat Fluid Flow* **2001**, *22*, 381–392.
247. Abdol-Hamid, K.S.; Pao, S.P.; Massey, S.J.; Elmiligui, A. Temperature corrected turbulence model for high temperature jet flow. *J. Fluids Eng.* **2004**, *126*, 844–850. [CrossRef]
248. Wang, H.; Zhu, R.; Gu, Y.L.; Wang, C.J. Behaviours of supersonic oxygen jet injected from four-hole lance during top-blown converter steelmaking process. *Canad. Metall. Quart.* **2014**, *53*, 367–380. [CrossRef]
249. Sumi, I.; Kishimoto, Y.; Kikuchi, Y.; Igarashi, H. Effect of high-temperature field on supersonic oxygen jet behavior. *ISIJ Int.* **2006**, *46*, 1312–1317. [CrossRef]
250. Lebon, G.S.B.; Patel, M.K.; Djambazov, G.; Pericleous, K.A. Mathematical modelling of a compressible oxygen jet entering a hot environment using a pressure-based finite volume code. *Comput. Fluids* **2012**, *59*, 91–100. [CrossRef]
251. Bodony, D.J.; Lele, S.K. On using large-eddy simulation for the prediction of noise from cold and heated turbulent jets. *Phys. Fluids* **2005**, *17*, 085103. [CrossRef]
252. Wang, M.; Freund, J.B.; Lele, S.K. Computational prediction of flow-generated sound. *Annu. Rev. Fluid Mech.* **2006**, *38*, 483–512. [CrossRef]
253. Bodony, D.J.; Lele, S.K. Current status of jet noise predictions using large-eddy simulation. *AIAA J.* **2008**, *46*, 364–380. [CrossRef]
254. Takatani, K. Effects of electromagnetic brake and meniscus electromagnetic stirrer on transient molten steel flow at meniscus in a continuous casting mold. *ISIJ Int.* **2003**, *43*, 915–922. [CrossRef]
255. Qian, Z.; Wu, Y. Large eddy simulation of turbulent flow with the effects of DC magnetic field and vortex brake application in continuous casting. *ISIJ Int.* **2004**, *44*, 100–107. [CrossRef]
256. Wang, S.; de Toledo, G.A.; Välimaa, K.; Louhenkilpi, S. Magnetohydrodynamic phenomena, fluid control and computational modeling in the continuous casting of billet and bloom. *ISIJ Int.* **2014**, *54*, 2273–2282. [CrossRef]
257. Chaudhary, R.; Thomas, B.G.; Vanka, S.P. Effect of electromagnetic ruler braking (EMBr) on transient turbulent flow in continuous slab casting using large eddy simulations. *Metall. Mater. Trans.* **2012**, *43B*, 532–553. [CrossRef]

258. Vakhrushev, A.; Kharicha, A.; Liu, Z.; Wu, M.; Ludwig, A.; Nitzl, G.; Tang, Y.; Hackl, G.; Watzinger, J. Electric current distribution during electromagnetic braking in continuous casting. *Metall. Mater. Trans.* **2020**, *51*, 2811–2828. [CrossRef]
259. Ji, H.-C.; Gardner, R.A. Numerical analysis of turbulent pipe flow in a transverse magnetic field. *Int. J. Heat Mass Transf.* **1997**, *40*, 1839–1851. [CrossRef]
260. Kenjereš, S.; Hanjalić, K. On the implementation of effects of Lorentz force in turbulence closure models. *Int. J. Heat Fluid Flow* **2000**, *21*, 329–337. [CrossRef]
261. Widlund, O.; Zahrai, S.; Bark, F.H. Development of a Reynolds stress closure for modeling of homogeneous MHD turbulence. *Phys. Fluids* **1998**, *10*, 1987–1996. [CrossRef]
262. Shimomura, Y. Large eddy simulation of magnetohydrodynamic turbulent channel flows under a uniform magnetic field. *Phys. Fluids* **1991**, *3*, 3098–3106. [CrossRef]
263. Kobayashi, H. Large eddy simulation of magnetohydrodynamic turbulent channel flows with local subgrid-scale model based on coherent structures. *Phys. Fluids* **2006**, *18*, 045107. [CrossRef]
264. Miao, X.; Timmel, K.; Lucas, D.; Ren, Z.; Eckert, S.; Gerbeth, G. Effect of an electromagnetic brake on the turbulent melt flow in a continuous-casting mold. *Metall. Mater. Trans.* **2012**, *43*, 954–972. [CrossRef]
265. Singh, R.; Thomas, B.G.; Vanka, S.P. Effects of a magnetic field on turbulent flow in the mold region of a steel caster. *Metall. Mater. Trans.* **2013**, *44*, 1201–1221. [CrossRef]
266. Singh, R.; Thomas, B.G.; Vanka, S.P. Large eddy simulations of double-ruler electromagnetic field effect on transient flow during continuous casting. *Metall. Mater. Trans.* **2014**, *45B*, 1098–1115. [CrossRef]
267. Chaudhary, R.; Vanka, S.P.; Thomas, B.G. Direct numerical simulations of magnetic field effects on turbulent flow in a square duct. *Phys. Fluids* **2010**, *22*, 075102. [CrossRef]
268. Chaudhary, R.; Shinn, A.F.; Vanka, S.P.; Thomas, B.G. Direct numerical simulations of transverse and spanwise magnetic field effects on turbulent flow in a 2:1 aspect ratio rectangular duct. *Comput. Fluids* **2011**, *51*, 100–114. [CrossRef]

metals

MDPI

Review

A Review of Bubble Dynamics in Liquid Metals

Tim Haas *, Christian Schubert, Moritz Eickhoff and Herbert Pfeifer

Department for Industrial Furnaces and Heat Engineering, RWTH Aachen University, Kopernikusstraße 10, D-52074 Aachen, Germany; schubert@iob.rwth-aachen.de (C.S.); eickhoff@iob.rwth-aachen.de (M.E.); pfeifer@iob.rwth-aachen.de (H.P.)
* Correspondence: haas@iob.rwth-aachen.de

Abstract: Gas bubbles are of major importance in most metallurgical processes. They promote chemical reactions, homogenize the melt, or float inclusions. Thus, their dynamics are of crucial interest for the optimization of metallurgical processes. In this work, the state of knowledge of bubble dynamics at the bubble scale in liquid metals is reviewed. Measurement methods, with emphasis on liquid metals, are presented, and difficulties and shortcomings are analyzed. The bubble formation mechanism at nozzles and purging plugs is discussed. The uncertainty regarding the prediction of the bubble size distribution in real processes is demonstrated using the example of the steel casting ladle. Finally, the state of knowledge on bubble deformation and interfacial forces is summarized and the scalability of existing correlations to liquid metals is critically discussed. It is shown that the dynamics of bubbles, especially in liquid metals, are far from understood. While the drag force can be predicted reasonably well, there are large uncertainties regarding the bubble size distribution, deformation, and lift force. In particular, the influence of contaminants, which cannot yet be quantified in real processes, complicates the discussion and the comparability of experimental measurements. Further open questions are discussed and possible solutions are proposed.

Keywords: liquid metals; bubble generation; bubble size distribution; porous plugs; bubble deformation; drag force; lift force

Citation: Haas, T.; Schubert, C.; Eickhoff, M.; Pfeifer, H. A Review of Bubble Dynamics in Liquid Metals. *Metals* **2021**, *11*, 664. https://doi.org/10.3390/met11040664

Academic Editors: Fernando Castro and Gunter Gerbeth

Received: 19 March 2021
Accepted: 14 April 2021
Published: 19 April 2021

1. Introduction

Gas injection and stirring is an integral part of most metallurgical processes. In some cases, the gas even determines metallurgical tasks. Examples are the removal of hydrogen in the degassing of aluminum or the removal of non-metallic inclusions from steel in the ladle or tundish. Thus, metallurgical bubble column reactors are of major importance for the process industry. Their main advantages are a simple construction, low maintenance costs, good mass, and heat transfer [1] and applicability in case mechanic stirring is prohibited by high temperature or reactive flows.

During the buoyancy-driven rise of bubbles, momentum is exchanged between the fluid and bubbles, which promotes a stirring effect. Characteristic of bubbles are their non-rigid phase boundaries. Thus, bubbles react to pressure gradients with a deformation. This leads to fascinating phenomena such as shape and path oscillation. These were already described by Leonardo DaVinci, which is why it is sometimes referred to as Leonardo's paradox [2]. On the other hand, this additional degree of freedom also leads to a very high complexity of the interaction between bubbles and fluid. Therefore, despite many years of extensive research, a comprehensive description of all phenomena is not yet possible. However, fundamental understanding of the bubble properties, shapes, rising velocities, or interfacial forces is crucial for the understanding, modelling, and optimization of bubble column reactors [3]. In metallurgy, the optimization of processes is often carried out using computational fluid dynamics (CFD). However, gaps in the knowledge of the behavior of gas bubbles in metals lead to considerable model uncertainties.

Bubble reactors are a multi-scale problem. It can be roughly divided into four different scales, though not all phenomena can be sharply assigned to one scale. On the microscale, the mass and heat transfer between fluid and bubble takes place. The bubble scale determines the formation mechanism of bubbles, bubble deformation, and interfacial forces. The swarm scale deals with the interaction of bubbles, as well as the characterization of the bubble column. At the macro or reactor scale, the stirring effect generated by the bubbles is considered. In order to limit the length of this paper, this review focuses on the bubble scale, largely omitting the other scales.

Different dimensionless numbers are used to describe bubble dynamics or bubble generation. The most important dimensionless numbers are the Reynolds number (Re), Eötvös number (Eo), Morton number (Mo), Weber number (We), and Froude number (Fr):

$$\begin{aligned} \mathrm{Re} &= \frac{\text{Inertial force}}{\text{Viscous force}} = \frac{u_C L_C \rho_l}{\mu_l} \\ \mathrm{Eo} &= \frac{\text{Buoyancy force}}{\text{Surface tension force}} = \frac{g(\rho_l - \rho_b) L_C^2}{\sigma} \\ \mathrm{Mo} &= \frac{g \mu_l^4 (\rho_l - \rho_b)}{\rho_l^2 \sigma^3} \\ \mathrm{We} &= \frac{\text{Inertial force}}{\text{Surface tension force}} = \frac{\rho_l u_C^2 L_C}{\sigma} \\ \mathrm{Fr} &= \frac{\text{Inertial force}}{\text{Gravity force}} = \frac{u_C^2}{g L_C} \end{aligned} \tag{1}$$

where u_C and L_C are the characteristic velocity and length, respectively. For this, different values are employed, though for bubbly flows, the bubble size, expressed as the diameter of a sphere with the same volume and the relative velocity between bubble and liquid, is commonly utilized. If other quantities are used, this is explicitly mentioned in the text.

In the following, different analytical methods and measurement techniques are discussed, and specific advantages, disadvantages, and restrictions are analyzed. Afterwards, different bubble scale phenomena are discussed. This review covers the bubble formation mechanism at nozzles and purging plugs, bubble deformation, and finally the interfacial forces between bubbles and fluid. Due to the small number of studies in liquid metals, conclusions drawn from aqueous systems are summarized first. In the second step, their transferability to liquid metals is discussed. This allows guidelines to be derived as to which models should currently be used for CFD, and the model uncertainty can be analyzed in more detail. Finally, open questions are identified and possible solution strategies are proposed.

2. Measurement Methods

Fundamental knowledge about the dynamics of bubbles can be achieved theoretically by analytical approximations, by experimental measurements, or by means of direct numerical simulation (DNS). The advantages and drawbacks of the different approaches are outlined briefly below.

2.1. Theoretical Investigations

Theoretical work is based on the analytical solution of the flow equations with certain boundary conditions at the phase boundary. However, this is only possible for special cases where certain simplifications can be made. Usually this is the assumption of spherical bubbles and inviscid ($Re \rightarrow \infty$) or creeping flows ($Re \rightarrow 0$). Due to these limitations, theoretical work is usually of limited use for industrial applications. However, these theoretical results serve as benchmarks for numerical models, the foundation for semi-empirical models, or reveal functional relationships.

2.2. Experimental Measurements

In experimental measurements, bubble characteristics are measured directly or indirectly and phenomenological conclusions can be drawn. However, there are numerous difficulties with measurements. First, the experimental boundary conditions are almost impossible to control. In some cases, influencing factors are unknown or cannot be determined. One example is the influence of contamination. It has been known for a long time that contamination has a significant influence on the bubble deformation, the rising velocity, and the lateral movement of the bubbles. Therefore, distilled water or deionized water has often been used for measurements. However, some studies [4,5] show that with these apparently pure systems, the influence of contamination can still be significant. In liquid metals, contamination by oxidation can hardly be avoided. Another example is the influence of injection, which often affects the bubble shape or movement some distance above the injector. As these effects cannot be fully quantified yet, this results in a large measurement uncertainty and similar studies are often not directly comparable. This applies in particular to liquid metals where the experimental setups are usually very small.

For the actual measurements, different methods can be found in the literature, which can be categorized into intrusive and non-intrusive methods. Intrusive methods, such as conductivity probes or fiber optic probes, interact to some extent with the flow and are usually restricted to single or few bubbles. Thus, the results may be biased by the applied measurement technique. On the other hand, intrusive methods can be applied in a broader range of setups and with different liquids. Non-intrusive methods do not affect the flow but have more constrains regarding their applicability. Amongst the non-intrusive methods, imaging methods, either based on a multi-stage image processing [6–10] or on machine learning [11,12], are the most frequently used for non-opaque systems because multiple objects can be observed simultaneously and the equipment is comparably cheap and flexible.

A major difficulty is that a large number of different phenomena must be considered for a comprehensive analysis of rising gas bubbles. These takes place on different temporal and spatial scales and require different measurement techniques:

Bubble path and velocity: For a statistically significant path, recordings on a decimeter or better a meter scale are necessary. The frame rate should be at least 200 frames per second (FPS), whereas the internal memory of high-speed cameras limits the duration of the recording. To reconstruct the three-dimensional rise, which can be zigzagging or spiraling, at least two cameras are required.

Bubble shape: In order to analyze the bubble oscillation, recording must be made in the millimeter range with high frame rates. The inaccuracy of the reconstruction of the three-dimensional shape of non-spherical bubbles from two-dimensional projections depends strongly on the number of cameras. Fu and Liu [13] showed that the bubble volume error of a single bubble is about 25% for a one-camera system and decreases to about 10% for a two-camera system.

Velocity distribution of the fluid: For a comprehensive understanding, not only the bubble but also the flow field around the bubble must be measured. For this purpose, the fluid must be seeded with some tracer, which in turn is likely to influence the flow. Moreover, the wake can stretch from a millimeter to a few centimeters.

Bubble swarms: A particular challenge for postprocessing are bubble swarms with high void fractions. Bubbles may overlap on the two-dimensional projection and are visible as a cluster of bubbles. Their segmentation is usually achieved by multi-staged image processing techniques [6–10] or machine learning [11,12]. The accuracy of the different approaches are discussed in [14]. However, no generally applicable workflow was proposed yet. Thus, experimental measurements are often restricted to relative dilute bubble swarms.

A simultaneous investigation of all phenomena is therefore only possible with many cameras, additional equipment, and a sophisticated triggering and could not be realized so far. Different approaches for non-opaque systems are discussed by Bröder and Sommerfeld [7].

In liquid metals, where the fluid is opaque and high temperatures and corrosivity prevent most measurement techniques, measurements are particularly cumbersome. Most early studies employed indirect measurement methods, for example by analyzing the pressure fluctuation [15,16] or the noise [17,18] when a bubble was released or by measuring the time between the generation of the bubble and the breaking of the bath surface [19]. However, these measurements are characterized by a high measurement uncertainty and only allow a macroscopic analysis of most phenomena. Furthermore, their applicability is limited to a few bubbles or low gas flow rates. Therefore, subsequent studies (e.g., [16,20,21]) used resistivity measurements. However, this technique also has disadvantages. First, the method is intrusive and provides only local information [22]. Second, the sensors have a short lifetime, especially at high temperatures [21]. Third, it is difficult to establish the functional relationship between the measured chord (or pierced) length and the actual bubble size [23]. On the other hand, this method enables the investigation of significantly higher flow rates. In addition, multi-needle sensors, like those developed by Iguchi et al. [24], allow a reconstruction of the bubble shape. X-ray measurements (e.g., [25–27]) are non-intrusive and allow an image-based analysis like described above. However, given the high absorption of the liquid metals, measurements are restricted to quasi two-dimensional experiments where the setup's thickness is not more than 12 mm [27]. Neutron radiography (e.g., [28,29]) employs a similar technique but allows thicker experimental setups up to 25 mm [30]. On the other hand, the resulting images are noisier and have less contrast than X-ray images, making subsequent image processing more difficult. Sarma et al. [29] showed that neutron radiography is capable of measuring the motion of tracers in the fluid, which would allow simultaneous measurements of fluid velocity and bubble velocity. This method, called NeuPIV, could provide interesting insights into bubble dynamics in liquid metals in the future. Other non-intrusive measurement techniques are ultrasound Doppler velocimetry (e.g., [31,32]) and inductive methods (e.g., [33,34]). However, they can only be used to measure the bubble velocity and the evaluation of the measurement signals can be very difficult with a higher number of bubbles. A problem with many measurement techniques is that sensors must be placed in close vicinity of the experimental setup, which prevents high temperature measurements. Therefore, measurements are often limited to Galinstan (GaInSn) or Mercury (Hg), as it is already liquid at room temperature.

Due to the described measurement difficulties in liquid metal, the majority of studies have been performed in aqueous systems. However, even for these systems, a comprehensive understanding of the complex bubble–fluid interaction by experimental measurements alone is almost impossible.

2.3. Direct Numerical Simulation

Due to the constant increase in the computing capacity of modern high-performance clusters, the possibility of an analysis by means of DNS is opening up. Here the phase boundary between fluid and gas bubble is directly modelled by means of a suitable interface tracking method. Thus, many phenomena can be analyzed far better than with experimental measurements, as quantities are directly accessible. Furthermore, the fluid properties can be easily varied, so that fluids can be examined, which cannot by experiments. However, DNS is a mathematical model that is affected by modelling and numerical errors. In particular, the interface tracking method and the treatment of surface tension are still being continuously developed. Therefore, existing results from DNS calculations must be critically examined for the influence of these models. They are therefore largely omitted in this review. Furthermore, the analysis is still limited by the computing capacity. According to Kolmogorov's length scales, the number of necessary cells increases with $Re^{9/4}$. This implies that computations are limited to relatively small computational areas with low turbulence (low Re) and only a few bubbles.

3. Bubble Formation Mechanisms

In most metallurgical processes, gas bubbles are generated by single nozzles (e.g., regular nozzles, open lances, Laval nozzles, impellers) or purging plugs. Gas bubbles generated by metallurgical reactions may be assigned to the micro scale and are omitted in this review. The bubble size distribution is an important parameter of the metallurgical process. It directly affects the removal efficiency of non-metallic inclusions [35] and degassing reactions [36] and can influence the fluid flow [37,38]. Generally, small bubbles and a narrow size distribution are desirable. In addition, the bubble size distribution is an important boundary condition of numerical models. Despite that importance, the bubble size distribution in actual processes is still largely unknown and comprehensive theoretical models have not been derived yet. A straightforward approach would be to assume bubbles in the model as well as in the melt in pressure equilibrium. A scaling of the bubble size from a physical model (M) to the real process (R) could then be maintained by [39]:

$$d_{b,R} = d_{b,M}\sqrt{\frac{\sigma_R(\rho_l-\rho_b)_M}{\sigma_M(\rho_l-\rho_b)_R}} \tag{2}$$

However, this approach implies that the bubble size is independent of the injector, which contradicts most experimental results for relevant flow rates.

Since the generation of gas bubbles by injectors plays an important role in a variety of processes, there is a great deal of research on the generation of gas bubbles in water or aqueous systems. These were discussed in the review by Kulkarni and Joshi [40]. The majority of these studies deal with the generation of gas bubbles at nozzles. Studies on the generation of gas bubbles on porous plugs or slot plugs are scarce. Very few studies, summarized in Table 1, investigated the generation of gas bubbles in molten metals, since measurements are cumbersome. These studies have common ground in the fact that they investigate the generation of bubbles at single nozzles. To the best of the author's knowledge, there are no studies on the generation of gas bubbles at purging plugs in liquid metals.

3.1. Bubble Generation at Single Nozzles

Before discussing the results in liquid metals, some important observations and correlations from water models are introduced, as these are often used to check the scalability of water experiment correlations to liquid metals or serve as foundation for correlations for liquid metals.

In an early and widely recognized study, Leibson et al. [41] correlated the bubble size with the orifice Reynolds number. They found two regimes. At laminar flow conditions, the bubble size increases with increasing orifice Reynolds number Re_{or}, where the nozzle diameter and the exit velocity are the characteristic scales, and depend on the orifice inner diameter d_{ni}. In contrast to that, at turbulent outflow conditions, the bubble size is considerably smaller and almost independent of the Reynolds number and the orifice diameter.

$$d_b = 0.029 d_{ni}^{0.5} Re_{or}^{0.33}\ Re_{or} < 2100 d_{b,32} = 0.0071 Re_{or}^{-0.05}\ Re_{or} > 10000 \tag{3}$$

However, all comparisons with measurements in liquid metals [17,42] showed poor agreement with Equation (3), so that it can be concluded that a scaling by the Reynolds number alone is insufficient.

Numerous studies on the generation of gas bubbles at nozzles have shown that a distinction can be made between the constant volume regime and the constant frequency regime [40]. In the former, an increase of the flow rate leads to an increase in bubble frequency, but the bubble size remains constant. In that regime, the bubble size is given by Tate's law [43]:

$$d_b = \left(\frac{6\sigma d_{ni}}{\rho_l g}\right)^{1/3} \tag{4}$$

With a further increase beyond a critical flow rate, the bubble detachment frequency remains constant but the bubble size increases. For that regime, Davidson and Amick [44] derived a correlation for intermediate gas Q_g flow rates up to 250 cm^3/s:

$$d_b = 0.0054\left(10^7 Q_g d_{ni}^{0.5}\right)^{0.289} \tag{5}$$

Mersmann [45] derived a correlation which covers both regimes:

$$d_b = 0.01 \cdot \left[\frac{3 \cdot 10^6 \sigma d_{ni}}{\rho_l g} + \left(\left(\frac{3 \cdot 10^6 \sigma d_{ni}}{\rho_l g} \right)^2 + K \frac{10^6 Q_g d_{ni}}{g} \right)^{0.5} \right]^{0.33} \tag{6}$$

where K is an empirical constant.

Investigations on gas bubble generation in liquid metals, as well as their boundary conditions and the most important results, are summarized in Table 1. Hg [15,20,42], silver (Ag) [15], iron (Fe) [18,46], copper (Cu), lead (Pb), tin (Sn) [17], Wood's metal [47], $Li_{17}Pb_{83}$ [16], GaInSn [22], or aluminum (Al) [48] were investigated. Thus, a broad range of physical properties is covered.

The earliest studies used indirect measurement methods. Here, either the pressure pulse in the gas supply was measured, which results from the rupture of the gas bubble [15,16], or the resulting vibration [17,18]. The bubble volume was calculated from the detachment frequency and the flow rate. Due to this indirect measurement technique, however, the gas flow rate was limited to rather small values, so that the rupture signals of individual gas bubbles could be distinguished.

Sano and Mori [15] compared the bubble size in liquid silver and mercury with theoretical correlations (Equations (4)–(6)) derived in aqueous systems. They reported good agreement between their measurements and theory. However, they pointed out that there is a difference in the wettability of the injector. Therefore, the inner diameter of the injector d_{ni} in Equations (4)–(6) has to be replaced by the outer diameter d_{no} in non-wetting systems. Irons and Guthrie [18] made similar observations, but found less satisfying agreement with the correlation of Davidson and Amick (Equation (5)) or Mersmann (Equation (6)). This was attributed to the fact that the experimental setup, in particular the chamber volume of the injector, differed between the experiments, which had a large influence on the resulting bubble size. Interestingly, both measurements confirm the observations from the water experiments that the bubble size becomes independent of the fluid's material properties for sufficiently large flow rates. Thus, this observation seems to be valid for systems with significantly higher surface tensions and density differences like molten metals. Based on the correlation of Mersmann (Equation (6)), Mori et al. [46] derived a first correlation of the bubble size in liquid metals (note that this correlation was first published in 1977 in Japanese by Sano and Mori and is therefore often called Sano and Mori equation):

$$d_b = 0.01 \cdot \left[\left(10^6 \frac{6\sigma \cdot d_{no}}{\rho_l g} \right)^2 + 0.0242 \left(10^7 Q_g d_{no}^{0.5} \right)^{1.734} \right]^{1/6} \tag{7}$$

Note that this correlation was originally derived using cm and dyn. The use of SI units can lead to deviations of the order of ±5%. This equation was derived for liquid iron and validated with values from previous measurements for Hg, Ag, and H_2O. It should be noted, however, that this correlation was derived for very small flow rates $Q_g \leq 70\ cm^3/s$. This is primarily due to the indirect measurement technique. This has the additional disadvantage that no conclusions can be drawn whether the bubbles disintegrate during injection or on their rising path. Therefore, subsequent studies used resistivity measurements.

The validity of Equation (7) has been critically reviewed in several subsequent studies, where different results were found. Mori et al. [20] found good agreement, even for higher flow rates. Irons and Guthrie [18] found larger gas bubbles than predicted by Equation (7), but this was attributed to differences in chamber volume, as mentioned above. A similar result was found by Iguchi et al. [21]. The reasons for that can only be speculated, since some details describing the experimental setup are not given in their paper. Tschuchiya et al. [16] also found different values. Zhang et al. [49] compared the values of Equation (7) with measured values of bubbles generated with a porous plug in a water model. For that, the diameter in Equation (7) was replaced by the pore size.

However, poor agreement was found. The gas bubbles generated by the porous plug were significantly smaller than predicted by Equation (7).

Moreover, there are also studies that contradict the observation that the initial bubble size is independent of the fluid properties. Tsuchiya et al. [16] argues that the fitting parameters of Equation (7) depend on the fluid properties and the nozzle orientation. However, it should be noted that the number of measurements was so small that these parameters might be overfitted. Andreini et al. [17] found relatively poor agreement with water model correlations, in particular with Equation (3). They argued that the bubble size should not be scaled with the orifice Reynolds number but with the orifice Froude and Weber number, where the characteristic length and velocity are the nozzle diameter and the exit velocity of the gas. They fitted their data to:

$$d_b = d_{ni} Fr_{or}^{0.224} We_{or}^{-0.109} \tag{8}$$

However, Irons and Guthrie [18] point out that the poor agreement with the water model correlations may also be due to different nozzles. Unfortunately, these parameters are not given in [14], so this cannot be further verified. Iguchi et al. [26] measured the bubble size by X-ray imaging. The results of their study suggest that the bubble size depends not only on gas flow rate but also on gas and fluid properties. However, the latter point is contradicted by the measurements of Xie et al. [47].

Resistivity measurements for higher flow rates suggest that the detachment bubble size can be reasonably predicted by Equation (5) [42]. However, the actual bubble size is strongly determined by bubble breakup and coalescence [42,47]. Therefore, Sano and Mori [42] correlated the actual bubble size with a proportional factor that takes into account bubble breakup and coalescence and fitted their data to:

$$d_{b,32} = 0.00091 \left(10^6 \frac{\sigma}{\rho_l}\right)^{0.5} \left(\frac{200\, Q_g}{\pi d_{ni}^2}\right)^{0.44} \tag{9}$$

It should be noted that this correlation was derived for the Sauter mean diameter. However, as shown in Figure 1, the deviation is small in case the bubble distortion is small.

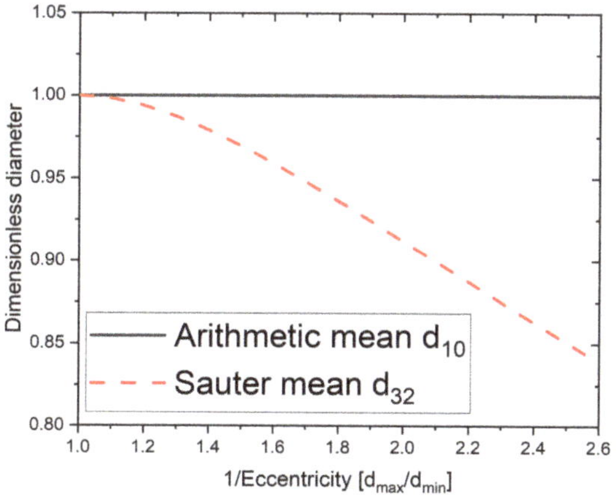

Figure 1. Arithmetic mean and Sauter mean diameter as a function of bubble aspect ratio.

Xie et al. [47] investigated bubble swarms in Wood's metal and reported that the bubble size distribution followed a log-normal distribution. For the mean bubble diameter, they correlated:

$$d_b = 0.000146 \left[10^4 \frac{Q_g}{z + H_0}\right]^{0.1} \tag{10}$$

where H_0 is the distance between the nozzle and the mathematical origin of the plume in m (correlation given in [47]).

In their study, no significant influence of the gas properties or the nozzle size was found. Using a similar experimental setup, Iguchi et al. [50] confirmed the applicability of Equation (10). However, a major criticism of Equation (10) is that it was only derived for one fluid and certainly overfits this system. Fluid property dependent bubble coalescence or breakup, which was found to be important, cannot be considered. Thus, Equation (10) is unlikely to generalize to other fluids.

The studies carried out so far on the formation of gas bubbles in liquid metals show that the theory established for water models for the formation at single nozzles at low gas flow rates also applies to liquid metals as long as the different wettability of the nozzles is considered. At very low flow rates, bubbles of identical size are formed, depending on the fluid properties and the injector. At slightly higher flow rates, however, the bubble size depends primarily on the flow rate and not on the injector or the fluid properties. Because of the difficult measurement conditions, especially at high flow rates, this observation refers to very low flow rates, and for more relevant high flow rates there are very few measurements [42,47]. At these high flow rates, the bubble size distribution seems to be largely dominated by bubble coalescence and breakup. However, there are too few studies in this area to evaluate whether existing models are applicable to liquid metals. Although both studies derived an empirical equation to determine the bubble size, it can be assumed that these correlations significantly overfit the measured data and that their scaling to other fluid systems or to higher flow rates is not valid. At very high flow rates, single nozzles go into jetting mode and the bubble size distribution results from breakup of the jet and sometimes subsequent coalescence [51].

Table 1. Studies on bubble generation in liquid metals (Pres Pul = indirect pressure pulse method, Ac = indirect acoustic method, Res = Electric resistivity, d_{no} = Outer diameter of the nozzle, d_{ni} = Inner diameter of the nozzle, BSD = Bubble size distribution).

Ref.	Year	Method	Nozzle [cm]	Flowrate [cm^3/s]	Fluid	Result
Sano & Mori [15]	1976	Pres Pul	d_{no}: 0.22–0.82 d_{ni}: 0.1–0.3	0.0167–70	Hg/Ag	Correlations from water can be used in case the O.D. instead of the I.D is used (non-wettability)
Andreini et al. [17]	1977	Ac	d_{ni}: 0.015–0.1		Cu/Pb/Sn	Bubble size depend on We and Fr, not on Re. Water model correlations cannot be used
Irons & Guthrie [18]	1978	Ac	d_{no}: 0.64–5.1, d_{ni}: 0.16–0.64	0.5–1000	Fe	Use of O.D because of non-wettability, BSD becomes independent of fluid properties at higher flow rates
Mori et al. [46]	1979	Pres. Pul	d_{no}: 0.32–0.68 d_{ni}: 0.12–0.33	0.1–36	Fe	Empirical Equation (7) for various liquids. At high flow rates, BSD becomes independent of the fluid
Sano & Mori [42]	1980	Res	d_{no}: 0.4–1.0 d_{ni}: 0.2–0.6	50–1330	Hg	Good agreement with water model correlation, BSD is independent of fluid properties, empirical correlation Equation (9)
Mori et al. [20]	1982	Res	d_{ni}: 0.1–0.4	0.05–4500	Hg	Good agreement with Equation (6) at higher flow rates
Xie et al. [47]	1992	Res	d_{ni}: 0.2–0.5	100–1200	Woods	BSD follows log-normal distribution, BSD becomes independent of nozzle, BSD is result of breakup
Tsuchiya et al. [16]	1993	Pres Pul	d_{no}: 0.2–0.64 d_{ni}: 0.1–0.4	0.5–10	H_20/Hg CH_3OH $Li_{17}Pb_{83}$	Fitting parameters of Equation (7) depend on nozzle orientation and fluid properties
Iguchi et al. [21]	1995	Res	d_{no}: 0.6 d_{ni}: 0.1	50–100	Fe	Larger bubbles than predicted by Equation (7)
Iguchi et al. [26]	1995	X-ray	d_{no}: 0.13–0.45 d_{ni}: 0.09–0.19	20–413	Fe	At high flow rates, BSD depend on I.D, not O.D. BSD depends on fluid and gas properties
Iguchi et al. [50]	1997	Res	d_{ni} = 0.005	60	Woods	Confirmed Equation (10)
Gnyloskurenko & Nakamura [48]	2003	X-ray	d_{no}: 0.2–1 d_{ni}: 0.1–0.4	0.43–12	Al	Non-wettability of the nozzle increases bubble size

3.2. Bubble Generation from Porous Plugs

So far, bubble formation in liquid metals at single nozzles has been discussed. However, transferring conclusions or even correlations to industrial vessels in which purging plugs are employed is difficult. It is known from water model experiments that from single nozzles and purging plugs, completely different bubble size distributions are generated, especially at high purging rates. Therefore, bubble formation by purging plugs is discussed in the following. The results presented are all based on observations in aqueous systems or theoretical considerations. Thus, the difficulty of scaling those results to other material systems arises. To the best of the authors' knowledge, measurements in liquid metals that could verify the scalability of existing models are not available.

Early studies [52] observed that at low flow rates, more pores become activated with increasing flow rate while the bubble size remains approximately constant. When all pores are activated, a high void fraction occurs in the direct vicinity of the porous plug. At a certain flow rate, the bubbles start to coagulate to form larger bubbles. A descriptive model for low flow rates was developed by Loimer et al. [53]. Here, a wetted sieve tray is considered first and the theory is then modified to represent the more complex case of a wetted and later a non-wetted porous plug. The model assumes a chamber below the sieve tray in which a certain isostatic pressure prevails, which is dependent on the gas flow rate. As soon as this pressure exceeds the capillary pressure in one of the orifices of the sieve tray, the first orifice gets activated and gas flows through it. If the volume flow is sufficient to lower the pressure in the chamber, the volume flow stops and the pressure increases again. If the volume flow is not sufficient to lower the pressure, successive orifices are activated. In mathematical terms, the minimum flow rate is given by:

$$Q_{min} = 2\pi\sqrt{\frac{\sigma r_{or}^3}{\rho_g}} \tag{11}$$

The number of activated orifices is given by:

$$n = \frac{Q_g}{Q_{min}} \tag{12}$$

For a porous plug, the assumption of an isostatic pressure cannot be made. Instead, it is assumed that the pressure around an activated pore decreases with $1/R$ [53]. Furthermore, the capillary pressure in a pore is different from those in a simple orifice. With these considerations, Equation (11) becomes:

$$Q_{min} = \frac{4\pi\sigma\kappa}{\mu_g} \tag{13}$$

where κ is the permeability of the porous plug. For non-wetted porous plugs, the bubble formation mechanism is different as shown in Figure 2. Bubbles can migrate vertically over the porous plug [54] and coagulate on the plug surface already [55], which results in the release of larger bubbles. Equation (13) has to be extended by a factor y_{max}/r_{or}, where y_{max} is the maximum radius of the contact line, which depends on the wetting angle θ. The contact line is the place at which the plug, the bubble, and the liquid are in three-phase contact. The relation between y_{max} and θ is summarized in [53].

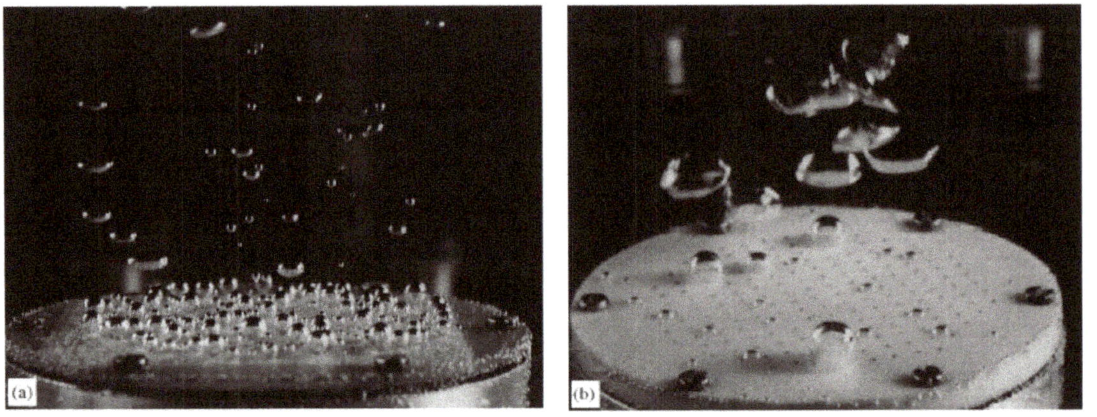

Figure 2. Bubble generation from a wetted (**a**) and non-wetted (**b**) sieve tray (Reproduced from [53], with permission from Elsevier, 2021).

Applying their theories to some typical plugs, Loimer et al. [53] concluded that, similar to the bubble formation at single nozzles, the bubbles at porous plugs are first formed in a quasi-steady regime at which the bubble size is given by Tate's law (Equation (4)). In case the superficial velocity, that is to say the quotient of flow rate and active plug area, exceeds a critical value, the bubble size starts to increase with the flow rate.

However, a rather mediocre agreement between experiments and theory was found. This was explained by the anisotopic permeability of the medium, which can cause pressure gradients in the chamber and in the porous material, respectively. Nevertheless, this model provides a descriptive idea of bubble formation at low flow rates. For higher flow rates, however, bubble coalescence in the vicinity of the porous plug gains importance. According to the theory of Koide et al. [56], all bubbles are formed with a similar size. However, in the vicinity of the plug, the void fraction is high and increases with the flow rate. At the same time, bubble velocity is low since bubbles have not reached their terminal rising velocity. This promotes coalescence of bubbles, which becomes more significant with increasing flow rates.

Since different physical mechanisms are dominant, it is useful to divide the bubble formation into different regimes depending on the gas flow rate. A first distinction [57] divided bubble formation into four different regimes: Quiescent column of discrete bubbles, pulsating column, onset of coalescence, and 'blanketing' at which the entire plug is covered by bubbles and coalescence is dominant. Images of the different regimes are shown in Figure 3.

The transition between the regimes was attributed to the superficial gas velocity. The transition from the quiescent to the pulsating column regime was found at 16 cm/s [57], 15 cm/s [59], or 14 cm/s [58]. The onset of coalescence was found between 30 and 38 cm/s [57], above 25 cm/s [59], or 40 cm/s [58]. 'Blanketing' was found above 67 cm/s [57]. Furthermore, it was shown that the superficial velocities at which transitions take place depend on the pore size, while the ambient pressure or the plug diameter had little effect [60]. For comparison, the superficial gas velocity in liquid steel during soft bubbling is approximately 90 cm/s. However, these studies investigated the same fluid system. Therefore, the transferability of these results to other fluids should be made rather critical. Furthermore, the differences between the various results, especially for the onset of coalescence, suggest that other factors besides the superficial gas velocity need to be considered.

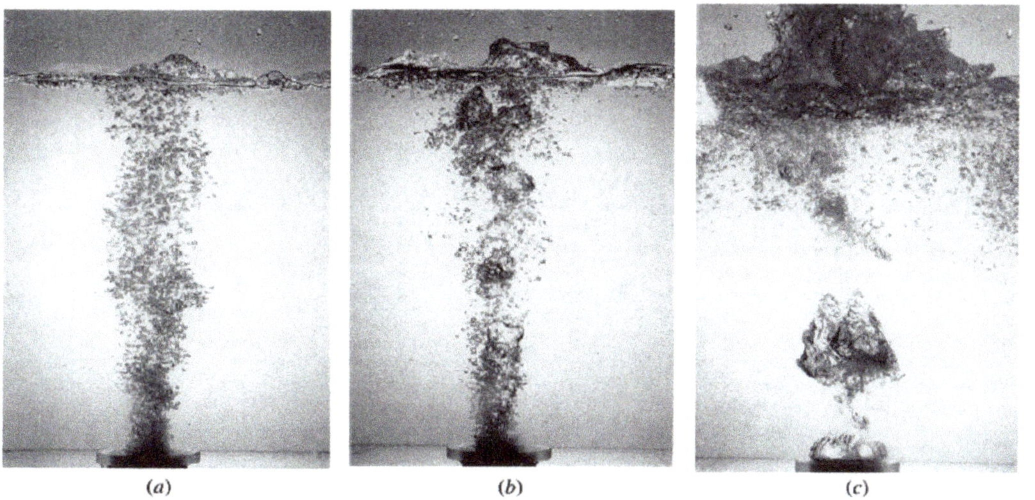

(*a*) (*b*) (*c*)

Figure 3. Modes of gas dispersion in porous plug injection: (**a**) Quiescent column of discrete bubbles, (**b**) onset of coalescence, and (**c**) blanketing (Reproduced from [58], with permission from Springer, 2021).

In more recent investigations, the subdivision into the homogeneous and heterogeneous bubble regime became established. The homogeneous regime is characterized by a 'non-coalescence induced' [3], approximately gaussian bubble size distribution [61] and an approximately linear dependence of the void fraction on the superficial gas velocity. In contrast, the heterogeneous regime is characterized by a 'coalescence-induced' [3] bubble size distribution, which follows a log-normal distribution [61]. The void fraction no longer follows a linear dependence of the superficial gas velocity.

The transition between the homogeneous and heterogeneous regime has been studied by Kazakis et al. [61] and Mohagheghian et al. [62]. In addition, the influence of numerous parameters on the transition between these regimes was discussed in the review by Besagni et al. [3]. It was found that the transition is independent of the type of gas [3,52,56]. A smaller average pore size of the porous plug seems to favor the heterogenous regime, thus shifting the transition to lower superficial velocities [61]. A reduction of the porous plug diameter increases the local void fraction whereby the transition maintains at lower flow rates [61]. Higher fluid flow velocity in the vicinity of the plug destabilize the homogeneous regime and the transition takes place at lower superficial gas velocities [3]. Smaller bubbles seem to stabilize the homogeneous regime [63].

The viscosity of the fluid does not seem to have an influence [61], whereas Mohagheghian et al. [62] reported an influence in high viscosity systems as the rising velocity of bubbles decreases, which increases the likelihood of coalescence. A higher surface tension is advantageous for coalescence, thus it destabilizes the homogenous regime [61].

Higher fluid temperature or higher pressure reduce the coalescence and favor breakup. Thus, the transition takes place at increased superficial gas velocities [3].

The composition of the fluid, in particular organic or inorganic contaminants or the addition of acids, is of great importance. Those suppress coalescence so that the homogeneous regime is stabilized, causing the transition to be shifted to higher superficial gas velocities [3,56].

An empirical equation to determine the transition superficial gas velocity was provided by Kazakis et al. [61] based on the Froude and Bond (equivalent to the Eötvös number) number. However, according to this correlation, the transition is mainly influenced by the ratio of porous plug to column diameter. In the experiments, that ratio was in the order of 0.1, which is much smaller than in most metallurgical processes. For example, it is 0.03 in the ladle. Nonetheless, the transition superficial velocity was typically reported in an order of a few cm/s while it is about 90 cm/s in a ladle. Therefore, it seems reasonable to assume that the bubble column in the ladle is in the heterogenous regime. Similar considerations can be made for most metallurgical processes. However, it should be noted that the influencing factors described above were derived for aqueous systems with much lower pressure, temperature, surface tension, and density.

The bubble size and its dependence on different parameters also depend on the regime of the bubble column. For very small flow rates, the bubble size remains constant and only depends on the plug's parameters [53]. An empirical correlation was proposed by Koide et al. [64] for pure and fully contaminated systems:

$$d_b = \hat{y} \cdot \left(\frac{d_{po} \sigma}{\rho_l g} \right)^{1/3} \tag{14}$$

where $\hat{y}$ is an empirical fitting parameter, ranging from equivalent to 1.47 for pure to 1.61 for fully contaminated systems, which is similar to Tate's law (Equation (4)). Similar results were reported by Miyahara and Tanaka [65].

For slightly higher flow rates, which still belong to the homogeneous regime, the bubble size increases proportionally to the flow rate. An empirical correlation for the whole homogenous regime was derived by Mohagheghian et al. [62]:

$$d_{b,32} = d_{po} \cdot 6.4 \left(Fr_b^{1.8} We_b^{1.7} Re_b^{0.7} \right)^{-0.132} \tag{15}$$

where the characteristic length and velocity of the dimensionless groups are the bubble size $d_{b,32}$ and the superficial gas velocity u_{sg}. In the heterogeneous regime, there are inconsistent results, but most studies found that the bubble size increases with increasing flow rate. In this regime, the bubble size is mainly determined by coalescence and breakup, which is influenced by many different parameters [3]. For the heterogenous regime, Mohagheghian et al. [62] argued that the Ohnesorg number, Oh, is a function of the Morton and the Capillary number, Ca:

$$\begin{aligned} Oh &= \frac{\text{Viscous force}}{\sqrt{\text{Inertial force} \cdot \text{Surface tension force}}} = \frac{\mu_l}{\sqrt{\rho_l \sigma d_b}} \\ Ca &= \frac{\text{Viscous force}}{\text{Surface tension force}} = \frac{\mu_l u_{sg}}{\sigma} \end{aligned} \tag{16}$$

They derived:

$$d_{b,32} = 0.2477 \frac{\mu_l^2}{\rho_l \sigma} \left(\frac{g u_{sg} \mu_l^5}{\rho_l \sigma^4} \right)^{-0.4} \tag{17}$$

For intermediate flow rates (up to 10 cm/s), Koide et al. [64] derived for organic liquids:

$$d_b = 0.0729 \cdot \left(10^6 \cdot \frac{d_{po} \sigma}{\rho_l g} \right)^{1/3} \left(\frac{\rho_l g d_{po}^2}{\sigma} \right)^{0.345} \left(\frac{Fr_{pl}}{\sqrt{We_{pl}}} \right)^{0.132 \left(\frac{\rho_l g d_{po}^2}{\sigma} \right)^{-0.139}} \tag{18}$$

For pure liquids, slightly different fitting parameters were found [64]:

$$d_b = 0.0703 \cdot \left(10^6 \cdot \frac{d_{po} \sigma}{\rho_l g} \right)^{1/3} \left(\frac{\rho_l g d_{po}^2}{\sigma} \right)^{0.364} \left(\frac{Fr_{pl}}{\sqrt{We_{pl}}} \right)^{0.133 \left(\frac{\rho_l g d_{po}^2}{\sigma} \right)^{-0.14}} \tag{19}$$

where We_{pl} is the plug Weber number ($u^2_{sg} d_{po} \rho_l / \varepsilon^2 \sigma$) and Fr_{pl} is the plug Froude number ($u^2_{sg} / \varepsilon^2 g d_{po}$). Another correlation for intermediately contaminated systems was proposed as well. However, since the influence of contaminations in liquid metals have not been studied yet, it is not given here.

The discussion of bubble generation on porous plugs shows that bubble formation at very low flow rates follows similar mechanisms as at single nozzles. However, the situation is somewhat more complex, since different bubble formation sites have to be considered. At higher flow rates, the bubble formation mechanism is quite different from single nozzles. Thus, it can be concluded that correlations for single nozzles at high flow rates cannot be applied to porous plugs. In addition, the literature review shows that a variety of parameters influence the bubble size distribution. However, to apply the correlations, a distinction must be made between different regimes. Due to the relatively small number of studies and the high number of influencing parameters, a comprehensive understanding of the bubble size distribution is not available, especially at higher gas flow rates. This is partly due to the difficulty of bubble size distribution measurements at high void fractions. Furthermore, all observations so far have been set up in aqueous systems; in those, the density and surface tension especially can only be varied to a small extent. Therefore, a simple transfer of those correlations to predict the bubble size in liquid metals has to be questioned critically. In addition, all correlations have been established for wettable purging plugs. However, Wang et al. [55] showed that non-wettability promotes coalescence directly at the plug surface, thus releasing larger bubbles. However, it is not clear whether this effect is also relevant for larger gas flow rates.

3.3. *Bubble Generation from Slot Plugs*

Slot plugs or hybrid plugs are sometimes used instead of porous plugs. However, there are only very few studies on the formation of bubbles at wetted [66–69] and non-wetted [70,71] slots.

Bubbles form at slots at a finite number of bubble sites. At each of these sites, gas bubbles are formed with a mechanism similar to that of single nozzles. Again, different regimes can be distinguished depending on the gas rate as shown in Figure 4. At low gas flow rates, the number of bubble formation sites increases with increasing gas rate (regime I). As the gas flow rate continues to increase, the number of bubble formation sites becomes so large that they touch each other and coagulate to form a larger bubble formation site (regime II). The number of gas bubble formation sites decreases with flow rate in this regime. At very high flow rates, a linear gas blanket forms (regime III) [70]. Harris et al. [69] showed that the transition between regime I and II depends on the slot Froude number.

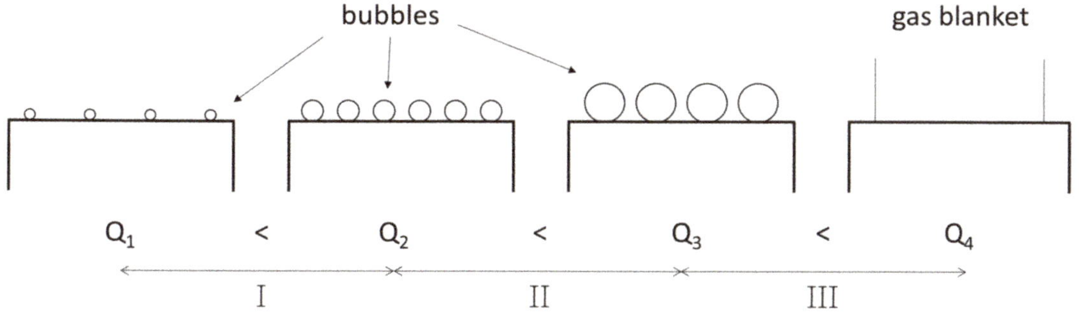

Figure 4. Regimes of bubble formation at a slot nozzle (Adapted from [70], with permission from Elsevier, 2021).

In regime I, Okumura et al. [71] reported that the number of active bubble formation sites on non-wetted slots depends on the shape of the slot, but not on the slot width.

In contrast to that, Li et al. [67] reported a slot width dependency for wetted slots and correlated Equation (20) for the distance between active formation sites:

$$\lambda_{sl} = 17.2 w_{sl} \left(\frac{\rho_l}{\rho_g} \right)^{0.16} We_{sl}^{-0.25} \tag{20}$$

where We_{sl} is the slot Weber number ($u^2_{sg} w_{sl} \rho_l / \sigma$).

As with porous plugs and single nozzles, wettability plays a major role. At wettable slots, significantly more bubble formation sites are formed, especially at low flow rates (about factor 5 [71]), and smaller bubbles are formed. At higher flow rates, the influence of wettability decreases significantly. In regime II, the distance between active formation sites on a non-wetted slot can be approximated by [71]:

$$\lambda_{sl} = 6.5 w_{sl} We_{sl}^{0.466} \tag{21}$$

In regime I, the bubble size does not depend on the gas rate but is given by Tate's law (Equation (4)). For a combination of regime I and II, a high agreement with the empirical correlation of Mori et al. [46] (Equation (7)) was found, assuming that the total flow rate is evenly distributed among the individual bubble formation sites and that they behave like single nozzles [71]. However, given the small number of studies, the transferability of this conclusion to other fluid properties cannot be verified.

3.4. Bubble Size in a Steel Casting Ladle

In the previous sections, the bubble formation mechanisms at individual nozzles, porous plugs, and slot plugs were described, and existing correlations were analyzed. The applicability of those correlations to industrial processes will be discussed in the following using the example of a steel casting ladle during the process stage of soft bubbling.

Commonly, purging plugs are used in ladle. However, different purging rates are used in different process stages. According to Trummer et al. [72], porous plugs are advantageous for soft bubbling and produce many small bubbles. At high flow rates, however, their wear is very high. Slot plugs are disadvantageous for soft bubbling as they produce much larger bubbles and low flow rates are difficult to control. However, they show significantly less wear for high flow rates. Therefore, there are also hybrid plugs, which have both a porous plug and a slot part.

In Table 2, the various correlations derived are summarized and the bubble size extrapolated to an industrial ladle are given. As shown, a wide range of mean bubble sizes is predicted, ranging from less than 1 mm up to 100 mm. This generally reveals that the extrapolation of existing correlations to another injection system or another fluid system is highly unreliable. In most numerical models, the bubble size, extrapolated from correlations for individual nozzles, is used as a boundary condition. However, it appears that those correlations probably overestimate the bubble size compared to porous plugs. The most likely scenario is that the bubble size distribution is determined partly by the injector and partly by coalescence and breakup. For example, Polli et al. [73] showed that an influence of the injector is even present in a considerable height above the injector itself, even if coalescence and breakup are the predominant effect. This is because coalescence and breakup are dynamic processes that require some time to reach an equilibrium state.

Table 2. Bubble size correlations and their corresponding value in soft bubbling in a 185 t ladle (Q_g = 200Nl/min, T = 1600 °C, H_{Fill} = 3.23 m, D_{pl} = 0.1 m).

System	Ref.	Equation	$d_{b,ladle}$
Analytical	[39]	$d_{b,R} = d_{b,M}\sqrt{\frac{\sigma_R(\rho_l-\rho_b)_M}{\sigma_M(\rho_l-\rho_b)_R}}$	8.5 mm
Single Nozzles	[47]	$d_b = 0.000146\left[10^4\frac{Q_g}{z+H_0}\right]^{0.1}$	15.8 mm
	[15]	$d_b = 0.01\cdot\left[\left(10^6\cdot\frac{6\sigma\cdot d_{no}}{\rho_l g}\right)^2 + 0.0242\left(10^7 Q_g d_{no}^{0.5}\right)^{1.734}\right]^{1/6}$	31.2 mm
	[42]	$d_{b,32} = 0.00091\left(10^6\frac{\sigma}{\rho_l}\right)^{0.5}\left(\frac{200Q_g}{\pi d_{ni}^2}\right)^{0.44}$	96.8 mm
Porous Plugs (heterogenous regime)	[64]	$d_b = 0.0703\cdot\left(10^6\cdot\frac{d_{po}\sigma}{\rho_l g}\right)^{1/3}\left(\frac{\rho_l g d_{po}^2}{\sigma}\right)^{0.364}\left(\frac{Fr_{pl}}{\sqrt{We_{pl}}}\right)^{0.133\left(\frac{\rho_l g d_{po}^2}{\sigma}\right)^{-0.14}}$	13.5 mm
	[62]	$d_{b,32} = 0.2477\frac{\mu_l^2}{\rho_l\sigma}\left(\frac{g u_{sg}\mu_l^5}{\rho_l\sigma^4}\right)^{-0.4}$	0.65 mm
Slot plug	[71]	$\lambda_{sl} = 6.5 w_{sl} We_{sl}^{0.466}$ $d_b = 0.01\cdot\left[\left(10^6\cdot\frac{6\sigma\cdot d_{no}}{\rho_l g}\right)^2 + 0.0242\left(10^7 Q_g d_{no}^{0.5}\right)^{1.734}\right]^{1/6}$	13.2 mm

A promising method to numerically determine the real bubble size distribution in the steel ladle is population balance models (PBM). Here, additional transport equations are solved for the particle density of different particle sizes:

$$\frac{\partial n(V_i, x, t)}{\partial t} + \nabla(u(V_i, x, t)\cdot n(V_i, x, t)) = S_{br} + S_{co} \tag{22}$$

For breakup, there are four different mechanisms that have to be considered: Turbulent fluctuation and collision, viscous shear stress, shearing-off, and phase boundary instability. These mechanisms differ in the number and size of the resulting daughter bubbles. That is, whether a bubble disintegrates into two bubbles of approximately the same size or two or more bubbles of different sizes. Different models for the breakup probability as well as the daughter bubble size are discussed in the review by Liao and Lucas [74]. This review show that a complete understanding of the breakup phenomenon, even for much-studied aqueous systems, is still lacking. Break up in liquid metals has so far only been investigated by Keplinger et al. [75] using X-ray in a GaInSn melt. It was shown that breakup is mainly caused by the interaction with the preceding bubble, either by the turbulence of their wake or by collision of the two bubbles. However, the X-ray measurement technique does not allow direct measurement of the mean flow not to speak of turbulent structures like wakes. Therefore, the effect cannot be quantified further. For small void fractions, breakup due to viscous shear stress is also relevant. In the investigations in the GaInSn melt, the onset of the effect was found at slightly higher Reynolds numbers than in water. However, the derived conditions seem to be a necessary, rather than a sufficient, condition. In contrast to aqueous systems, breakup results in two bubbles of different sizes.

The coalescence rate is primarily determined by the collision frequency of the bubbles. Collision is evoked by turbulent fluctuations or velocity gradients in the fluid, different bubble velocities, bubble capture in eddies, or bubble wake interaction. However, not all collisions lead to coalescence. Therefore, there is also the concept of coalescence efficiency, which specifies how likely coalescence is in the event of a collision. Existing models for the collision frequency as well as for coalescence efficiency are discussed in the review by Liao and Lucas [76]. Again, there is only one study for liquid metals by Keplinger et al. [75]. In an investigation of collision in a GaInSn melt, good qualitative agreement with the phenomena observed in aqueous systems was reported. Bubble collision seems to be primarily caused by wake capture of the trailing bubble. However, it is important to note that a bubble chain was studied, which favors this effect. For the coalescence efficiency, a strong dependence on the turbulence in the vicinity of the bubbles was suspected. However, this effect cannot be quantified due to the limited measurability. These results show that PBMs have a potential for deriving the actual bubble size in the ladle. However, there is a high degree of uncertainty due to incomplete knowledge of the governing phenomena. In particular, it is critical that the differences between aqueous systems and liquid metals cannot yet be quantified.

Another difficulty is that hybrid purging plugs are often used in the industry i.e., two different bubble formation mechanisms have to be considered. However, it is not possible to evaluate when and to what extent the different mechanisms are active in the real process, since no studies are available apart from the water model study by Trummer et al. [72]. Sahai and Guthrie [77] argue that a precise knowledge of the bubble size is not essential, since it can be assumed that the bubbles are large enough to be of a cap shape where the drag coefficient c_d can be assumed to be 8/3. However, this reasoning is based on extrapolation of correlations for single nozzles, so the bubble size is probably overestimated by the authors. For the most probable range, the bubbles are mostly ellipsoids, so the drag coefficient cannot be assumed to be constant as discussed in the subsequent sections.

4. Bubble Deformation

In contrast to solid particles, droplet and bubble's phase boundary may deform due to local pressure gradients. This leads to a complex interaction between the phase boundary and the surrounding flow. Bubble deformation influences the interfacial forces and heat and mass transfer, and vice versa. For the drag force, the deformation is usually considered implicitly in the drag coefficient though some exceptions exists (e.g., [78,79]). For the modelling of the lift force, the bubble shape has to be considered explicitly by the modified Eötvös number as discussed below.

An overview on different bubble shapes is provided in Figure 5. For small bubbles, the surface tension dominates, and the bubbles are of spherical shape (a). For medium-sized bubbles, the surface tension, viscous friction, and the inertia of the surrounding fluid are relevant for the bubble's shape. If there is a relative velocity between bubble and fluid, the flow stops at the front stagnation point and the pressure increases towards the inside of the bubble. The surrounding streamlines bend to form a curve around the bubble. Due to the incompressibility of the fluid, an acceleration along the streamline must therefore take place, which reduces the local pressure. This effect results in the formation of an ellipsoidal bubble (b). At even higher relative velocities, cap or mushroom bubbles (c) are formed, which are caused by vortices in the wake behind the bubble [80]. Subsequently, the discussion is limited to spherical and ellipsoidal bubbles, largely omitting cap bubbles since the influence of deformation on the interfacial forces is assumed to be less pronounced for that case.

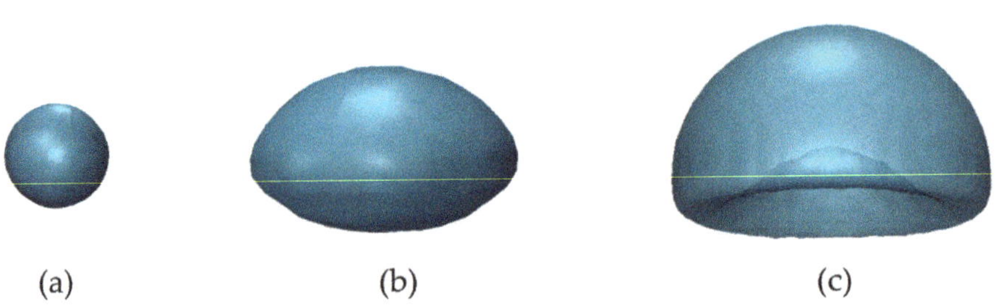

Figure 5. Bubble shapes: (**a**) Spherical, (**b**) ellipsoidal, (**c**) mushroom.

The deformation of ellipsoidal bubbles is usually described by either the eccentricity E or the shape factor φ given by Equation (23). In the stationary shape, oscillation is usually not considered, except for some measurements that concentrate on the interaction between wake, shape, and path oscillation.

$$\text{Eccentricity E} = \frac{d_{min}}{d_{max}} \text{Shape factor } \varphi = \frac{d_{eq}}{d_{max}} = E^{1/3} \tag{23}$$

where the equivalent diameter, d_{eq}, is defined as the diameter of a sphere with the same volume as the distorted bubble. A simplified analytical solution for inviscid flows was derived by Moore [78] by balancing the normal stresses caused by dynamic pressure and surface tension at the stagnation points and on the equatorial plane. This analysis yields:

$$E = \frac{1}{1 + \frac{9}{64}\text{We} + O(\text{We}^2)} \tag{24}$$

The third term (O notification) indicates that this theory losses validity for higher Weber numbers, that is to say for stronger deviation from the spherical shape. Thus, this theory is only applicable for the limited range of slightly distorted ellipsoid bubbles in pure liquids. Therefore, dimensionless numbers describing the system are derived and coefficients for different f-functions (hypothesis) are established by fitting experimental datasets. Wellek et al. [81] discussed that a droplet's shape depends on eight parameters: Density and viscosity of the gas and the surrounding fluid, the volume equivalent bubble diameter, surface tension, gravity, and the relative bubble velocity in steady state. By dimensional analysis, they derived five dimensionless numbers:

$$E = f\left(\frac{d_{eq}u^2\rho_l}{\sigma}, \frac{d_{eq}u\rho_l}{\mu_l}, \frac{\left(\rho_l - \rho_g\right)gd_{eq}^2}{\sigma}, \frac{u^2}{gd_{eq}}, \frac{\mu_l}{\mu_g}\right) = f(\text{We}, \text{Re}, \text{Eo}, \text{Fr}, N_{\mu r}) \tag{25}$$

Haberman and Morton (1953) found the same parameters for bubbles, though they derived the drag coefficient c_D, the bubble Reynolds number, the Morton number, the Reynolds number inside the bubble, and the density ratio as dimensionless groups. If the density and viscosity of the gas is considered to be negligible, the number of dimensionless groups can be further reduced to three. Nowadays the bubble Reynolds number, Eötvös number, and Morton number are commonly used, although some correlations have been derived for the Weber number.

Grace et al. (original publication in French, found in [82]) presented the bubble shape as a function of these three dimensional groups in a diagram (Figure 6), often named the Grace diagram. This allows a qualitative estimate of the bubble shape. It can be seen that the deformation increases with increasing Eötvös and Reynolds number and decreasing Morton number. Furthermore, some shapes may only be present in certain Morton number ranges, although it should be mentioned that the boundaries are less strict in reality than shown in the diagram. Using neutron radiography in lead bismuth ($\log_{10}$(Mo)~−13), Mishima et al. [28] found a good qualitative agreement between the predicted and observed bubble shapes.

Quantitative estimates are obtained by regression analysis of experimental data, mostly from experiments with single bubbles. Different f-functions exists. Wellek et al. [81] proposed:

$$\frac{1}{E} = 1 + \alpha_0 \prod_{i=1}^{n_{tot}} \theta_i^{\alpha_i} \tag{26}$$

where θ is a dimensionless group and α are the regression coefficients. A more complex f-function was proposed by Aoyama et al. [83]:

$$\frac{1}{E} = \left[1 + \alpha_0 \prod_{i=1}^{n_{tot}} \theta_i^{\alpha_i}\right]^{\alpha_{exp}} \tag{27}$$

Equation (26) might miss some higher-order correlations between the dimensionless groups while Equation (27) has a stronger tendency to overfit the experimental data.

Using Equation (26), Wellek et al. [81] correlated:

$$E = \frac{1}{1 + 0.163Eo^{0.757}} \tag{28}$$

Equation (28) was originally derived for liquid droplets, but it is now widely accepted that it also applies to contaminated bubbles [84]. For super purified fluids, Sanada et al. [5] used the same approach but correlated:

$$E = \frac{1}{1 + 6.5Eo^{1.925}} \tag{29}$$

This shows that the deformation is significantly reduced by contaminants. However, it should be mentioned that the measurements of Sanada et al. [5] were limited to small bubbles.

Correlations that only consider the Eötvös number have a limited ability to generalize. For instance, Wellek et al. [81] showed that Equation (28) is only applicable in a limited range of Morton numbers and is inaccurate for high viscous systems. They proposed that, instead, a correlation with the Weber number, similar to the analytical solution of the Moore (Equation (24)), should be used to cover a broader range of fluid properties. Similar results were found by Besagni and Deen [84]. Furthermore, the Eötvös number does not capture all dynamic effects of the hydrodynamic system, as shown in different studies. For example, Tomiyama et al. [85] observed an influence of the bubble release mechanism on the deformation, which cannot be captured by the Eötvös number. On the other hand, the Eötvös number is independent of the relative velocity, which makes it easily applicable in bubble swarms.

Tadaki and Maeda (1961, original publication in Japanese, found in [82]) experimentally found a dependency on the Reynolds and Morton number, which was later termed the Tadaki number Ta:

$$Ta = ReMo^{0.23} \tag{30}$$

For bubble deformation, expressed by the shape factor φ, they derived:

$$\varphi = E^{\frac{1}{3}} = \begin{matrix} 1 & Ta < 2\ (\text{spherical}) \\ 1.14Ta^{-0.176} & 2 < Ta < 6\ (\text{ellipsoid}) \\ 1.36Ta^{-0.28} & 6 < Ta < 16.5\ (\text{ellipsoid}) \\ 0.62 & 16.5 < Ta\ (\text{spherical cap}) \end{matrix} \tag{31}$$

Fan and Tsuchiya [80] compared this correlation with literature data for purified liquids and determined that it is applicable for contaminated fluids. They generalized an existing correlation by Vakhrushev and Efremov [86] to be used for pure and contaminated liquids:

$$\frac{1}{E} = \begin{matrix} 1 & Ta \le Ta_1 \\ \{c_1 + c_2 \tanh[c_3(c_4 - \log(Ta))]\}^m & Ta_1 \le Ta \le Ta_2 \\ (h/b)_{cap} & Ta_2 \le Ta \end{matrix} \tag{32}$$

The fitting parameters of this correlation are summarized in Table 3.

Table 3. Parameters for Equation (32) from Vakhrushev and Efremov [86] and Fan and Tsuchiya [80].

System	m	$(h/b)_{cap}$	Ta_1	Ta_2	c_1	c_2	c_3	c_4
3Dcontaminated	3	0.24	1	40	0.81	0.2	2.0	0.8
3Dpure	3	0.24	0.3	20	0.81	0.2	1.8	0.4

Aoyama et al. [87] used a more complex f-function given in Equation (27) and found a correlation that takes the bubble Reynolds and Eötvös numbers into account.

$$E = \frac{1}{\left[1 + 0.016\mathrm{Eo}^{1.12}\mathrm{Re}\right]^{0.388}} \tag{33}$$

All correlations presented up to this point have common ground in the fact that they are based on measurements of single bubbles in stagnant liquids. However, it is likely that bubble deformation is affected by the presence of other bubbles in a bubble swarm. This subject was investigated by Ziegenhein and Lucas [88] using data from six different experimental setups with pure water. The dataset included experiments with single bubbles in linear shear flows and different bubble columns. Because not all instantaneous local quantities could be measured simultaneously, averaged values for the turbulence, dynamic pressure, and bulk flow field were used. Their measurements showed that deformation is largely independent of the flow rate, i.e., void fraction and fluid flow. The deformation found in single bubble experiments were less pronounced than in bubble swarms, but the difference is generally small. For smaller Eötvös numbers, this difference is more pronounced than for larger Eötvös numbers. A comparison with existing correlations showed that none of them reflect the data well over the whole range of values. For small Eötvös numbers, there is a good agreement with measurements in super purified liquids by Sanada et al. (Equation (29)) [5]. For larger Eötvös numbers, however, this correlation clearly overestimates the deformation. Compared with Tadaki number correlations, there is good agreement with the correlation of Fan and Tsuchiya (Equation (32) and Table 3 up to Ta < 3. Beyond that, significantly less deformation was found. From these results, the authors concluded that for a final evaluation of bubble deformation in swarms, the local flow conditions would have to be considered, which is currently not possible. For larger bubbles, however, there are very similar tendencies as in experiments with single bubbles.

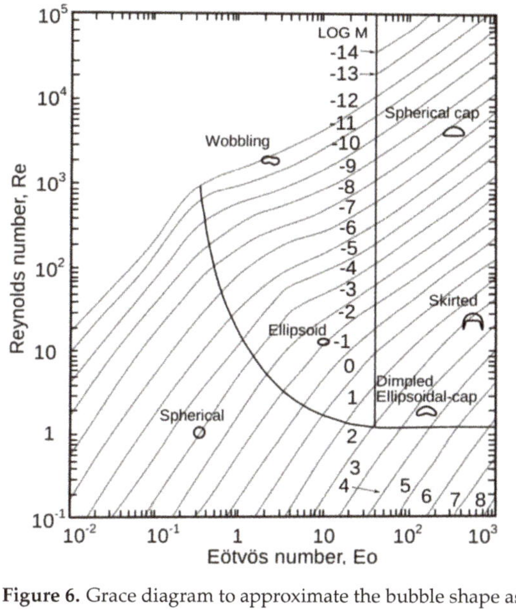

Figure 6. Grace diagram to approximate the bubble shape as a function of the Eötvös, Reynolds, and Morton number (original in French, reproduced from [88], with permission from Elsevier, 2021).

Besagni and Deen [84] evaluated a large dataset of measurements in pure and contaminated systems for single bubbles and swarms, ranging from $-10.8 < \log_{10}(\text{Mo}) < 2.3$, which was collected by different researchers. Using Equation (27) as f-function, they correlated:

$$E = \frac{1}{\left[1 + 0.4\text{Eo}^{1.19}\text{Re}^{1.05}\right]^{0.07}} \approx \frac{1}{[1 + 0.45\text{EoRe}]^{0.08}} \tag{34}$$

In addition, they compared existing correlations with their data and showed than none of them could predict the eccentricity very well in a wide range of conditions. This result indicates that a comprehensive prediction of the bubble deformation is still difficult. Though a large number of correlations have been proposed, none seems to be applicable for all experimental conditions. There are different reasons for that. First, the dimensionless numbers are on different scales. While Morton numbers are usually very small, the Reynolds number is usually some order of magnitudes larger. Feature scaling should be applied before fitting an f-function to avoid any scale effects. More importantly, all measurements suffer from a systematic bias by contaminants. Measurements by Sanada et al. [5] showed that even smallest amounts of organic substances, which cannot be avoided experimentally, have significant influence of the bubble rising behavior and its deformation. Since the purity of the investigated systems is usually provided qualitatively, the bias of contamination results in different measurement uncertainties. Finally, one has to bear in mind that the approximation of a symmetric ellipsoid becomes rather inaccurate for larger bubbles, particular in low Morton number systems and that shape oscillation is usually neglected. Thus, the experimental data, in particular for larger bubbles, scatter significantly. Even more difficult is the extrapolation of existing correlations to liquid metals. Iguchi et al. [24] developed a multi-needle resistivity probe to reconstruct the shape of gas bubbles. A cap or mushroom shape was found for approximately 25 mm bubbles in Wood's metal. A correlation was found between the averaged bubble Reynolds number (mean bubble size and velocity as characteristic length and velocity, respectively) and the modified Weber number (multiplied by the density ratio) [23]. However, the measurements were made just above the injector, so that a strong influence of the initial deformation due to the tearing off of the bubble can be assumed. There are very few studies investigating the deformation of gas bubbles in liquid metals quantitatively [22,26,27,31]. These studies are all based on X-ray measurements, which can be used to visualize the bubble. However, the deformation is only explicitly stated by Richter et al. [31] and Keplinger et al. [27]. For the other studies, the deformation must be roughly estimated from the published images. A comparison of the presented correlation with the measured values in liquid steel and GaInSn is shown in Figure 7. The bubble Reynolds number and Weber number were estimated for this purpose with the drag coefficient according to Dijkhuizen et al. [89] (Equation (45)). The physical properties employed for this estimate are listed in Table 4.

Table 4. Physical properties of the investigated liquid metals.

Fluid	Density [kg/m^3]	Viscosity [Ns/m^2]	Surface Tension [N/m]	log10 (Mo)
GaInSn	6330	0.00234	0.585	−12.6
Steel	6900	0.00506	1.5	−12.6
Mercury	13,550	0.00155	0.487	−13.4
Silver	9510	0.00389	0.92	−12.5

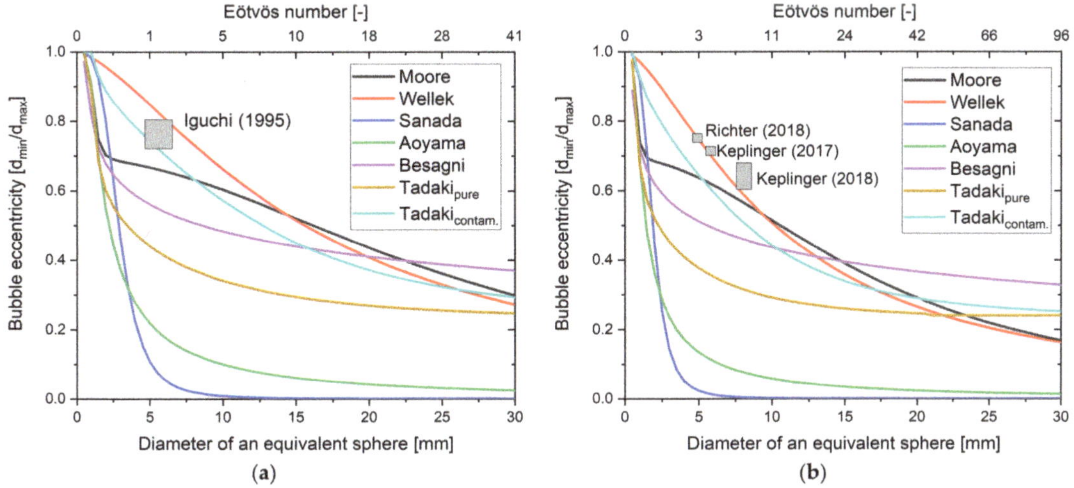

Figure 7. Bubble deformation in a steel melt (**a**) and GaInSn (**b**).

It can be seen that most correlations significantly overestimate the deformation. However, there is good agreement with the model of Wellek et al. [81]. This is surprising, since this model is only valid for contaminated bubbles in aqueous systems. It is not clear whether the lower deformation can be explained by the presence of contaminants or by the significantly higher surface tension of the liquid metals. A general problem of the used X-ray measurements is, however, that only the contour of the bubble is visible. Thus, it is difficult to distinguish between ellipsoids and spherical cap bubbles [24]. In addition, the high absorption of the liquid metals conditions that very narrow experimental setups have to be used. It is therefore quite possible that the proximity to the wall has an influence on the deformation. Numerically, the deformation of argon bubbles in steel was qualitatively investigated by Xu et al. [90] and Wang et al. [91]. Both use the volume of fluid surface tracking method. They reported that bubbles are spherical after injection but become wobbly ($d_b \leq 7$ mm) or ellipsoidal cap shaped ($d_b > 7$ mm), depending on their size and their Eötvös and Reynolds numbers. However, when evaluating these results, it should be noted that no turbulence model was used in both studies, but the mesh resolution used is too coarse for a real DNS. Therefore, the models can be classified as implicit large eddy turbulence model (LES) without a subgrid model, which underestimates the draining effect of the small-scale eddies on the energy cascade. Therefore, it is likely that the resolved turbulence is overestimated in both studies, which certainly has some influence on the bubble shape.

5. Interfacial Force Closure

According to the second Newtonian axiom, the acceleration of bubbles and thus their motion can be described by the sum of all forces acting on them. For bubbles, the buoyancy, the drag, the lift, and the virtual mass force are decisive. Thus, the velocity of a bubble is governed by:

$$\frac{\partial u_b}{\partial t} = \frac{1}{m_b} \sum_i F_i = gV_b(\rho_l - \rho_b) + F_D + F_L + F_{VM} \quad (35)$$

Subsequently, the different forces, shown in Figure 8, are discussed, and existing correlations are critically analyzed with regard to their application to liquid metals.

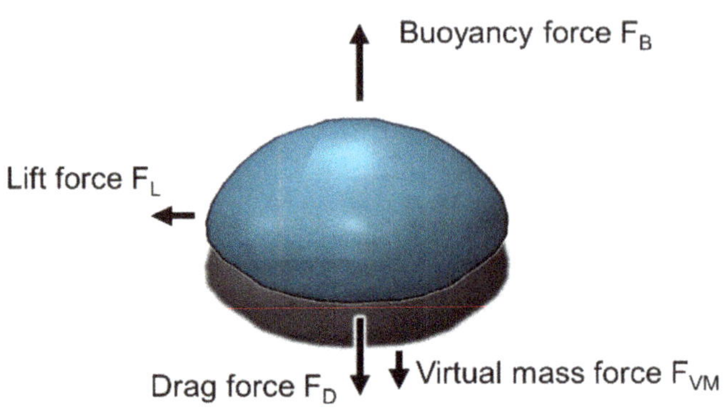

Figure 8. Forces on a bubble.

5.1. Drag Force

The drag force determines the bubble's rising velocity and contributes the main part of the momentum transfer from bubbles to liquid. It arises through a relative velocity between the bubble and the continuous phase. It comprises two phenomena, a viscous friction force and a force caused by a pressure gradient along the bubble surface in the direction of movement. The local pressure gradient is the result of the bubble wake and is therefore more pronounced in the case of high bubble Reynolds numbers. The drag force is determined by:

$$F_D = \frac{1}{2} c_D \rho_l \frac{\pi d_b^2}{4} (u_b - u_l)^2 \tag{36}$$

Equation (34) indicates, that the drag force is directly proportional to the dimensionless drag coefficient c_D, which depends on bubble properties and the flow conditions. The effect of bubble deformation is usually lumped into c_D as well. By dimensionless analysis, it can be shown that the drag coefficient depend on the bubble Reynolds number, the Eötvös number, and the Morton number [82].

Analytically, Levich (in [92]) derived inviscid flows, for which the influence of viscous friction can be neglected:

$$c_D = \frac{48}{Re} \tag{37}$$

For creeping flows, for which the pressure contribution by the wake is negligible, Hadamard (in [82]) derived:

$$c_D = \frac{16}{Re} \tag{38}$$

Mei and Klausner [93] proposed an analytical approximation for spherical bubbles at arbitrary Reynolds numbers, which fits Equations (37) and (38) asymptotically.

$$c_D = \frac{16}{Re}\left[1 + \left(\frac{8}{Re} + \frac{1}{2}\left(1 + 3.315 Re^{-0.5}\right)\right)^{-1}\right] \tag{39}$$

These analytical solutions are often used as benchmark for experimental or numerical approximations of c_D. However, they are not applicable for most industrial processes, since the bubbles are usually deformed and the fluid system is contaminated to some extent, which increase the real drag coefficient.

The second and most common approach to derive c_D are experimental measurements. Usually, the rise of single bubbles in stagnant ($u_l = 0$), mostly aqueous liquids, is studied.

After the bubble reaches its terminal rising velocity, the drag force is in equilibrium with the buoyancy force. The drag coefficient is then given by:

$$c_D = \frac{4}{3}\frac{(\rho_l - \rho_b)d_b g}{\rho_l u_b^2} \tag{40}$$

A large number of correlations based on this approach can be found in the literature. These are listed for example by Clift et al. [82] or Pang and Wei [94]. Here, only some correlations, which were found in close agreement with validation data in CFD studies, are repeated.

By reviewing existing correlations, Tomiyama et al. [95] derived three correlations in dependence on the degree of contamination of the system. For pure liquids, they derived:

$$c_D = \max\left[\min\left[\frac{16}{Re}\left(1 + 0.15Re^{0.687}\right), \frac{48}{Re}\right], \frac{8}{3}\frac{Eo}{Eo + 4}\right] \tag{41}$$

This correlation is often employed in numerical models. For example, Frank et al. [96] showed that this correlation has the best accordance with experimental validation data. Bröder and Sommerfeld [7] proved its validity by experimental measurements in (dilute) bubble swarms.

An entirely different approach to correlate the drag coefficient was proposed by Bozzano and Dente [79]. Because the bubble deformation and the wake structure are correlated, they argued that the drag may be described by the bubble deformation, which is usually implicitly considered:

$$c_D = f\left(\frac{a}{R_{eq}}\right)^2 \tag{42}$$

where f is a friction factor given by:

$$f = \frac{48}{Re}\left(\frac{1 + 12Mo^{0.33}}{1 + 36Mo^{0.33}}\right) + 0.9\frac{Eo^{1.5}}{1.4\left(1 + 30Mo^{0.167}\right) + Eo^{1.5}} \tag{43}$$

The second term on the right hand side is a deformation factor given by:

$$DEF = \left(\frac{2a}{d_{eq}}\right)^2 \cong \frac{10\left(1 + 1.3Mo^{0.167}\right) + 3.1Eo}{10\left(1 + 1.3Mo^{0.167}\right) + Eo} \tag{44}$$

However, experimental measurements have the disadvantage that it is very difficult to control all boundary conditions exactly. In particular, contaminants have a significant influence on the rising velocity as discussed in detail below.

A correlation for c_D, based on a DNS, was proposed by Dijkhuizen et al. [89]:

$$c_D = \sqrt{c_D(Re)^2 + c_D(Eo)^2} \tag{45}$$

where $c_D(Re)$ is the analytical solution by Mei and Klausner (Equation (39)) and $c_D(Eo)$ is given by:

$$c_D(Eo) = \frac{4Eo}{Eo + 9.5} \tag{46}$$

A comparison of the different drag correlations independent of the Eötvös number and the bubble equivalent diameter is provided in Figure 9. The physical properties of the liquid metals employed for this comparison are listed in Table 4. It can be seen that Equation (45) predicts higher rising velocities and lower drag coefficients, respectively, compared to Equations (41) and (42), indicating that the experimental correlations might be influenced by contaminants.

Those correlations show that the drag force at low Reynolds numbers is governed by the Reynolds number, but at sufficiently high Reynolds numbers, it is determined by the Eötvös number. Though some correlations explicitly include the Morton number [79,97], most studies agree that the Reynolds, Eötvös, and Morton numbers depend on one another when considering the drag force [95]. Thus, the system is fully described by the Reynolds and Eötvös number. However, it should be noted that almost all measurements were performed in aqueous systems where the density difference between fluid and bubble and the surface tension can be varied only to a small extent. Due to measurement difficulties in liquid metals, data from this fluid systems are very scarce. Usually, bubble chains were investigated, which has the disadvantage that the assumption of a stagnant fluid cannot be made. Thus, the bubble rising velocity is different from the relative velocity and a drag coefficient cannot be derived. There are only a few exceptions such as the measurements by Davenport [19], Mori et al. [98], Zhang et al. [99], and Wang et al. [32]. Davenport [19] performed measurements in mercury at room temperature and in liquid silver at 1000 °C. Bubbles were generated by a rotating cup. This cup initially pointed downwards, so that gas flowing in was collected in it. When the cup was turned, the collected gas rose as a single bubble. The rising velocity was derived by dividing the rising path by the time between bubble release and bubble's breaking through the bath surface. Mori et al. [98] measured the rising velocity by electric conductivity probes. Zhang et al. [99] and Wang et al. [32] both employed ultrasonic doppler velocimetry. A comparison between the different drag correlations described above and the measurements is provided in Figure 9. For mercury (Figure 9a), the measured rising velocities of small bubbles are slightly lower than predicted, but generally agree very well with the correlations. For larger bubbles, the experimental values lie between the predicted values by Tomiyama et al. [95] (Equation (41)) and Bozzano and Dente [79] (Equation (42)) and the numerical correlation by Dijkhuizen et al. [89] (Equation (45)). A possible explanation might be a slight contamination of the used mercury or the measurement uncertainty. Unfortunately, information about the purity of the mercury is neither provided by Davenport [19] nor by Mori et al. [98]. For intermediate Eötvös numbers, the measurements are in better accordance with the experimentally derived correlations, while for higher Eötvös numbers, they fit the numerically derived correlation better.

For liquid silver (Figure 9b), two measurement methods were used by Davenport [19], depending on whether the bubble's acceleration phase was taken into account or not. With the latter method (marked with an x), about 15% higher rising velocities were derived. This leads to an excellent agreement with the Dijkhuizen et al. [89] (Equation (45)) correlation. Again, no remark on the purity of the liquid silver was given. In GaInSn (Figure 9c), where measurements were made at relatively small Eötvös numbers, both authors found slightly lower rising velocities than predicted for pure systems. This could be an indication that contaminants influence the rising velocities.

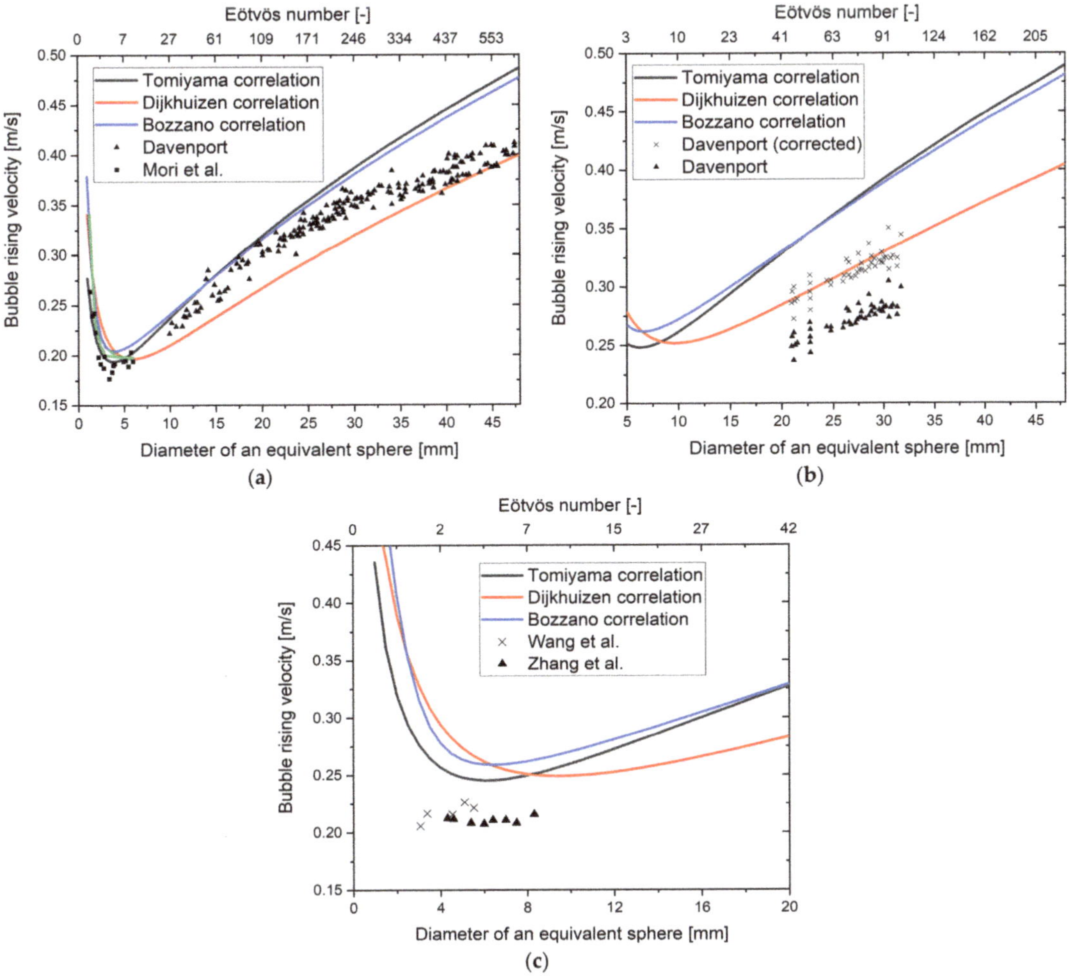

Figure 9. Comparison of drag models with measured bubble rising velocities in mercury (**a**), silver (**b**), and GaInSn (**c**) [19,32,98,99].

In addition, the bubble rising velocity of single bubbles in liquid steel was investigated numerically [90,91]. A comparison of the results with the correlations is shown in Figure 10. As with the experimental measurements, there is fairly good agreement with the correlations. However, no distinct tendency to which correlation fits better can be derived. As with the bubble deformation, however, it should be noted that the resolution of the numerical grid used was too coarse for a real DNS.

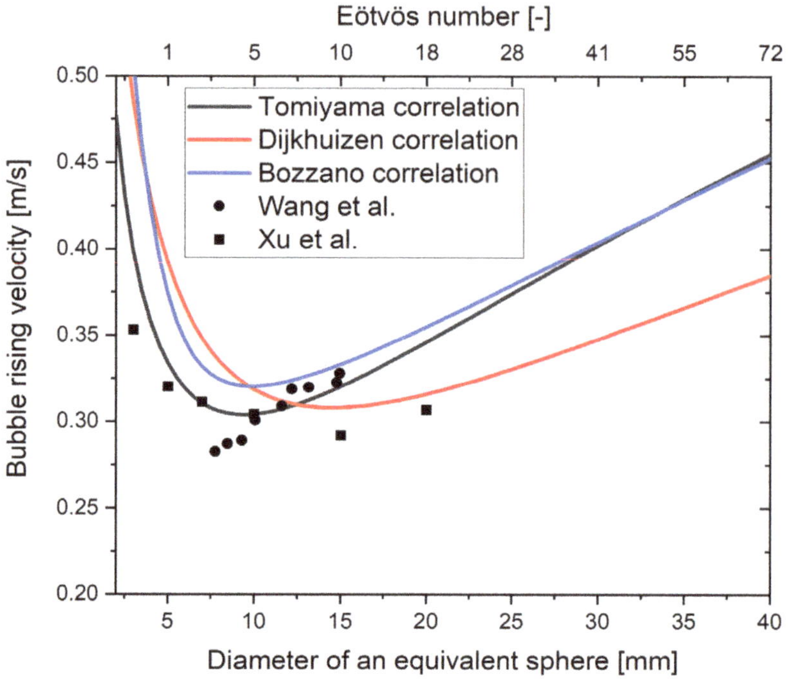

Figure 10. Comparison of drag models with numerically computed bubble rising velocities in steel [90,91].

Nevertheless, the comparison can be seen as an indication that the correlations derived from aqueous systems can be extrapolated to liquid metals. For small Eötvös numbers, there is a small uncertainty due to the influence of contaminants. For larger Eötvös numbers, the correlations are based on relatively few data, so there may be another uncertainty due to extrapolation of the correlations. For a final conclusion regarding which correlation should be used for molten metals, no conclusion can be drawn by now since the verification dataset is too small and its measurement uncertainty is too high. The same applies for the influence of contaminants in liquid metals.

5.1.1. Influence of Contaminants

Contaminants cause a reduction of the drag force due to the Marangoni effect or by a demobilization of the phase boundary. Because the viscosity of the fluid in the bubble is typically much smaller than that of the surrounding fluid, the boundary condition at the uncontaminated bubbles is typically a zero-shear-stress condition (Figure 11a). However, in contaminated systems, contaminants attached at the front of the bubble slide along the surface and accumulate at the bubble's rear (Figure 11b). This leads to gradients in surface tension along the bubble surface, causing a tangential shear force [100]. At sufficiently high contamination, complete demobilization of the bubble's surface occurs, and its flow condition changes from a no-shear to a no-slip condition (Figure 11c). This causes the bubbles to behave approximately like rigid particles.

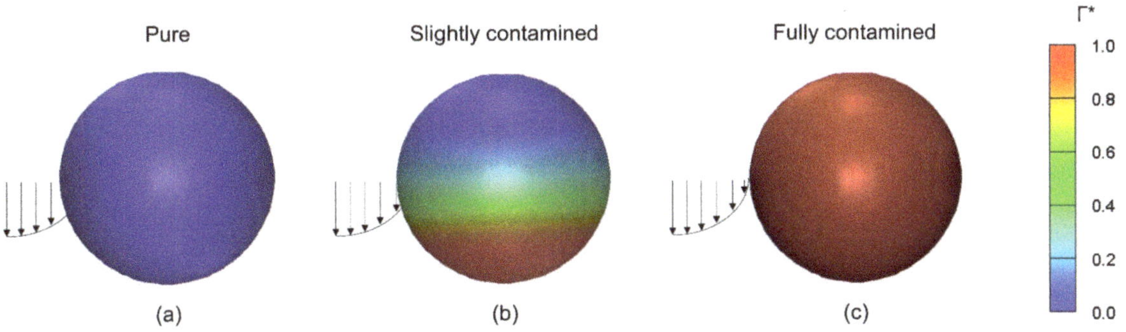

Figure 11. Schematic overview of the effect of contaminants at (**a**) pure, (**b**) slightly contaminated, and (**c**) fully contaminated systems (Γ* = dimensionless contaminant concentration).

Although this effect has been known for a long time, it is still not fully understood and can only be roughly quantified. A frequently employed approximation was derived by Tomiyama et al. [95]. Amongst a correlation for pure systems (Equation (41)), they derived correlations for slightly contaminated (Equation (47)) and fully contaminated (e.g., tap water) (Equation (48)) systems.

$$c_D = \max\left[\min\left[\frac{24}{Re}\left(1+0.15Re^{0.687}\right), \frac{72}{Re}\right], \frac{8}{3}\frac{Eo}{Eo+4}\right] \quad (47)$$

$$c_D = \max\left[\frac{24}{Re}\left(1+0.15Re^{0.687}\right), \frac{8}{3}\frac{Eo}{Eo+4}\right] \quad (48)$$

The different rising velocities derived from the three correlations are shown in Figure 12. It can be seen that, according to Tomiyama et al. [95], the contamination is only relevant for small bubbles.

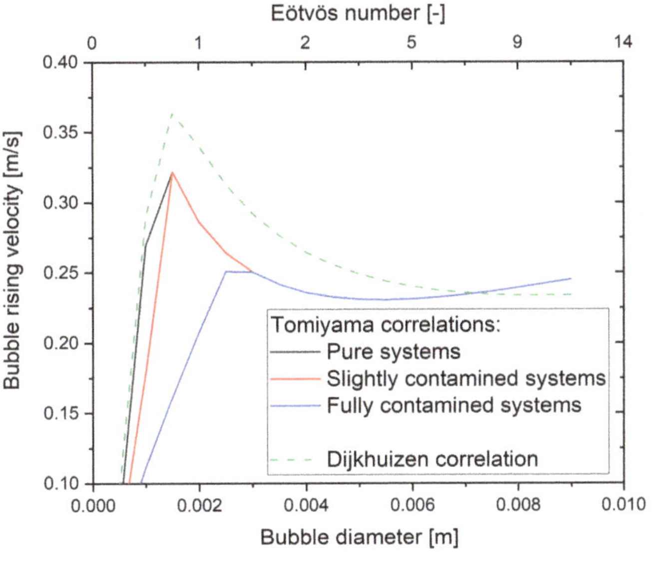

Figure 12. Bubble rising velocity of air in water in dependency of the degree of contamination of the fluid.

However, a comparison with the correlation by Dijkhuizen et al. [89] (Equation (45)) suggest that the measurement for all three correlations are influenced by contaminants and that the rising velocity of larger bubbles is influenced as well. Indeed, the correlation for pure systems was derived from measurements in distilled water, which is far from being pure. Therefore, the question of whether contamination is important for all bubble sizes cannot yet be answered. Some authors argue that for large Eötvös and Weber numbers, surface tension no longer plays a role for c_D and thus contaminants also no longer have an influence on the rising velocity. However, contaminants not only affect the flow boundary condition at the bubble surface, but also the deformation and vortex shedding. Therefore, Tasoglu et al. [101] found that contaminants are relevant for a broad range of Eötvös numbers, but the effect is much more pronounced for smaller bubbles. A problem here is that DNS studies usually investigate very small Reynolds number flows and large Morton numbers, since the computational costs increases with $Re^{9/4}$. Experimental studies, on the other hand, face the problem that it is practically impossible to generate completely uncontaminated systems. Therefore, sound studies on the effect of contaminates for large bubbles at relevant Reynolds numbers are largely lacking. A further critique of the Tomiyama et al. [95] approach is that for most engineering applications, there is no a priori definition of the degree of contamination of the system.

There are numerous attempts to quantify the range of intermediate contamination more precisely. The results are summarized by the reviews of Cuenot et al. [102] and Takagi and Matsumoto [100]. For low contaminant concentrations, Zhang and Finch [103] showed that bubbles may rise a few meters until they reach their terminal velocity. In that case, the effect is governed by the Hatta number, Ha, which is the ratio of adsorption velocity to the bubble rising velocity:

$$\mathrm{Ha} = \frac{k c_\infty d_b}{|u_d - u_l|} \tag{49}$$

Furthermore, the strength of the effect not only depends on the concentration of the contaminants, but also on the type of contaminants [104]. Thus, it was suggested that the effect should be quantified by the Langmuir number, La, which is defined as the ratio between the adsorption and desorption rate:

$$\mathrm{La} = \frac{c_\infty}{\beta} \tag{50}$$

The drag coefficient as a function of the Langmuir number is shown in Figure 13. It can be seen that the Langmuir number not only influences the viscous share of the drag force, but also the pressure share. It should be noted though that the results shown in Figure 13 were derived for low Re numbers. Thus, it is not clear whether the dependency of the pressure share on the Langmuir number can be extrapolated to higher Re numbers where the flow is mainly governed by inertia effects.

However, the relationships described above are still the subject of molecular dynamics research. They can hardly be used in engineering science, since information on concentrations and absorption and desorption kinetics is generally not available. This is particularly true for liquid metal systems. Therefore, it is not possible to determine whether contaminants, for example alloying elements, trace elements, or non-metallic inclusions, have an influence on bubble behavior in metallurgical processes.

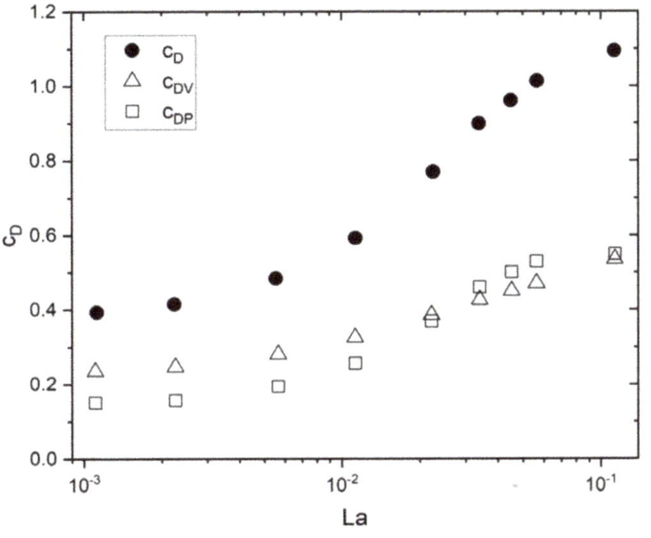

Figure 13. Drag coefficient as a function of the Langmuir number (adapted from [105] with permission from AIP publishing, 2021).

5.1.2. Influence of Surrounding Bubbles

The studies and correlations presented so far all dealt with single bubbles in stagnant liquids. However, some authors, summarized by Simonnet et al. [106], showed that these correlations are no longer valid when the bubble is surrounded by other bubbles. A similar effect for the sedimentation of solid particles was found by Richardson and Zaki [107]. In particular, they found a correlation between the sedimentation velocity of single particles, the sedimentation velocity of particles in suspensions, and the suspension density (equivalent to the void fraction in bubbly flows):

$$u_{\text{rel, swarm}} = u_{\text{rel, single}}\left(1 - \alpha_{\text{glob}}\right)^{n} \tag{51}$$

where the exponent n was coined Richardson and Zaki exponent by Simonnet et al. [106]. Because the particle velocity and the drag force are correlated, the effect of particle swarms on the drag force can be taken into account by multiplying a correction factor to the drag coefficient of single particles:

$$c_{D,\text{swarm}} = c_{D,\text{single}} \cdot f(\alpha, \ldots) \tag{52}$$

Most early studies on bubble swarms used Equation (52) and fitted the Richardson and Zaki exponent to their experimental data [106]. However, in bubble swarms, the situation is more complex since the possible degree of freedom for bubbles is greater than those of solid particles. Thus, Lockett and Kirkpatrick [108] added an additional factor that takes into account bubble deformation.

These correlations used a global void fraction. However, Garnier et al. [109] argued that the change of drag is caused by local, not global, phenomena. Experimentally, the local void fraction was determined by the quotient of time in which a bubble was detected in the measurement probe and the total measurement time [110].

A correlation based on the local void fraction was established by Simonnet et al. [106]:

$$f(\alpha) = (1 - \alpha_{\text{loc}})\left[(1 - \alpha_{\text{loc}})^{m} + \left(4.8\frac{\alpha_{\text{loc}}}{1 - \alpha_{\text{loc}}}\right)^{m}\right]^{-\frac{2}{m}} \tag{53}$$

where m is a smoothing parameter, estimated with a value of 25 by experimental data. Simonnet et al. [106] observed that the increase of the drag coefficient with increasing void fraction, which was reported by all previous studies, is only valid for small bubbles ($d_b \leq 7$ mm). For larger bubbles, the drag increases up to a critical void fraction of approximately 15% and decreases afterwards. Thus, Equation (53) is only valid for an air–water system, where bubble diameters are between 5 and 10 mm, with the restriction that for void fractions above 15%, the bubble diameter has to be larger than 7 mm. Simonnet et al. [106] assumed that the increase is caused by a hindrance effect by surrounding bubbles while the decrease might be explained by the aspiration in the bubble wakes of preceding bubbles.

Roghair et al. [111] investigated the effect of surrounding gas bubbles on the drag coefficient by DNS. In contrast to previous studies, they found a linear relationship with the void fraction as well as a dependence on the Eötvös number:

$$f(\alpha) = (1 - \alpha_{loc})\left(1 + \left(\frac{18}{Eo}\right)\alpha\right) \tag{54}$$

Equation (54) was established for $1 \leq Eo \leq 5$ and local void fractions up to 45%. However, due to the larger bubble size and density gradients, the Eötvös range is much larger in molten metals.

The reasons for the differences in the derived functional relationship can only be speculated. For example, Roghair et al. [111] suggested the influence of contaminants in experimental measurements as a possible explanation. In addition, an accurate comparison is difficult because some researchers used the global void fraction, while others used the local void fraction, and there is no direct connection between the two.

A fundamental problem is that, with the exception of the numerical study by Roghair et al. [111], the effect has only been studied for water–air systems. Roghair et al. [111] observed that the Morton number does not seem to have any effect. It should be noted, though, that in their study, it was only slightly varied.

Using numerical methods, different results were found. While Grienberger and Hofmann [112] did not find any significant improvement by including the swarm effect, Simonnet et al. [113] found that the transition between homogeneous and heterogeneous regimes was significantly better reproduced when the swarm effect was included with Equation (53). Lau et al. [114] found a significantly improved agreement between validation experiments and numerical results regarding fluid velocity and turbulence quantities when the swarm effect was taken into account by Equation (54). However, one difficulty is the determination of the local void fraction when using the Lagrangian particle tracking model. In the study of Simonnet et al. [113], it was assumed that bubbles are small compared to the numerical grid. Therefore, the local void fraction could be estimated as the concentration of bubbles in each cell. However, the grid spacing Δx was 10 mm while the bubble size d_b was 8.5 mm. Thus, this assumption seems to be oversimplified. Lau et al. [114] used a mapping technique to determine the local void fraction. However, within this approach, the result were dependent on the size of the mapping window, n. For sizes $n > d_b$, however, this effect became negligible.

5.2. Lift Force

In bubble column reactors, a lateral spreading of the bubble column can be observed. This behavior cannot be explained by the drag force alone. Instead, this phenomenon is attributed to the lift force, causing a lateral motion of the bubbles. Though other formulations for the lift force exists, the most frequently used form nowadays is the shear-induced lift model proposed by Zun [115]:

$$F_L = c_L \frac{\pi d_b^3}{6} \rho_l (u_b - u_l) \times (\nabla \times u_l) \tag{55}$$

where c_L is the lift coefficient. The first analytical solution for the lift coefficient was derived by Saffman [116] for rotating, solid spheres in creeping flows (Re→0) and infinite shear rates, which could later be extended numerically by McLaughlin [117] to arbitrary shear rates:

$$c_L = \frac{1.37J(Re, Sr)}{\sqrt{Re \cdot Sr}} \tag{56}$$

where J is a function of the bubble Reynolds number and the dimensionless shear rate Sr:

$$J(Re, Sr) = \frac{2.255}{\left(1 + 0.2\frac{Re}{Sr}\right)^{3/2}} \text{ for } Re << 1 \tag{57}$$

$$Sr = \frac{\omega d_b}{|u_b - u_l|} \tag{58}$$

Legendre and Magnaudet [118] showed, based on a first attempt by Mei and Klausner [119], that the lift coefficient for uncontaminated, spherical bubbles is 4/9 of those for solid spheres:

$$c_L = \frac{6J(Re, Sr)}{\pi^2\sqrt{Re \cdot Sr}} \tag{59}$$

For inviscid flows (Re→∞), Auton [120] showed analytically that the lift coefficient for spherical particles or bubbles is 0.5.

The first correlation for arbitrary Reynolds numbers, approaching both solutions asymptotically, was numerically derived by Legendre and Magnaudet [121]:

$$c_L = \sqrt{\left(\frac{6J(Re, Sr)}{\pi^2\sqrt{Re \cdot Sr}}\right)^2 + \left(\frac{1Re + 16}{2Re + 29}\right)^2} \text{ for } Sr << 1 \tag{60}$$

As shown in Figure 14, the correlation predicts that the lift coefficient is determined by the dimensionless shear rate for low Reynolds numbers and approaches 0.5. Furthermore, the correlation predicts an independence of the shear rate for intermediate and high Reynolds numbers. In addition, all lift coefficients are positive.

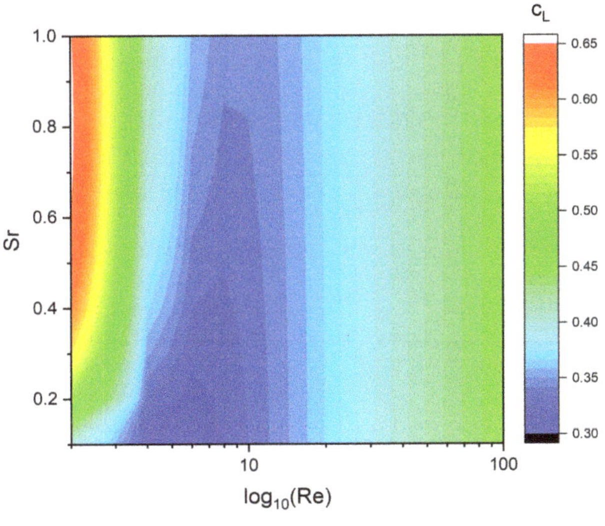

Figure 14. Lift coefficient of a spherical bubble in dependence of the dimensionless shear rate (Sr) and the Reynolds number (Re) according to the model of Legendre and Magnaudet [121].

However, Legendre and Magnaudet [121] already pointed out that real conditions, that is to say ellipsoid bubbles with inner circulation in turbulent flows with arbitrary shear rates, are far more complex. Indeed, the theoretical model was not able to describe all phenomena observed in experiments. Experimentally, lift coefficients between 0.25 and 0.3 have been found for small bubbles in air–water systems [115,122], which is about half the value proposed by Equation (60). More importantly, Zun [115] reported that in a pipe flow, larger bubbles move towards the pipe center while smaller bubbles migrate to the walls. This indicates a change of sign of the lift coefficient, which depends on the bubble diameter. For air–water systems, Liu [123] could determine the change of sign for equivalent bubble diameters of about 5 mm in a pipe. Naciri (in [124], original in French) analytically showed that this effect is not solely caused by bubble deformation. Instead, the reason for the sign change is that the lift force, similar to the drag force, composes two different effects. By reviewing existing experimental data, Serizawa und Kataoka [125] came to the conclusion that the lateral motion is caused by unsteady asymmetries in the bubble wake and a shear flow around the bubble. This theory was later confirmed in a simplified numerical simulation by Tomiyama et al. [126] and experimentally by Brücker [127]. Hibiki and Ishii [124] speculated that the bubble orientation in the shear flow, the wake structure modification, and the bubble shape may cause the sign change. This was proved by Adoua et al. [128], who refined the theory by numerically showing that the change of sign is caused by the generation of counter rotating streamwise vorticity at the bubble surface and its complex interaction with the shear flow. Experimental evidence for that was provided by Aoyama et al. [129], which found that the sign change and the onset of path instabilities, which is also linked to the generation of vorticities, follow similar patterns.

However, in mathematical models of industrial scale, the flow field in the close vicinity of the bubble practically cannot be resolved. Instead, semi-empirical models are used for the macroscopic description of the bubble behavior. Therefore, the cause of the lift force plays only a minor role outside of fundamental research. Tomiyama et al. [85] suggested that the superimposed effect of shear and wake have similar mathematical forms and can be jointly incorporated into the lift coefficient c_L so that Equation (55), which was originally derived for shear-induced lift forced only, can be employed to describe both effects.

A systematic experimental investigation on c_L was carried out in Tomiyama's much-acclaimed work [85]. A rotating belt was used to induce a laminar shear flow in distilled water–glycerol mixtures of different viscosities. In this shear flow, the rise of bubbles with different diameters was analyzed in the ranges of $-5.5 \leq \log_{10}(\mathrm{Mo}) \leq -2.8$, $1.39 \leq \mathrm{Eo} \leq 5.75$, and $0 \leq \omega \leq 8.3\ \mathrm{s}^{-1}$. Measurement in systems with lower viscosity were not feasible, because the rotating belt requires a sufficient viscosity to induce a linear shear flow. The measurements yield the following empirical correlation:

$$c_L = \begin{cases} \min\left[0.288\tanh\left(0.121\mathrm{Re}; f(\mathrm{Eo_d})\right)\right] & \mathrm{Eo_d} \leq 4 \\ 0.00105\mathrm{Eo_d^3} - 0.0159\mathrm{Eo_d^2} - 0.0204\mathrm{Eo_d} + 0.474 & 4 < \mathrm{Eo_d} \leq 10.7 \end{cases} \quad (61)$$

where $\mathrm{Eo_d}$ is the modified Eötvös number, using the major axis of deformed bubbles as characteristic length:

$$\mathrm{Eo_d} = \frac{g(\rho_l - \rho_b)d_{max}^2}{\sigma} \quad (62)$$

Later, Frank et al. [96] added:

$$c_L = -0.27 \quad \mathrm{Eo_d} > 10 \quad (63)$$

to ensure that the lift coefficient c_L is a monotonic function of the modified Eötvös number. However, it is questionable whether it is physically justified to extrapolate the correlation for very large bubbles, because the bubble size has a significant influence on the wake structure.

In the experiments, a Reynolds number dependency of the lift coefficient could only be found for very small bubbles. For larger bubbles, a dependence on the modified Eötvös number was found, but the shear rate had practically no influence on the lift coefficient. It should be noted that $\log_{10}(\mathrm{Mo})$ for water at room temperature is about -10.6. Nonetheless, though the study was conducted for higher viscous systems ($-5.5 \leq \log_{10}(\mathrm{Mo}) \leq -2.8$), a very good agreement with the experimental measurements in air–water systems [115,122,123] and Equation (61) was found. Because of that, the authors concluded that their correlation can be extrapolated to air–water systems.

The correlation of Tomiyama et al. [85] has become popular. It has been used in many industrial scales in CFD models [130] to date and is the most widespread lift coefficient model nowadays. However, it should be kept in mind that it is purely empirical and was derived in a limited range of fluid properties. Extrapolation from this range can be error prone. For example, numerical investigations by Bothe et al. [131] found slightly lower c_L values and a limit of 0.5 for small Eötvös numbers. For a range of $-10.8 < \log_{10}(\mathrm{Mo}) < -1.8$, $0 < \mathrm{Eo_d} < 20$, $0 < \mathrm{Eo} < 12$, $0 < \mathrm{Re} < 1500$, Dijkhuizen et al. [132] found, via numerical simulation, that Equation (61) is only applicable in the range that the measurements actually took place. They showed that for systems of higher viscosity or for very small Reynolds numbers ($\mathrm{Re} < 10$ and $E < 0.95$), the predicted values can vary quite significantly from the simulated ones. For the latter case of small Reynolds numbers, Dijkhuizen et al. [132] found good agreement with the correlation of Legende and Magnaudet [121] (Equation (60)) rather than the plateau proposed by Tomiyama et al. [85]. On the other hand, they found that Equation (60) overestimates c_L significantly in case the bubble shape deviates from a sphere. Based on their results, Dijkhuizen et al. [132] proposed:

$$c_L = \min\left(\sqrt{\left(\frac{6J(\mathrm{Re},\mathrm{Sr})}{\pi^2\sqrt{\mathrm{Re}\cdot\mathrm{Sr}}}\right)^2 + \left(\frac{1\mathrm{Re}+16}{2\mathrm{Re}+29}\right)^2}, 0.5 - 0.11\mathrm{Eo_d} + 0.002\mathrm{Eo_d^2}\right) \tag{64}$$

Experimentally, the measurement range was expanded ($1.9 \times 10^{-2} < \mathrm{Re} < 1.2 \times 10^2$, $-6.6 \leq \log_{10}(\mathrm{Mo}) \leq -3.2$, $2.2 \times 10^{-2} < \mathrm{Eo} < 5.0$, $3.4 \times 10^{-2} < \mathrm{Sr} < 3.5$) by Aoyama et al. [129] using a similar rotating belt system as Tomiyama et al. [85]. Similar to Dijkhuizen et al. [132], it was found that Equation (61) is only applicable in a limited range. Moreover, it was found that none of the tested dimensionless numbers (Re, Eo, We, $\mathrm{Eo_d}$, Ca) alone can be used to correlate an accurate equation for the lift coefficient. The most promising attempts in this direction were made with the Reynolds and modified Eötvös numbers, though for both approaches, the Morton number has to be taken into account, too.

The first systematic measurements for air–water systems was achieved by Ziegenhein et al. [133] in a different experimental setup. Instead of inducing the linear shear by a rotating belt, which limits the measurement range to highly viscous systems, the shear was generated by a bubble column. An additional advantage of this system is that no moving parts are used, which allows a better control of the contamination level of the system. On the other hand, the bubbles introduce some turbulence [134] that makes the evaluation more challenging. Their measurements showed that the modified Eötvös number is a better choice than the Reynolds number for correlating c_L for a broader range of Morton numbers. In addition, it was found that the instantaneous lift force can vary quite significantly in water systems. Combining their results with those of Aoyama et al. [129] suggests that there is always a sign change and a linear behavior of c_L around the sign change, as shown in Figure 15. However, the modified Eötvös number at which the sign change occurs as well as the slope of the linearity around the sign change depend on the Morton number. The sign change occurs at smaller modified Eötvös numbers with a decreasing Morton number. However, this trend reverses at a Morton number not exactly known yet. In addition, it was found that c_L asymptotically approaches an upper and a lower limit for high and low modified Eötvös numbers, respectively. These limits again seem to depend on the Morton number, too.

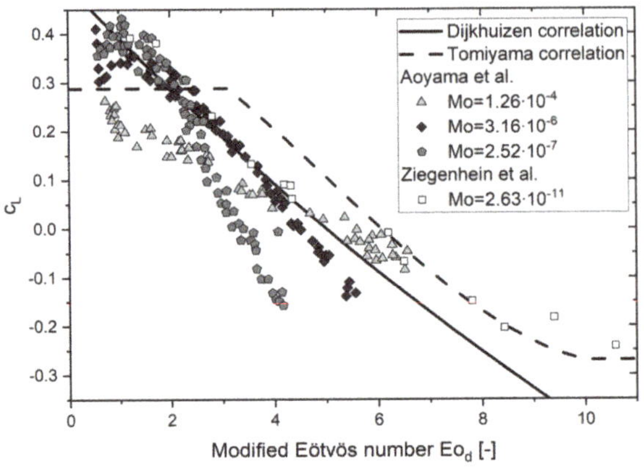

Figure 15. Experimental lift coefficients as a function of the modified Eötvös number (adapted from [133], with permission from Elsevier, 2021).

In the study by Ziegenhein et al. [133] on an air–water system, an asymptotical behavior was found for smaller modified Eötvös numbers than predicted by Tomiyama et al. [85]. Moreover, a higher asymptotic limit was found. For the lower limit, it can be assumed that c_L is increasing in case of very small Re numbers. However, these small Reynolds numbers could not be produced in an air–water system. For the investigated air–water system, Ziegenhein et al. [133] correlated, via second-order polynomial regression, their results:

$$c_L = 0.5 - 0.1\mathrm{Eo_d} + 0.002\mathrm{Eo_d^2},\ 1.2 < \mathrm{Eo_d} < 10.5 \tag{65}$$

which is quite close to Equation (64).

The range of the most important studies is given in Figure 16. There is a clear gap of experimental studies in the range of $-10.5 < \log_{10}(\mathrm{Mo}) < -6.6$ and $-10.5 > \log_{10}(\mathrm{Mo})$, which should be complemented to make correlations more applicable to a wider range. For Morton numbers corresponding to the range of liquid metals, studies are missing.

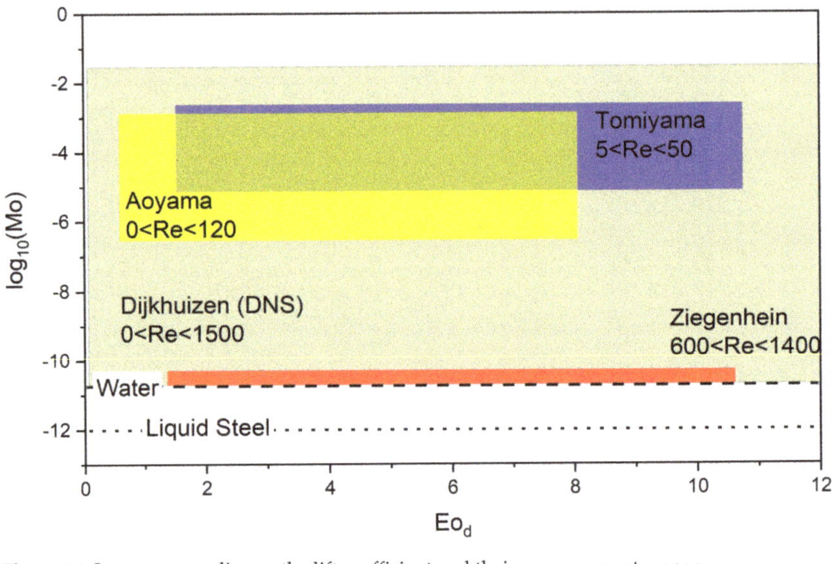

Figure 16. Important studies on the lift coefficient and their measurements range.

The correlations for deformed bubbles described so far were purely empirical. To compensate for this disadvantage, Hibiki and Ishii [124] used the experimental data from Tomiyama et al. [85] to extend the correlation of Legendre and Magnaudet [121] by an additional amplitude factor. The aim of this ambitious attempt was to couple the change of the lift coefficient with the different ranges of the drag coefficient, thus giving the lift coefficient a more physical background:

$$c_L = \left(2 - \exp(0.136Eo^{1.11})\right)\sqrt{\left(\frac{6J(Re, Sr)}{\pi^2\sqrt{Re \cdot Sr}}\right)^2 + \left(\frac{1Re + 16}{2Re + 29}\right)^2} \tag{66}$$

However, Dijkhuizen et al. [132] showed that Equation (66) is unable to predict c_L accurately for a wider range of physical properties.

Influence of Contamination

The influence of contaminants and solid particles on the lift force has not been included to the discussion of the lift coefficient in the previous paragraph. However, it is known that their impact can be quite significant. In the above-discussed studies, the authors used purified water to minimize this effect. However, the pureness of this liquids varied and contamination sources, like moving parts in the rotating belt setup used by Tomiyama et al. [85], Dijkhuizen et al. [132], and Aoyama et al. [129], cause some level of uncertainty. As described in the subsequent section, this uncertainty complicates the comparability of the experimental studies, but may also explain some discrepancy of their results.

Ogasawara [135] experimentally observed that the tendency of small bubbles to cluster in the vicinity of the wall was reduced in case surfactants were added. Dijkhuizen et al. [132] used the rotating belt setup with tab water–glycerin mixtures to validate their numerical results. Here, they measured significantly lower lift coefficients for larger bubbles, but larger lift coefficients for smaller bubbles than predicted by the numerical results. In addition, they observed a shear flow dependency, which was not found numerically for pure liquids.

Influencing mechanisms of contaminations on the lift were also studied via numerical simulation. Fukuta et al. [105] found that the lift coefficient for a contaminated spherical bubble ($Re = 100$, $Sr = 0.2$) is significantly reduced and can become negative. To quantify the surfactant effects, they used the Langmuir number. As shown in Figure 17, it was found that in pure liquids, the total lift force (c_L) on small bubbles is mainly driven by pressure components (c_{LP}), while increasing surfactant levels decrease its impact until it disappears entirely, so that the small negative lift force is due to viscous stress (c_{LV}). Overall, the effect of surfactants was attributed to the Marangoni effect.

Hayashi and Tomiyama [136] expanded the range to larger bubbles, larger Langmuir numbers, and larger Hatta numbers ($2 < Re < 70$, $0.6 < Eo < 5$, $-6 < \log_{10}(Mo) < -4$, $0 < Sr < 1$, $1.38 \le La \le 13.8$, $0 < Ha < 41$). Similarly, they found that the lift coefficient decreases with increasing Langmuir number. The main effect, however, depends on the Hatta number. In case of large Hatta numbers, which means that the absorption of contaminants is much faster than the bubble rising, the decrease of the lift coefficient is attributed to a decrease of the effective surface tension by:

$$\sigma(\Gamma) = \sigma_0\left[1 + \frac{R_g T\Gamma_{max}}{\sigma_0}\ln(1 - \frac{\Gamma}{\Gamma_{max}})\right] \tag{67}$$

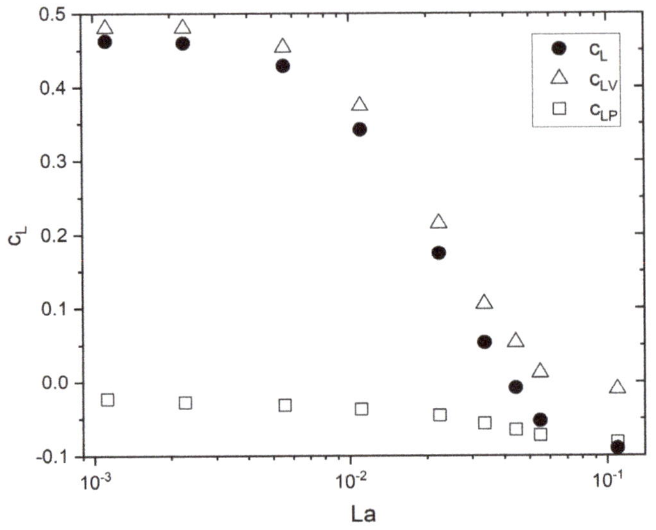

Figure 17. Effect of the Langmuir number on the lift coefficient (c_L) and its pressure (c_{LP}) and viscosity (c_{VL}) components (adapted from [105] with permission from AIP publishing, 2021).

In that case, the Marangoni effect is negligible. The lift coefficient can be correlated with the modified Eötvös number, employing the effective surface tension. In case of smaller Hatta numbers, the Marangoni effect gains importance. They also found an effect of the shear rate, which they attributed to a slight flow induced inclination of the bubble for contaminated bubbles.

Rastello et al. [137] investigated the impact on hydrophobic tracer particles, that attach to the bubble, on the lift coefficient in a rotating flow in a cylinder (D = 100mm). They observed that clean bubbles have an unseparated wake while the wake of contaminated bubbles separates. In addition, the tracer induces a surface spinning, which results in a Magnus effect. In fact, fully contaminated bubbles rotated like solid particles. Thus, c_L increases due to this additional contribution of the Magnus effect, which itself depends on the shear rate. In contrast, Hessenkemper et al. [138] found no influence of tracer particles in a linear shear flow. Until further research has been conducted, one can only speculate about this discrepancy. Reasons can either be that the dimensionless shear rates investigated by Rastello were quite high (Sr > 0.2), the flow was rotating, the number of tracers attached to the bubble in the experiment by Hessenkemper et al. [138] were too small, or the tracer sizes were different and therefore the surface tension interaction of the particles with the bubbles was significantly different. Hessenkemper et al. [138] investigates the influence of inorganic surfactants. In contrast to organic surfactance, it was found that they increase the lift coefficient.

These studies reveal that the impact of surfactants and solid particles is far from being understood. It is likely that the different effects superimpose on each other, which complicates analysis further. In addition, the number of studies and their range is too limited to derive a comprehensive theory or a quantification of the effects to a wider range of physical conditions. Furthermore, there is currently no link between the described phenomena and measurable variables, so that a quantitative estimation of some of the mentioned influences in experimental measurements would be possible. On the other hand, the influence of contaminations might explain the good agreement with macroscale CFD models employing the Tomiyama correlation (Equation (61)) and measurements in water models. Because usually these models are filled with tab water and sometimes tracer particles are added, it might be that the extrapolation error is coincidentally mitigated by the contamination effect. This would also explain the discrepancy between early measurements of the lift coefficient in tap water [115,122] and more recent measurement in deionized water [133].

Unfortunately, measurements in liquid metals are lacking entirely. Because there is no comprehensive understanding of the lift force in pure liquids yet, the extrapolation of existing correlations to very low Morton number systems, like liquid metals, suffers from a very high degree of uncertainty. The study by Aoyama et al. [129] suggests that both the asymptotic limits of the lift coefficient for large and small bubbles, as well as the sign change and the slope around it, depend on the Morton number. In liquid metals, the Eötvös number is usually significantly larger than in the systems investigated. Therefore, existing correlations would predict negative lift coefficients even for small, approximately spherical bubbles. However, this contradicts its physical justification and is therefore very unlikely. For larger, presumably ellipsoidal bubbles, large negative values for c_L would be predicted, which is likely to lead to unphysical behavior and stability problems in numerical models. Therefore, in particular, the negative asymptotical limit of c_L for large bubble in liquid metals is of great interest. The discussion is further complicated by the large uncertainty of the prediction of the bubble's deformation analyzed above, which is necessary for the calculation of the modified Eötvös number.

Another problem arises from the high surface tension of liquid metals. All experimental studies were made in systems with very low surface tension. The only evidence that the surface tension effect is entirely captured in the Morton number and modified Eötvös number was provided numerically by Bothe et al. [131], varying the surface tension from 0.1 to 0.8 N/m for pure liquids. However, in their study, systems with a significantly larger Morton number were investigated. Whether this relationship applies for higher surface tensions or if it can be reproduced experimentally cannot be predicted with the current knowledge. Even more difficult is the impact of surfactants and solid particles. It can be assumed that alloying elements and non-metallic inclusions affect the lift coefficient. However, it is likely that the strength of these effects is different from those observed in aqueous solutions because of the much higher surface tension. Finally, the lift coefficient may even change on the rising path of the bubbles due to the adherence of inclusions. Therefore, it is currently not possible to make reliable estimates about the lift force in metallurgical processes.

5.3. *Virtual Mass Force*

When a bubble rises through a liquid, some of the surrounding liquid is carried by the bubble. The virtual mass force is the force arising from the acceleration of the surrounding fluid. Since it virtually increases the mass of the bubble, it is called virtual mass force. It is given by [92]:

$$F_{VM} = c_{VM} \rho_l V_b \left(\frac{du_l}{dt} - \frac{du_b}{dt} \right) \tag{68}$$

The virtual mass force limits the bubble acceleration. Thus, it is important to stabilize numerical calculations [139]. Like the drag force and the lift force, the virtual mass force is proportional to a coefficient c_{VM}, by which all influences on the force are represented. If c_{VM} is too small, the bubble acceleration may become too large, which can cause numerical instability depending on the solution process and settings. If c_{VM} is too large, the acceleration phase becomes unphysically long and smaller oscillations in the bubble's rising path may be suppressed. For a spherical bubble in stagnant liquids, an analytical value of c_{VM} = 0.5 was derived [140]. This value is also used in almost all numerical studies. There are only a small number of quantitative studies on the exact value of c_{VM} in real flow conditions. These suggest that the value of c_{VM} also depends on the void fraction [141,142] and the bubble deformation ([143] found in [144]). However, if acceleration effects are not significant in the flow, then the virtual mass force is not important for the flow either. This applies, for example, to bubble reactors [142]. Therefore, the exact knowledge of c_{VM} plays a minor role in this type of flow.

6. Discussion

Despite numerous studies and many years of research, the dynamics of gas bubbles are not yet fully understood. This applies in particular to liquid metals, where even experiments to verify existing correlations are difficult, if possible at all, and place the highest demands on measurement techniques. In this review, the transferability of correlations from more accessible systems to liquid metals was analyzed. In this concluding discussion, the results are briefly summarized and approaches to resolve existing uncertainties are proposed.

The same formation regimes as in water models had been observed on single nozzles in liquid metals. However, the mechanisms differ with respect to wettability of the nozzle. For higher, more industrially relevant flow rates, this effect seems to decrease in relevance. Instead, the bubble size distribution is mainly determined by coalescence and especially breakup. The bubble formation mechanism is different at purging plugs than at single nozzles and presumably other bubble size distributions are generated. However, coalescence, rather than breakup, has a decisive role here. Extrapolating observations made in aqueous systems, it can be assumed that the bubble column in most processes is in the heterogeneous regime and the bubble size distribution follows a log-normal distribution. However, a verification of this assumption in liquid metals is difficult because relatively large experimental setups are needed, which exceed the limits of most available measurement techniques today. The only exception so far are resistivity probes, which are intrusive and consequently introduce a bias into the measurements. Therefore, the real bubble size distribution in industrial plants can only be predicted with considerable uncertainty, as the example of a ladle shows. For specific applications, such as the ladle, resistivity probe measurements in the real process can provide a reasonable estimate of the bubble size distribution. However, these measurements will be associated with difficulties such as the measurement bias and the short lifetime of the sensors at high temperatures [21]. More importantly, the bubble size distribution depends on local flow conditions and the type of gas injection, so that results of such measurements can hardly be generalized. Therefore, to predict the bubble size distribution in different processes, it is essential to develop a more detailed understanding of the coalescence and breakup mechanisms. This will be a major experimental challenge even in aqueous systems, since the instantaneous flow field and the behavior of bubbles have to be measured simultaneously in three dimensions. In liquid metals, this is currently not possible, although there is a first approach to simultaneously measure the flow and bubbles in two dimensions [29]. Studies in liquid metals like those by Keplinger et al. [22,75] can be used to critically discuss the scalability of mechanisms investigated in aqueous models, even though the physical restrictions of imaging techniques in liquid metals allow for a significantly lower degree of detail. Due to the difficulty of the experiments, DNS could be a useful alternative to develop fundamental knowledge of coalescence and breakup mechanisms. Nevertheless, this requires further development of phase boundary modeling, and these calculations will be highly computationally expensive. Once the coalescence and breakup mechanisms are sufficiently understood, these fundamental insights can be transferred to the reactor scale using PBMs. Since the first experiments in liquid metals suggest that turbulence effects are inherently important, at least an LES approach should be chosen. For a precise validation of such PBMs, however, exact experimental benchmark cases are necessary. For these benchmark cases it is important that all boundary conditions are reported in detail and the measurement uncertainty can be quantified [14].

A comparison of different bubble deformation models and experimental measurements in liquid metals revealed that these models, derived for aqueous systems, significantly overestimate the deformation. The best agreement was found with the model of Wellek et al. [81] for contaminated systems. This result can be interpreted in different ways. First, the less pronounced deformation could indicate that the examined liquid metal systems were contaminated, since, for example, they were partially oxidized. Another explanation is that the deformation has a dependence on the fluid properties that is not yet fully understood. Indeed, there are few measurements on systems with Morton numbers smaller than water, which approximately correspond to the values of liquid metals. A final assessment of which explanation applies cannot be made due to the small number of experimental studies. Experiments with high-purity metals in an inert gas atmosphere could provide new insights here.

Besides the influence of contaminants, other factors can introduce an uncertainty to the measurement of bubble deformation. These can partly explain the distinct differences of existing models and should be taken into account in the design and evaluation of future measurements of bubble deformation in liquid metals. Especially in shallow experimental setups, there may be an influence of the injector. When the bubble is released from the injector, a strong deformation occurs, which is subsequently damped. If the measuring point is too close to the injector, the results may be biased. This can be a particular problem for measurements in liquid metals, where the measurement volumes are often quite small. Therefore, a sufficient filling height must be ensured and an influence of the injector should be examined, for example by using different injectors. Other problems arise from the physical restrictions of imaging techniques such as X-ray or neutron radiography measurements. An influence of the walls on the deformation may arise due to the small thickness of the measurement volume. Furthermore, this limits the measurements to relatively small bubbles, so extrapolation of deformation models to larger bubbles will remain a problem. Finally, it should be considered that a measurement uncertainty arises from the reconstruction of the three-dimensional bubble shape by a single two-dimensional projection [13]. On the other hand, the alternative usage of multi-needle resistivity probes has the drawback of being intrusive.

Even though the number of experimental measurements is too small for a conclusive assessment, the drag force seems to be predictable employing existing correlations. The frequently used correlation according to Tomiyama et al. [95] generally provides reasonable accordance with measurements in liquid metals. For smaller bubbles, there is some uncertainty due to the influence of contaminants. [19] Additionally, the influence of surrounding bubbles in a bubble swarm on the drag coefficient in liquid metals cannot be quantified yet. Nevertheless, employing existing measurement methods, it should be possible to further refine the knowledge of the drag force in liquid metals. For this purpose, bubble swarms as well as single bubbles of different size should be examined. One challenge is the generation of single large bubbles, which is difficult to control by nozzles. For this purpose, a rotating cup as proposed by Davenport [19] might be used.

For the lift force there is no comprehensive understanding even for aqueous systems. For liquid metals, experimental data are lacking entirely. From a numerical point of view, it seems reasonable to define an asymptotic limit for negative values similar to Frank et al. [96], to prevent numerical instability. However, these values are currently highly uncertain for liquid metal systems and there is probably a dependency of the lift force on the Morton number, which is not yet quantifiable. Experimentally, it will be very difficult to determine the lift coefficient for liquid metals since an exactly defined shear flow must be generated, which is difficult to control in liquid metal systems. A possible solution might be a bubble column induced shear flow proposed by Ziegenhein et al. [133]. This shear flow could be quantified by NeuPIV [29], simultaneously measuring the rising path of a single bubble. On the other hand, given that the experimental setup has to the thin to employ neutron radiography, the influence of the walls might prevent a defined shear flow.

When considering metallurgical reactors, the virtual mass force seems mostly important for numerical stability and has little effect on the flow. The currently most frequently used value of 0.5 for the virtual mass coefficient seems to be sufficiently accurate in almost all cases.

A fundamental problem in the scaling of all described phenomena is that the dynamics of bubbles are strongly influenced by contaminants. However, with the current state of knowledge, it is not possible to conclusively assess whether contaminants are relevant in industrial processes, too. It can even be assumed that the influence changes during the process. It can be speculated whether contaminants have an influence on the conductivity of the melt and can therefore be quantified. However, this remains speculative at the present state of knowledge. Moreover, it is unknown whether contaminants have the same effect in liquid metals as in aqueous systems. Fundamental research in the field of molecular dynamics could provide new insights into this topic.

In summary, the behavior of gas bubbles in liquid metals is still far from being fully understood. This leads to a considerable uncertainty of numerical models of metallurgical reactors involving bubble flows. However, some direct measurement techniques have been developed in recent years that allow a more detailed analysis of different phenomena. Using these methods, further measurements should be carried out to reduce the lack of experimental data in liquid metals. With more experimental data, especially for value ranges and phenomena that have not yet been sufficiently analyzed, the discussion of the scalability of existing models can be significantly improved. This could help to reduce many of the remaining knowledge gaps in the future.

Author Contributions: Conceptualization, T.H., C.S., M.E. and H.P.; methodology, T.H.; software, T.H. and C.S.; validation, T.H., C.S., M.E. and H.P.; formal analysis, T.H.; investigation, T.H.; resources, T.H.; data curation, T.H.; writing—original draft preparation, T.H.; writing—review and editing, T.H.; visualization, T.H.; supervision, M.E. and H.P.; project administration, T.H., M.E. and H.P.; funding acquisition, H.P. All authors have read and agreed to the published version of the manuscript.

Funding: This research was funded by German Research Foundation (Deutsche Forschungsgemeinschaft (DFG)), grant number RU 2050/2-1.

Institutional Review Board Statement: Not applicable.

Informed Consent Statement: Not applicable.

Data Availability Statement: Data sharing not applicable.

Conflicts of Interest: The authors declare no conflict of interest.

Abbreviations

List of Symbols (Equations have been adjusted to SI Units).

Symbol	Description
c_D	Drag coefficient
$c_{1\text{–}4}$	Fitting parameter
c_L	Lift coefficient
c_{VM}	Virtual mass coefficient
c_∞	Far-field concentration of contaminants, mol/m^3
d_b	Arithmetic mean equivalent bubble diameter, m
$d_{b,32}$	Bubble Sauter mean diameter, m
d_{eq}	Equivalent bubble diameter, m
d_{max}	Length of bubble major axis, m
d_{min}	Length of bubble minor axis, m
d_{ni}	Orifice inner diameter, m
d_{no}	Orifice outer diameter, m
d_{po}	Pore diameter
E	Bubble eccentricity
Eo	Eötvös number
Eo_d	Modified Eötvös number
Fr	Froude number
F_i	Force i on bubble, N
F_D	Drag force, N
F_L	Lift force, N
F_{VM}	Virtual mass force, N
f	Friction factor in Bozzano drag model

Symbol	Description
g	Gravity, 9.81 m/s^2
H_0	Distance between nozzle and the mathematical origin of the nozzle, m
H_{Fill}	Filling height, m
K	Empirical constant in Mersmann correlation
k	Adsorption rate, $m^3/mol \cdot s$
Lc	Characteristic length, m
Mo	Morton number
m	Fitting parameter
m_B	Bubble mass, kg
n	Number of activated orifices
Q_g	Gas flow rate in m^3/s
Q_{min}	Minimum gas flow rate at which an orifice gets activated, m^3/s
R_g	Universal gas constant, 8.3145 $J/mol \cdot K$
Re	Reynolds number
r_{or}	Orifice radius, m
Sr	Dimensionless shear rate
S_{br}	Breakup rate
S_c	Coalescence rate
T	Temperature, K
Ta	Tadaki number
u_c	Characteristic velocity, m/s
u_{sg}	Superficial gas velocity
u_b	Bubble velocity, m/s
u_l	Liquid velocity, m/s
V_b	Bubble volume, m^3
We	Weber number
w_{sl}	Slot width, m
$\hat{y}$	Fitting parameter
z	Height, m
α	Global void fraction
α_i	Fitting parameter
α_{loc}	Local void fraction
β	Desorption rate, mol/m^3
Γ	Contaminant concentration, mol/m^2
ε	Porosity of a porous plate
θ	Fitting parameter
κ	Permeability of a porous plug
λ_{sl}	Distance between active bubble formation sites on slot nozzles, m
μ_g	Gas viscosity, Ns/m^2
μ_l	Liquid viscosity, Ns/m^2
ρ_b	Bubble density, kg/m^3
ρ_l	Liquid density, kg/m^3
σ	Surface tension, N/m
φ	Shape factor
ω	Shear rate, 1/s

References

1. Deckwer, W.-D.; Field, R.W. *Bubble Column Reactors*; Wiley: New York, NY, USA, 1992; Volume 200.
2. Prosperetti, A. Bubbles. *Phys. Fluids* **2004**, *16*, 1852–1865. [CrossRef]
3. Besagni, G.; Inzoli, F.; Ziegenhein, T. Two-phase bubble columns: A comprehensive review. *ChemEngineering* **2018**, *2*, 13. [CrossRef]
4. Veldhuis, C.; Biesheuvel, A.; Van Wijngaarden, L. Shape oscillations on bubbles rising in clean and in tap water. *Phys. Fluids* **2008**, *20*, 040705. [CrossRef]
5. Sanada, T.; Sugihara, K.; Shirota, M.; Watanabe, M. Motion and drag of a single bubble in super-purified water. *Fluid Dynam. Res.* **2008**, *40*, 534. [CrossRef]
6. Fu, Y.; Liu, Y. Development of a robust image processing technique for bubbly flow measurement in a narrow rectangular channel. *Int. J. Multiph. Flow* **2016**, *84*, 217–228. [CrossRef]

7. Bröder, D.; Sommerfeld, M. Planar shadow image velocimetry for the analysis of the hydrodynamics in bubbly flows. *Meas. Sci. Technol.* **2007**, *18*, 2513. [CrossRef]
8. Ferreira, A.; Pereira, G.; Teixeira, J.; Rocha, F. Statistical tool combined with image analysis to characterize hydrodynamics and mass transfer in a bubble column. *Chem. Eng. J.* **2012**, *180*, 216–228. [CrossRef]
9. Karn, A.; Ellis, C.; Arndt, R.; Hong, J. An integrative image measurement technique for dense bubbly flows with a wide size distribution. *Chem. Eng. Sci.* **2015**, *122*, 240–249. [CrossRef]
10. Lau, Y.; Sujatha, K.T.; Gaeini, M.; Deen, N.; Kuipers, J. Experimental study of the bubble size distribution in a pseudo-2D bubble column. *Chem. Eng. Sci.* **2013**, *98*, 203–211. [CrossRef]
11. Poletaev, I.; Pervunin, K.; Tokarev, M. Artificial neural network for bubbles pattern recognition on the images. *J. Phys. Conf. Ser.* **2016**, *754*, 072002. [CrossRef]
12. Haas, T.; Schubert, C.; Eickhoff, M.; Pfeifer, H. BubCNN: Bubble detection using Faster RCNN and shape regression network. *Chem. Eng. Sci.* **2020**. [CrossRef]
13. Fu, Y.; Liu, Y. 3D bubble reconstruction using multiple cameras and space carving method. *Meas. Sci. Technol.* **2018**, *29*, 075206. [CrossRef]
14. Haas, T.; Suarez, A.L.; Eickhoff, M.; Pfeifer, H. Toward a Strong-Sense Validation Benchmark Database for Numerical Ladle Flow Models. *Metal. Mater. Trans. B* **2021**, *52*, 199–222. [CrossRef]
15. Sano, M.; Mori, K. Bubble formation from single nozzles in liquid metals. *Trans. Jpn. Instit. Metals* **1976**, *17*, 344–352. [CrossRef]
16. Tsuchiya, K.; Aida, M.; Fujii, Y.; Okamoto, M. Bubble formation from a single nozzle in moderator liquids and liquid metal for tritium breeding. *J. Nucl. Mater.* **1993**, *207*, 123–129. [CrossRef]
17. Andreini, R.; Foster, J.; Callen, R. Characterization of gas bubbles injected into molten metals under laminar flow conditions. *Metal. Trans. B* **1977**, *8*, 625–631. [CrossRef]
18. Irons, G.; Guthrie, R. Bubble formation at nozzles in pig iron. *Metal. Trans. B* **1978**, *9*, 101–110. [CrossRef]
19. Davenport, W.G. *Bubbles in Liquids Including Liquid Metals*; The Royal School of Mines: London, UK, 1964.
20. Mori, K.; Ozawa, Y.; Sano, M. Characterization of gas jet behavior at a submerged orifice in liquid metal. *Trans. Iron Steel Inst. Jpn.* **1982**, *22*, 377–384. [CrossRef]
21. Iguchi, M.; Kawabata, H.; Nakajima, K.; Morita, Z.-I. Measurement of bubble characteristics in a molten iron bath at 1600 °C using an electroresistivity probe. *Metal. Mater. Trans. B* **1995**, *26*, 67–74. [CrossRef]
22. Keplinger, O.; Shevchenko, N.; Eckert, S. Visualization of bubble coalescence in bubble chains rising in a liquid metal. *Int. J. Multiph. Flow* **2018**, *105*, 159–169. [CrossRef]
23. Iguchi, M.; Nakatani, T.; Tokunaga, H. The shape of bubbles rising near the nozzle exit in molten metal baths. *Metal. Mater. Trans. B* **1997**, *28*, 417–423. [CrossRef]
24. Iguchi, M.; Nakatani, T.; Kawabata, H. Development of a multineedle electroresistivity probe for measuring bubble characteristics in molten metal baths. *Metal. Mater. Trans. B* **1997**, *28*, 409–416. [CrossRef]
25. Davis, K.G.; Irons, G.A.; Guthrie, R.I.L. X-ray cinematographic observations of gas injection into liquid metals. *Metal. Mater. Trans. B* **1978**, *9*, 721–722. [CrossRef]
26. Iguchi, M.; Chihara, T.; Takanashi, N.; Ogawa, Y.; Tokumitsu, N.; Morita, Z.-I. X-ray fluoroscopic observation of bubble characteristics in a molten iron bath. *ISIJ Int.* **1995**, *35*, 1354–1361. [CrossRef]
27. Keplinger, O.; Shevchenko, N.; Eckert, S. Validation of X-ray radiography for characterization of gas bubbles in liquid metals. *IOP Conf. Ser. Mater. Sci. Eng.* **2017**, *228*, 012009. [CrossRef]
28. Mishima, K.; Hibiki, T.; Saito, Y.; Nishihara, H.; Tobita, Y.; Konishi, K.; Matsubayashi, M. Visualization and measurement of gas–liquid metal two-phase flow with large density difference using thermal neutrons as microscopic probes. *Nucl. Instrum. Methods Phys. Res. Sect. A Accel. Spectromet. Detect. Assoc. Equip.* **1999**, *424*, 229–234. [CrossRef]
29. Sarma, M.; Ščepanskis, M.; Jakovičs, A.; Thomsen, K.; Nikoluškins, R.; Vontobel, P.; Beinerts, T.; Bojarevičs, A.; Platacis, E. Neutron radiography visualization of solid particles in stirring liquid metal. *Phys. Proc.* **2015**, *69*, 457–463. [CrossRef]
30. Baake, E.; Fehling, T.; Musaeva, D.; Steinberg, T. Neutron radiography for visualization of liquid metal processes: Bubbly flow for CO2 free production of Hydrogen and solidification processes in EM field. *IOP Conf. Ser. Mater. Sci. Eng.* **2017**, *228*, 12026. [CrossRef]
31. Richter, T.; Keplinger, O.; Strumpf, E.; Wondrak, T.; Eckert, K.; Eckert, S.; Odenbach, S. Measurements of the Diameter of Rising Gas Bubbles by Means of the Ultrasound Transit Time Technique. *Magnetohydrodynamics (0024-998X)* **2017**, *53*, 383–392. [CrossRef]
32. Wang, Z.; Wang, S.; Meng, X.; Ni, M. UDV measurements of single bubble rising in a liquid metal Galinstan with a transverse magnetic field. *Int. J. Multiph. Flow* **2017**, *94*, 201–208. [CrossRef]
33. Gundrum, T.; Büttner, P.; Dekdouk, B.; Peyton, A.; Wondrak, T.; Galindo, V.; Eckert, S. Contactless inductive bubble detection in a liquid metal flow. *Sensors* **2016**, *16*, 63. [CrossRef]
34. Karcher, C.; Lyu, Z.; Boeck, T.; Tran, N.; Lüdtke, U. Experimental and numerical investigation on particle-induced liquid metal flow using Lorentz force velocimetry. *IOP Conf. Ser. Mater. Sci. Eng.* **2018**, *424*, 1. [CrossRef]
35. Zhang, L.; Taniguchi, S. Fundamentals of inclusion removal from liquid steel by bubble flotation. *Int. Mater. Rev.* **2000**, *45*, 59–82. [CrossRef]
36. Stapurewicz, T.; Themelis, N.J. Mixing and mass transfer phenomena in bottom-injected gas–liquid reactors. *Can. Metal. Quart.* **1987**, *26*, 123–128. [CrossRef]

37. Owusu, K.B.; Haas, T.; Gajjar, P.; Eickhoff, M.; Kowitwarangkul, P.; Pfeifer, H. Interaction of Injector Design, Bubble Size, Flow Structure, and Turbulence in Ladle Metallurgy. *Steel Res. Int.* **2019**, *90*, 1800346. [CrossRef]
38. Haas, T.; Schubert, C.; Eickhoff, M.; Pfeifer, H. Numerical modeling of the ladle flow by a LES-based Eulerian-Lagrange approach: A systematic survey. *Metal. Mater. Trans. B* **2021**, *52*, 903–921. [CrossRef]
39. Kraume, M.; Zehner, P. Modellierung der Fluiddynamik in Blasensäulen. *Chemieingenieurtechnik* **1989**, *61*, 332–333. [CrossRef]
40. Kulkarni, A.A.; Joshi, J.B. Bubble formation and bubble rise velocity in gas— liquid systems: A review. *Ind. Eng. Chem. Res.* **2005**, *44*, 5873–5931. [CrossRef]
41. Leibson, I.; Holcomb, E.G.; Cacoso, A.G.; Jacmic, J.J. Rate of flow and mechanics of bubble formation from single submerged orifices. I. Rate of flow studies. *AIChE J.* **1956**, *2*, 296–300. [CrossRef]
42. Sano, M.; Mori, K. Size of bubbles in energetic gas injection into liquid metal. *Trans. Iron Steel Inst. Jpn.* **1980**, *20*, 675–681. [CrossRef]
43. Tate, T. On the magnitude of a drop of liquid formed under different circumstances. *Lond. Edinb. Dublin Philos. Magaz. J. Sci.* **1864**, *27*, 176–180. [CrossRef]
44. Davidson, L.; Amick, E.H., Jr. Formation of gas bubbles at horizontal orifices. *AIChE J.* **1956**, *2*, 337–342. [CrossRef]
45. Mersmann, A. Druckverlust und Schaumhöhen von gasdurchströmten Flüssigkeitsschichten auf Siebböden. *VDI ForschHft B* **1962**, *28*, 144.
46. Mori, K.; Sano, M.; Sato, T. Size of bubbles formed at single nozzle immersed in molten iron. *Trans. Iron Steel Inst. Jpn.* **1979**, *19*, 553–558. [CrossRef]
47. Xie, Y.; Orsten, S.; Oeters, F. Behaviour of bubbles at gas blowing into liquid Wood's metal. *ISIJ Int.* **1992**, *32*, 66–75. [CrossRef]
48. Gnyloskurenko, S.V.; Nakamura, T. Wettability effect on bubble formation at nozzles in liquid aluminum. *Mater. Trans.* **2003**, *44*, 2298–2302. [CrossRef]
49. Zhang, L.; Taniguchi, S.; Matsumoto, K. Water model study on inclusion removal from liquid steel by bubble flotation under turbulent conditions. *Ironmak. Steelmak.* **2002**, *29*, 326–336. [CrossRef]
50. Iguchi, M.; Tokunaga, H.; Tatemichi, H. Bubble and liquid flow characteristics in a wood's metal bath stirred by bottom helium gas injection. *Metal. Mater. Trans. B* **1997**, *28*, 1053–1061. [CrossRef]
51. Castillejos, A.; Brimacombe, J. Physical characteristics of gas jets injected vertically upward into liquid metal. *Metal. Trans. B* **1989**, *20*, 595–601. [CrossRef]
52. Houghton, G.; McLean, A.; Ritchie, P. Mechanism of formation of gas bubble-beds. *Chem. Eng. Sci.* **1957**, *7*, 40–50. [CrossRef]
53. Loimer, T.; Machu, G.; Schaflinger, U. Inviscid bubble formation on porous plates and sieve plates. *Chem. Eng. Sci.* **2004**, *59*, 809–818. [CrossRef]
54. Mirsandi, H.; Baltussen, M.; Peters, E.; van Odyck, D.; van Oord, J.; van der Plas, D.; Kuipers, J. Numerical simulations of bubble formation in liquid metal. *Int. J. Multiph. Flow* **2020**, *131*, 103363. [CrossRef]
55. Wang, Z.; Mukai, K.; Izu, D. Influence of wettability on the behavior of argon bubbles and fluid flow inside the nozzle and mold. *ISIJ Int.* **1999**, *39*, 154–163. [CrossRef]
56. Koide, K.; Kato, S.; Tanaka, Y.; Kubota, H. Bubbles generated from porous plate. *J. Chem. Eng. Jpn.* **1968**, *1*, 51–56. [CrossRef]
57. Baxter, R.; Wraith, A. Transitions in the bubble formation mode of a submerged porous disc. *Chem. Eng. Sci.* **1970**, *25*, 1244–1247. [CrossRef]
58. Anagbo, P.; Brimacombe, J. Plume characteristics and liquid circulation in gas injection through a porous plug. *Metal. Trans. B* **1990**, *21*, 637–648. [CrossRef]
59. Goyal, P.; Themelis, N.; Zanchuk, W.A. Gaseous refining of anode copper. *JOM* **1982**, *34*, 22–28. [CrossRef]
60. Iguchi, M.; Kaji, M.; Morita, Z.-I. Effects of pore diameter, bath surface pressure, and nozzle diameter on the bubble formation from a porous nozzle. *Metal. Mater. Trans. B* **1998**, *29*, 1209–1218. [CrossRef]
61. Kazakis, N.; Papadopoulos, I.; Mouza, A. Bubble columns with fine pore sparger operating in the pseudo-homogeneous regime: Gas hold up prediction and a criterion for the transition to the heterogeneous regime. *Chem. Eng. Sci.* **2007**, *62*, 3092–3103. [CrossRef]
62. Mohagheghian, S.; Ghajar, A.J.; Elbing, B. Effect of operation regime on bubble size and void fraction in a bubble column with porous sparger. *Authorea Prepr.* **2020**. [CrossRef]
63. Lucas, D.; Ziegenhein, T. Influence of the bubble size distribution on the bubble column flow regime. *Int. J. Multiph. Flow* **2019**, *120*, 103092. [CrossRef]
64. Koide, K.; Hayashi, T.; Noro, M.; Takemura, Y.; Kawamata, N.; Kubota, H. Bubbles generated from porous plate in pure liquid and in aqueous solutions of inorganic electrolytes. *J. Chem. Eng. Jpn.* **1972**, *5*, 236–242. [CrossRef]
65. Miyahara, T.; Tanaka, A. Size of bubbles generated from porous plates. *J. Chem. Eng. Jpn.* **1997**, *30*, 353–355. [CrossRef]
66. Li, R.-Q.; Harris, R. Bubble formation from a very narrow slot. *Can. Metal. Quart.* **1993**, *32*, 31–37. [CrossRef]
67. Li, R.-Q.; Wraith, A.; Harris, R. Gas dispersion phenomena at a narrow slot submerged in a liquid. *Chem. Eng. Sci.* **1994**, *49*, 531–540. [CrossRef]
68. Wraith, A.; Li, R.-Q.; Harris, R. Gas bubble volume at a narrow slot nozzle in a liquid. *Chem. Eng. Sci.* **1995**, *50*, 1057–1058. [CrossRef]
69. Harris, R.; Ng, K.; Wraith, A. Spargers for controlled bubble size by means of the multiple slot disperser (MSD). *Chem. Eng. Sci.* **2005**, *60*, 3111–3115. [CrossRef]

70. Okumura, K.; Harrist, R.; Sanot, M. Bubble formation from nonwetted slot nozzles. *Can. Metal. Quart.* **1998**, *37*, 49–56. [CrossRef]
71. Okumura, K.; Komarov, S.V.; Sano, M. Gas injection from slot nozzles with various shapes in water. *ISIJ Int.* **2000**, *40*, 544–548. [CrossRef]
72. Trummer, B.; Fellner, W.; Viertauer, A.; Kneis, L.; Hackl, G. A Water Modelling Comparison of Hybrid Plug, Slot Plug and Porous Plug Designs. *RHI Worldwide* **2016**, *1*, 35.
73. Polli, M.; Di Stanislao, M.; Bagatin, R.; Bakr, E.A.; Masi, M. Bubble size distribution in the sparger region of bubble columns. *Chem. Eng. Sci.* **2002**, *57*, 197–205. [CrossRef]
74. Liao, Y.; Lucas, D. A literature review of theoretical models for drop and bubble breakup in turbulent dispersions. *Chem. Eng. Sci.* **2009**, *64*, 3389–3406. [CrossRef]
75. Keplinger, O.; Shevchenko, N.; Eckert, S. Experimental investigation of bubble breakup in bubble chains rising in a liquid metal. *Int. J. Multiph. Flow* **2019**, *116*, 39–50. [CrossRef]
76. Liao, Y.; Lucas, D. A literature review on mechanisms and models for the coalescence process of fluid particles. *Chem. Eng. Sci.* **2010**, *65*, 2851–2864. [CrossRef]
77. Sahai, Y.; Guthrie, R. Hydrodynamics of gas stirred melts: Part I. Gas/liquid coupling. *Metal. Trans. B* **1982**, *13*, 193–202. [CrossRef]
78. Moore, D. The velocity of rise of distorted gas bubbles in a liquid of small viscosity. *J. Fluid Mech.* **1965**, *23*, 749–766. [CrossRef]
79. Bozzano, G.; Dente, M. Shape and terminal velocity of single bubble motion: A novel approach. *Comp. Chem. Eng.* **2001**, *25*, 571–576. [CrossRef]
80. Liang-Shih, F.; Tsuchiya, K. *Bubble Wake Dynamics in Liquids and Liquid-Solid Suspensions*; Butterworth-Heinemann: Oxford, UK, 2013.
81. Wellek, R.; Agrawal, A.; Skelland, A. Shape of liquid drops moving in liquid media. *AIChE J.* **1966**, *12*, 854–862. [CrossRef]
82. Clift, R.; Grace, J.R.; Weber, M.E. *Bubbles, Drops, and Particles*; Courier Corporation: Chelmsford, MA, USA, 2005.
83. Aoyama, S.; Hayashi, K.; Hosokawa, S.; Tomiyama, A. Shapes of single bubbles in infinite stagnant liquids contaminated with surfactant. *Exp. Therm. Fluid Sci.* **2018**, *96*, 460–469. [CrossRef]
84. Besagni, G.; Deen, N.G. Aspect ratio of bubbles in different liquid media: A novel correlation. *Chem. Eng. Sci.* **2020**, *215*, 115383. [CrossRef]
85. Tomiyama, A.; Tamai, H.; Zun, I.; Hosokawa, S. Transverse migration of single bubbles in simple shear flows. *Chem. Eng. Sci.* **2002**, *57*, 1849–1858. [CrossRef]
86. Vakhrushev, I.; Efremov, G. Interpolation formula for computing the velocities of single gas bubbles in liquids. *Chem. Technol. Fuels Oils* **1970**, *6*, 376–379. [CrossRef]
87. Aoyama, S.; Hayashi, K.; Hosokawa, S.; Tomiyama, A. Shapes of ellipsoidal bubbles in infinite stagnant liquids. *Int. J. Multiph. Flow* **2016**, *79*, 23–30. [CrossRef]
88. Ziegenhein, T.; Lucas, D. Observations on bubble shapes in bubble columns under different flow conditions. *Exp. Therm. Fluid Sci.* **2017**, *85*, 248–256. [CrossRef]
89. Dijkhuizen, W.; Roghair, I.; Annaland, M.V.S.; Kuipers, J. DNS of gas bubbles behaviour using an improved 3D front tracking model—Drag force on isolated bubbles and comparison with experiments. *Chem. Eng. Sci.* **2010**, *65*, 1415–1426. [CrossRef]
90. Xu, Y.; Ersson, M.; Jönsson, P. Numerical simulation of single argon bubble rising in molten metal under a laminar flow. *Steel Res. Int.* **2015**, *86*, 1289–1297. [CrossRef]
91. Wang, G.; Zhou, H.; Tian, Q.; Ai, X.; Zhang, L. The Motion of Single Bubble and Interactions between Two Bubbles in Liquid Steel. *ISIJ Int.* **2017**. [CrossRef]
92. Magnaudet, J.; Eames, I. The motion of high-Reynolds-number bubbles in inhomogeneous flows. *Annu. Rev. Fluid Mech.* **2000**, *32*, 659–708. [CrossRef]
93. Mei, R.; Klausner, J.F. Unsteady force on a spherical bubble at finite Reynolds number with small fluctuations in the free-stream velocity. *Phys. Fluids A Fluid Dynam.* **1992**, *4*, 63–70. [CrossRef]
94. Pang, M.J.; Wei, J.J. Analysis of drag and lift coefficient expressions of bubbly flow system for low to medium Reynolds number. *Nucl. Eng. Design* **2011**, *241*, 2204–2213. [CrossRef]
95. Tomiyama, A.; Kataoka, I.; Zun, I.; Sakuguchi, T. Drag Coefficients of Single Bubbles under Normal and Micro Gravity Conditions. *JSME Int. J. Ser. B Fluids Therm. Eng.* **1998**, *41*, 472–479. [CrossRef]
96. Frank, T.; Shi, J.; Burns, A. Validation of Eulerian multiphase flow models for nuclear safety applications. In Proceedings of the 3rd International Symposium on Two-Phase Flow Modelling and Experimentation, Pisa, Italy, 22–24 September 2014.
97. Peebles, F.; Garber, H. Studies on the Motion of Gas Bubbles in Liquids. *Chem. Eng. Progress* **1953**, *49*, 88–97.
98. Mori, Y.; Hijikata, K.; Kuriyama, I. Experimental study of bubble motion in mercury with and without a magnetic field. *J. Heat Transfer* **1977**, *99*, 404–410. [CrossRef]
99. Zhang, C.; Eckert, S.; Gerbeth, G. Experimental study of single bubble motion in a liquid metal column exposed to a DC magnetic field. *Int. J. Multiph. Flow* **2005**, *31*, 824–842. [CrossRef]
100. Takagi, S.; Matsumoto, Y. Surfactant effects on bubble motion and bubbly flows. *Annu. Rev. Fluid Mech.* **2011**, *43*, 615–636. [CrossRef]
101. Tasoglu, S.; Demirci, U.; Muradoglu, M. The effect of soluble surfactant on the transient motion of a buoyancy-driven bubble. *Phys. Fluids* **2008**, *20*, 040805. [CrossRef]

102. Cuenot, B.; Magnaudet, J.; Spennato, B. The effects of slightly soluble surfactants on the flow around a spherical bubble. *J. Fluid Mech.* **1997**, *339*, 25–53. [CrossRef]
103. Zhang, Y.; Finch, J. A note on single bubble motion in surfactant solutions. *J. Fluid Mech.* **2001**, *429*, 63. [CrossRef]
104. Takagi, S.; Ogasawara, T.; Matsumoto, Y. The effects of surfactant on the multiscale structure of bubbly flows. *Philos. Trans. R. Soc. A Math. Phys. Eng. Sci.* **2008**, *366*, 2117–2129. [CrossRef]
105. Fukuta, M.; Takagi, S.; Matsumoto, Y. Numerical study on the shear-induced lift force acting on a spherical bubble in aqueous surfactant solutions. *Phys. Fluids* **2008**, *20*, 040704. [CrossRef]
106. Simonnet, M.; Gentric, C.; Olmos, E.; Midoux, N. Experimental determination of the drag coefficient in a swarm of bubbles. *Chem. Eng. Sci.* **2007**, *62*, 858–866. [CrossRef]
107. Richardson, J.; Zaki, W. The sedimentation of a suspension of uniform spheres under conditions of viscous flow. *Chem. Eng. Sci.* **1954**, *3*, 65–73. [CrossRef]
108. Lockett, M.J.; Kirkpatrick, R.D. Ideal bubbly flow and actual flow in bubble columns. *Trans. Inst. Chem. Eng.* **1975**, *53*, 267–273.
109. Garnier, C.; Lance, M.; Marié, J. Measurement of local flow characteristics in buoyancy-driven bubbly flow at high void fraction. *Exp. Therm. Fluid Sci.* **2002**, *26*, 811–815. [CrossRef]
110. Guet, S.; Ooms, G.; Oliemans, R.; Mudde, R. Bubble size effect on low liquid input drift–flux parameters. *Chem. Eng. Sci.* **2004**, *59*, 3315–3329. [CrossRef]
111. Roghair, I.; Lau, Y.; Deen, N.; Slagter, H.; Baltussen, M.; Annaland, M.V.S.; Kuipers, J. On the drag force of bubbles in bubble swarms at intermediate and high Reynolds numbers. *Chem. Eng. Sci.* **2011**, *66*, 3204–3211. [CrossRef]
112. Grienberger, J.; Hofmann, H. Investigations and modelling of bubble columns. *Chem. Eng. Sci.* **1992**, *47*, 2215–2220. [CrossRef]
113. Simonnet, M.; Gentric, C.; Olmos, E.; Midoux, N. CFD simulation of the flow field in a bubble column reactor: Importance of the drag force formulation to describe regime transitions. *Chem. Eng. Proces. Process Intensif.* **2008**, *47*, 1726–1737. [CrossRef]
114. Lau, Y.; Roghair, I.; Deen, N.; van Sint Annaland, M.; Kuipers, J. Numerical investigation of the drag closure for bubbles in bubble swarms. *Chem. Eng. Sci.* **2011**, *66*, 3309–3316. [CrossRef]
115. Zun, I. The transverse migration of bubbles influenced by walls in vertical bubbly flow. *Int. J. Multiph. Flow* **1980**, *6*, 583–588. [CrossRef]
116. Saffman, P. The lift on a small sphere in a slow shear flow. *J. Fluid Mech.* **1965**, *22*, 385–400. [CrossRef]
117. McLaughlin, J.B. Inertial migration of a small sphere in linear shear flows. *J. Fluid Mech.* **1991**, *224*, 261–274. [CrossRef]
118. Legendre, D.; Magnaudet, J. A note on the lift force on a spherical bubble or drop in a low-Reynolds-number shear flow. *Phys. Fluids* **1997**, *9*, 3572–3574. [CrossRef]
119. Mei, R.; Klausner, J. Shear lift force on spherical bubbles. *Int. J. Heat Fluid Flow* **1994**, *15*, 62–65. [CrossRef]
120. Auton, T. The lift force on a spherical body in a rotational flow. *J. Fluid Mech.* **1987**, *183*, 199–218. [CrossRef]
121. Legendre, D.; Magnaudet, J. The lift force on a spherical bubble in a viscous linear shear flow. *J. Fluid Mech.* **1998**, *368*, 81–126. [CrossRef]
122. Lance, M.; De Bertodano, M.L. Phase distribution phenomena and wall effects in bubbly two-phase flows. *Multiph. Sci. Technol.* **1994**, *8*. [CrossRef]
123. Liu, T. Bubble size and entrance length effects on void development in a vertical channel. *Int. J. Multiph. Flow* **1993**, *19*, 99–113. [CrossRef]
124. Hibiki, T.; Ishii, M. Lift force in bubbly flow systems. *Chem. Eng. Sci.* **2007**, *62*, 6457–6474. [CrossRef]
125. Serizawa, A.; Kataoka, I. Dispersed flow-I. *Multiph. Sci. Technol.* **1994**, *8*. [CrossRef]
126. Tomiyama, A.; Zun, I.; Sou, A.; Sakaguchi, T. Numerical analysis of bubble motion with the VOF method. *Nucl. Eng. Design* **1993**, *141*, 69–82. [CrossRef]
127. Brücker, C. Structure and dynamics of the wake of bubbles and its relevance for bubble interaction. *Phys. Fluids* **1999**, *11*, 1781–1996. [CrossRef]
128. Adoua, R.; Legendre, D.; Magnaudet, J. Reversal of the lift force on an oblate bubble in a weakly viscous linear shear flow. *J. Fluid Mech.* **2009**, *628*, 23–41. [CrossRef]
129. Aoyama, S.; Hayashi, K.; Hosokawa, S.; Lucas, D.; Tomiyama, A. Lift force acting on single bubbles in linear shear flows. *Int. J. Multiph. Flow* **2017**, *96*, 113–122. [CrossRef]
130. Dhotre, M.; Deen, N.; Niceno, B.; Khan, Z.; Joshi, J. Large Eddy simulation for dispersed bubbly flows. *Int. J. Chem. Eng.* **2013**. [CrossRef]
131. Bothe, D.; Schmidtke, M.; Warnecke, H.J. VOF-simulation of the lift force for single bubbles in a simple shear flow. *Chem. Eng. Technol. Ind. Chem. Plant Equipm. Process Eng. Biotechnol.* **2006**, *29*, 1048–1053. [CrossRef]
132. Dijkhuizen, W.; van Sint Annaland, M.; Kuipers, J. Numerical and experimental investigation of the lift force on single bubbles. *Chem. Eng. Sci.* **2010**, *65*, 1274–1287. [CrossRef]
133. Ziegenhein, T.; Tomiyama, A.; Lucas, D. A new measuring concept to determine the lift force for distorted bubbles in low Morton number system: Results for air/water. *Int. J. Multiph. Flow* **2018**, *108*, 11–24. [CrossRef]
134. Ma, T.; Santarelli, C.; Ziegenhein, T.; Lucas, D.; Fröhlich, J. Direct numerical simulation–based Reynolds-averaged closure for bubble-induced turbulence. *Phys. Rev. Fluids* **2017**, *2*, 034301. [CrossRef]
135. Ogasawara, T. Surfactant effect on bubble migration toward the wall in upward bubbly flow. *Prog. Multiphase Flow Res.* **2006**, *1*, 9–16. [CrossRef]

136. Hayashi, K.; Tomiyama, A. Effects of surfactant on lift coefficients of bubbles in linear shear flows. *Int. J. Multiph. Flow* **2018**, *99*, 86–93. [CrossRef]
137. Rastello, M.; Marié, J.-L.; Lance, M. Clean versus contaminated bubbles in a solid-body rotating flow. *J. Fluid Mech.* **2017**, *831*, 592–617. [CrossRef]
138. Hessenkemper, H.; Ziegenhein, T.; Lucas, D. Contamination effects on the lift force of ellipsoidal air bubbles rising in saline water solutions. *Chem. Eng. J.* **2019**. [CrossRef]
139. Dhotre, M.T.; Niceno, B.; Smith, B.L.; Simiano, M. Large-eddy simulation (LES) of the large scale bubble plume. *Chem. Eng. Sci.* **2009**, *64*, 2692–2704. [CrossRef]
140. Auton, T.; Hunt, J.; Prud'Homme, M. The force exerted on a body in inviscid unsteady non-uniform rotational flow. *J. Fluid Mech.* **1988**, *197*, 241–257. [CrossRef]
141. Ishii, M.; Mishima, K. Two-fluid model and hydrodynamic constitutive relations. *Nucl. Eng. Design* **1984**, *82*, 107–126. [CrossRef]
142. Paladino, E.E.; Maliska, C.R. Virtual mass in accelerated bubbly flows. In Proceedings of the 4th European Thermal Sciences, National Exhibition Centre, Birmingham, UK, 29–31 March 2004.
143. Niemann, J.; Laurien, E. The Virtual Mass of Gas-Bubbles in a Liquid at High Void Fractions. In Proceedings of the Abstract submitted to: ERCOFTAC Conference on Particle-Laden Flows, Kappel, Switzerland, 3–5 June 2000.
144. Koebe, M. *Numerische Simulation Aufsteigender Blasen Mit und Ohne Stoffaustausch Mittels der Volume of Fluid (VOF) Methode*; University of Paderborn: Paderborn, Germany, 2004.

metals

MDPI

Article

Adaptation of the Rist Operating Diagram as a Graphical Tool for the Direct Reduction Shaft

Thibault Quatravaux [1,*,†], Jose Barros [2], Pascal Gardin [2] and Gabriel Lucena [3]

[1] Institut Jean Lamour, Campus Artem, UMR 7198 Université de Lorraine/CNRS, 2 Allée André Guinier-BP 50840, LabEx DAMAS, CEDEX, 54011 Nancy, France
[2] ArcelorMittal R&D Maizières Process, Voie Romaine, 57280 Maizières-lès-Metz, France; jose.barroslorenzo@arcelormittal.com (J.B.); pascal.gardin@arcelormittal.com (P.G.)
[3] ArcelorMittal Produits Longs Canada, 4000, Route des Aciéries, Contrecoeur, QC J0L 1C0, Canada; gabriel.lucena@arcelormittal.com
* Correspondence: thibault.quatravaux@univ-lorraine.fr
† formerly ArcelorMittal R&D Maizières Process.

Abstract: The blast-furnace operating diagram proposed by Rist was revised to direct reduction and was specifically applied to the Midrex NG^{TM} process. The use of this graphical tool in the study of an industrial process highlighted the staggered nature of the reduction in the shaft furnace with, in particular, the existence of a prereduction zone in the upper part where metallization is thermodynamically impossible. A sensitivity study also showed the impact of the in situ reforming rate on the ability of the gas to completely reduce iron oxides. Finally, we graphically defined the minimum quality required for the top gas to produce direct-reduced iron.

Keywords: direct reduction; Midrex; HYL; Rist diagram

Citation: Quatravaux, T.; Barros, J.; Gardin, P.; Lucena, G. Adaptation of the Rist Operating Diagram as a Graphical Tool for the Direct Reduction Shaft. *Metals* **2021**, *11*, 1953. https://doi.org/10.3390/met11121953

Received: 15 October 2021
Accepted: 2 December 2021
Published: 4 December 2021

1. Introduction

The production of direct-deduced Iron (DRI) is the main alternative route to blast furnace ironmaking. Over the last decade, this route has become increasingly popular and counts from 1.5% in the 1980s to more than 6% of the overall reduced-iron production nowadays [1,2]. The most widespread direct-reduction (DR) technology is the vertical shaft reactor fabricating reducing gas from natural gas (i.e., Midrex NG^{TM} and HYL processes). ArcelorMittal operates an installed capacity of 13 Mt/year including nine Midrex NG^{TM} modules and four HYL reactors.

Experimental and mathematical modeling of DR processes is important in terms of energy efficiency and productivity. Therefore, the development and application of tools to study the chemical reactions along the reduction shaft are necessary for the understanding and optimization of the process [3–5].

In this respect, it is interesting to draw a parallel with the blast furnace, an older process on which technical and scientific investigations are not comparable to the direct reduction processes. Numerous studies during the 1960s led to a more complete understanding of the blast furnace and a profound change in its operating point.

As such, Rist and Meyson have developed in the early 1960s a graphical tool to describe the operating point of the blast furnace [6–9]. The graphical approach also existed more widely in chemical engineering to describe the operating conditions of many industrial processes. The graphical tool they have developed allows to describe more specifically the ideal operating point of the blast furnace. It takes into account the thermal equilibria and some critical chemical equilibria, which limit the heat and material exchanges, respectively, in this process.

This graphical tool is based on the work of Kitaiev and Michard and, in particular, on the mathematical modeling of the blast furnace proposed by Michard [7]. The blast furnace is considered as a counter-current gas–solids reactor, and it behaves as an oxygen exchanger in the iron oxide reduction reaction.

The graphical representation of the blast furnace was particularly relevant from an educational point of view. Moreover, through the evaluation of material and heat balances within the blast furnace, the diagram has been used for years both to describe the operating point of a blast furnace and to establish prospective operating points for optimization [7,10–13].

The more-recently developed Midrex NG^{TM} and Tenova HYL processes have strong similarities with the blast furnace. The reduction shaft furnace can also be thought of as an oxygen exchanger in a counter-current gas–solids reactor configuration.

Rist proposed in 1992 an original graphic description of the Midrex NG^{TM} process [14], combining a diagram for the reformer with that of the shaft furnace. Nevertheless, it is an idealized and simplified description, with an educational vocation, which cannot be directly exploited to describe the real operations. For example, natural-gas injections downstream of the reformer are not taken into account for a fair description of the gas mass balance in the reduction zone. Moreover, in situ methane cracking or carburization of iron bearing materials are not taken into account.

It may therefore be interesting to adapt the original diagram proposed by Rist and Meysson for the blast furnace to this process, in order to identify the industrial operating point and study the prospects of such a graphical tool.

The adaptation of the Rist diagram to the Midrex NG^{TM} process, and to direct reduction processes in general, requires overcoming two difficulties:

- The gas composition is more complex than in the blast furnace. Particular attention will be paid to considering the hydrocarbons, including methane.
- The carburization phenomena in DRI will impact the operating line of the diagram.

This article reviews the formalism of the Rist approach in order to provide a comprehensive description of the Midrex NG^{TM} reduction shaft, compatible with the Rist diagram and applicable to direct-reduction processes. The approach developed on a Midrex NG^{TM} flowsheet is easily transferable to the HYL III process.

2. Modeling

2.1. Description of the Midrex Process

The Midrex NG^{TM} process is one of the most-spread technologies in the steel industry for the production of direct-reduced iron (DRI). The main differences with the conventional reduction route (blast furnace) are the use of natural gas as a reducing agent and as a heat source, allowing an almost complete reduction of iron oxides at a lower temperature (below 1000 °C) without any melting phenomenon. The process itself is based on the coupling of three main components, a heat recovery device, a reformer, and a shaft furnace, for an optimized use of the natural-gas consumption for the production of DRI. Figure 1 shows the operating principle of the Midrex NG^{TM} process.

Moreover, the compositions of the inlet and outlet gases measured in the Midrex shaft at the Gilmore plant [3] are presented as an illustration in Table 1. The reducing gas corresponds to the mixing of the bustle gas (**B**) with the natural gas injected into the transition zone.

Table 1. Compositions of inlet and outlet gases of the Midrex shaft in Gilmore plant. [3]

Gas Composition (% mol)	H_2	CO	H_2O	CO_2	$CH_4 + N_2$
Reducing gas	52.90	30.0	4.7	4.8	8.1
Top gas (**T**)	37.0	18.9	21.2	14.3	8.6

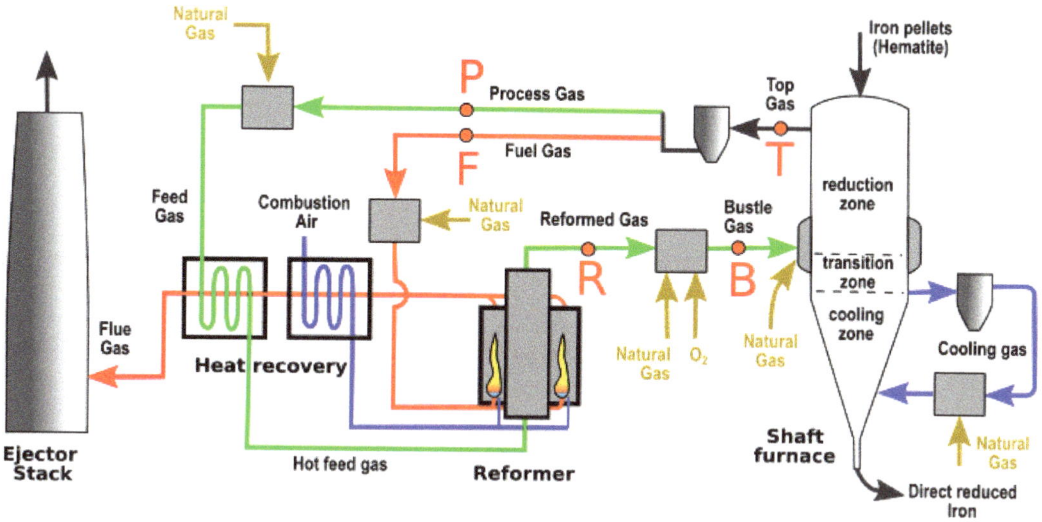

Figure 1. Operating principle of the Midrex NG^{TM} process.

The operating point of the process is summarized in the following overview:

- The shaft furnace is a vertical gas–solid countercurrent reactor with a downward flow of iron oxides and an upward flow of a hot reducing gas. The iron pellets, consisting mainly of hematite (%Fe_2O_3 > 95%), are then both reduced and carburized. The direct-reduced iron (DRI) leaving the shaft achieves a high rate of metallization (92–96% iron metal, with residual iron oxides being Wustite) and a moderate level of carburization (2–2.5% carbon in the total mass).
- The preparation of the reducing gas follows several steps. First, a fraction of the recycled top gas, named the process gas (**P**), is mixed with injected natural gas. The corresponding mixture, called the feed gas, is preheated in the recovery heat device and then injected into the reformer. Cracking occurs between CH_4, CO_2, and H_2O in the tubes of the reformer. The resulting reformed gas (**R**) is mainly composed of CO, CO_2, H_2, and H_2O, with a low rate of remaining CH_4 (few %).
- The reformed gas (**R**) is mixed with additional natural gas and pure oxygen and injected in the shaft furnace at a temperature around 950 °C. Additional natural gas is also injected in the shaft, in the bottom area, inside the loop of the cooling gas and in the transition zone with the bustle gas.

2.2. Description of the Local Mass Balance in the Shaft Furnace

In the frame of a counter-current gas–solid reaction, for steady-state conditions without any diffusion phenomena, the local balance law can be written in the following way:

$$\frac{\partial c_i}{\partial t} = \vec{\nabla} \cdot \vec{\phi}_i^{gas} + \vec{\nabla} \cdot \vec{\phi}_i^{s} = 0 \quad (1)$$

where c_i is the local molar concentration of atom i. $\vec{\phi}_i^{gas}$ and $\vec{\phi}_i^{s}$ are the local gaseous and solid molar fluxes in the shaft furnace, respectively. It can be applied to calculate any mass balance of O, H, C, and Fe.

We adopt the following formalism to describe the counter-current gas–solid reduction, as shown in Figure 2.

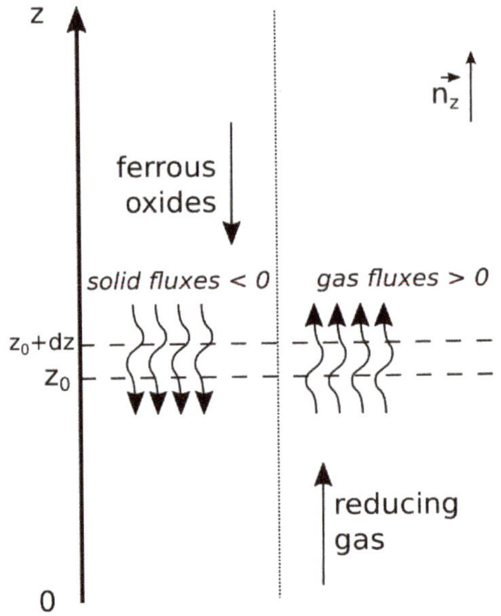

Figure 2. Formalization of the counter-current configuration of the reduction zone in the shaft furnace.

Each local mass balance equation is simplified according to this formalism:

$$\vec{\nabla} \cdot \vec{\phi}_i^{gas} = \frac{d\phi_i^{gas}}{dz}, \vec{\nabla} \cdot \vec{\phi}_i^{s} = -\frac{d\phi_i^{s}}{dz} \tag{2}$$

Finally, Equations (1) and (2) lead to :

$$\frac{d\phi_i^{gas}}{dz} = \frac{d\phi_i^{s}}{dz} \tag{3}$$

The molar gaseous flux of the atom i is calculated with the following relation:

$$\phi_i^{gas} = \frac{Q_v^{gas}}{V_m} \times \sum_{molecule\, j} a_j \cdot n_i^j \tag{4}$$

where Q_v^{gas} is the volumetric gas flow rate, V_m the molar volume of the gas, a_j the volumetric fraction of the molecule j in the gas, and n_i^j the number of atoms i in the molecule j.

ϕ_i^s, the solid molar flux of the atom i, provided by the iron bearing material, is obtained using Equation (5), according to the mass flow rate Q_m^s of the burden, the mass fraction w_i^s, and the molar mass M_i.

$$\phi_i^s = Q_m^s \times \frac{w_i^s}{M_i} \tag{5}$$

2.3. Presentation of the Operating Diagram

The operating diagram represents on the x-axis and y-axis the gas and burden oxidation degrees denoted by X^{gas} and y, respectively, and defined by the following relations:

$$X^{gas} = \frac{\phi_O^{gas} + \frac{1}{2}\phi_H^{gas}}{\phi_C^{gas} + \frac{1}{2}\phi_H^{gas}} \tag{6}$$

$$y = \frac{\phi_O^s + \frac{1}{2}\phi_H^s}{\phi_{Fe}^s} \tag{7}$$

In a counter-current configuration, according to a complete oxygen transfer from oxides to the gas all along the reduction path, the burden and gas oxidation degree variations are correlated. μ represents the specific consumption as the stoichiometric ratio between gas and solid, according the following relation:

$$\mu = \frac{\phi_C^{gas} + \frac{1}{2}\phi_H^{gas}}{\phi_{Fe}^{s}} \tag{8}$$

Rist also introduces a coupling with the Chaudron diagram. As a reminder, Chaudron has established the equilibrium conditions between the CO-CO_2 and H_2-H_2O gas mixtures and the various iron oxides, as a function of temperature. The diagram represents the thermodynamic equilibria gas–Wustite and gas–Magnetite as a function of x^{gas} on the abscissa, a derived definition of the gas oxidation degree, and temperature on the ordinate, with:

$$x^{gas} = X^{gas} - 1 = \frac{\phi_O^{gas} - \phi_C^{gas}}{\phi_C^{gas} + \frac{1}{2}\phi_H^{gas}} \tag{9}$$

Rist plots the points W and M for these equilibria. Abscissae are provided by the Chaudron diagram, and ordinates correspond to the oxidation degrees of these oxides (y_W = 1.056 and y_M = 1.33), as shown in Figure 3. He thus delimits the shape of a thermodynamic boundary excluding thermodynamically impossible situations, when the gas is not sufficiently reducing: the forbidden zone.

In the blast furnace, it has been shown that the main thermodynamic constraint is located in the reserve zone, halfway up the shaft, where coke gasification is preponderant, for a temperature around 950–1000 °C. Rist defines the ideal blast furnace operation when thermal and thermodynamic equilibria are reached in this reserve zone and the iron oxides are reduced to pure Wustite. Here, the operating line passes through the W-point, as shown in Figure 3.

Rist also establishes the existence of an invariant point P about which the operating line rotates when the operating point of the blast furnace is changed. This point is plotted on the basis of the heat balance calculated in the blast furnace elaboration zone below the reserve zone. The details of these calculations are described in [7], but this section does not apply to the Midrex NG^{TM} process. Therefore, it is not developed further in this document.

2.4. Application to the Description of the Shaft Furnace in the Midrex NG^{TM} Process

To use the Rist diagram in the Midrex NG^{TM} process, we need to consider the following phenomena that do not occur in the blast furnace:

- The more complex composition of the reducing gas, including the presence of hydrocarbons,
- the in situ reforming of hydrocarbons in the shaft furnace,
- the phenomena of carburization of DRI by gas, through the Boudouard and Beggs reactions.

We consider the following two situations, with an increasing degree of complexity.

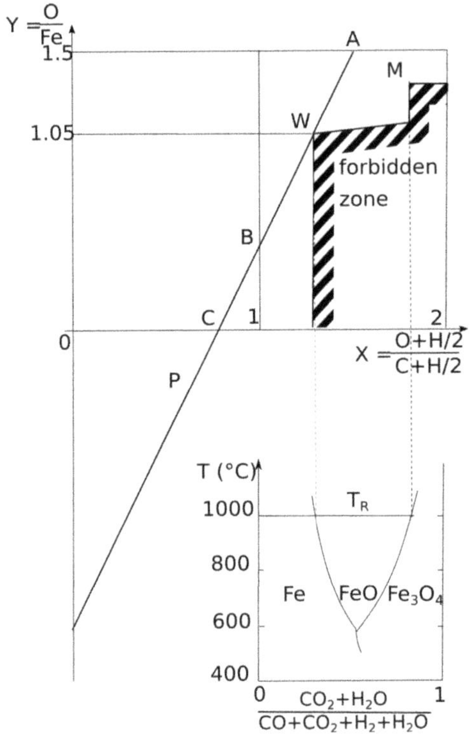

Figure 3. An illustration of the Rist diagram applied for the blast furnace, coupled with thermodynamics constraints from the Chaudron diagram.

2.4.1. Simplified Case: No Carburization Phenomena

In a simplified case, the carbon remains in a gaseous state, and the chemical reactor is considered to be an oxygen exchanger, similar to Rist's approach for the blast furnace. Therefore, the changes in oxidation states x and y are calculated from the following relationships:

$$\frac{dx^0}{dz} = \frac{dX^0}{dz} = \frac{1}{\phi_C^g + \frac{1}{2}\phi_H^g}\frac{\partial \phi_O^g}{\partial z} \tag{10}$$

$$\frac{dy}{dz} = \frac{1}{\phi_{Fe}^s}\frac{\partial \phi_O^s}{\partial z} \tag{11}$$

According to the local oxygen mass balance Equation (3), we deduced the simplified specific consumption:

$$\mu^0 = \frac{dy}{dX^0} = \frac{\phi_C^g + \frac{1}{2}\phi_H^g}{\phi_{Fe}^s} \tag{12}$$

In the shaft furnace of the Midrex NG^{TM} process, the reducing gas is composed of CO, CO_2, H_2, H_2O, N_2, and hydrocarbons denoted C_mH_n. The degree of oxidation of the gas and the specific consumption can then be deduced from these relationships:

$$x^0 = \frac{\%CO_2 + \%H_2O - \sum m\,\%C_mH_n}{\eta_{CH}} \tag{13}$$

$$\mu^0 = \frac{Q_v^g \cdot \eta_{CH}}{V_m} \cdot \frac{M_{Fe}}{Q_m^{DRI} \,.\, \%Fe^{DRI}} \tag{14}$$

where:

$$\eta_{CH} = \%CO + \%CO_2 + \%H_2 + \%H_2O + \sum \left(m + \frac{n}{2}\right) \%C_m H_n \tag{15}$$

2.4.2. Real Case: Carburization Is Taken into Account

We assumed that the carburization phenomena occur below the reduction zone in the shaft furnace. Therefore, the efficient carbon content of the reducing gas should exclude the carbon deposited on DRI:

$$\phi_C^g = \phi_C^{inlet} - \phi_C^s \tag{16}$$

We can thus complete the definition of specific consumption and oxidation degrees, on the basis of the simplified forms defined above.

The corresponding new specific consumption is deduced from Equations (14) and (16):

$$\mu = \frac{\phi_C^g + \frac{1}{2}\phi_H^g}{\phi_{Fe}^s} = \mu^0 - \frac{\phi_C^s}{\phi_{Fe}^s} \tag{17}$$

We defined μ_{carb} as the decrease in specific fuel consumption due to carburization:

$$\mu_{carb} = \frac{M_{Fe}}{M_C}\left(\frac{\%C}{\%Fe}\right)_{DRI} \tag{18}$$

We then introduced the correction factor r_μ to be applied to the specific consumption to take into account its decrease due to carburization:

$$r_\mu = \frac{\mu_0 - \mu_{carb}}{\mu_0} = 1 - \frac{1}{\mu_0}\frac{M_{Fe}}{M_C}\left(\frac{\%C}{\%Fe}\right)_{DRI} \tag{19}$$

Finally, the generalized specific consumption, considering carburization, was calculated from the following relationship:

$$\mu = \mu_0 \cdot r_\mu \tag{20}$$

By similar reasoning, the degree of oxidation was calculated from the correction coefficient r_μ:

$$X = \frac{X^0}{r_\mu} \tag{21}$$

We finally deduced:

$$x = \frac{x^0 + 1 - r_\mu}{r_\mu} \tag{22}$$

2.5. Plotting of the Thermodynamic Forbidden Zone

As previously explained, the Chaudron diagram determines the gas–solid equilibrium states between the different types of iron oxides and a reducing gas composed of H_2, H_2O, CO, and CO_2. The abscissa, denoted $x_{Chaudron}$, was calculated from the reactive component of the gas, according to the following relationship:

$$x_{Chaudron} = \frac{\%CO_2 + \%H_2O}{\%CO + \%CO_2 + \%H_2 + \%H_2O} \tag{23}$$

Contrary to the situation encountered for the blast furnace, the $x_{Chaudron}$ abscissa is not directly transferable from the Chaudron diagram to the Rist diagram. Consequently, we applied the following transformation to trace the critical points **W** and **M** in this generalized formalism, according to Equations (13) and (23):

$$x_{Rist} = (1-\alpha) \cdot x_{Chaudron} - \alpha \tag{24}$$

$$\alpha = \frac{\sum \left(m + \frac{n}{2}\right) \%C_mH_n}{\eta_{CH}} \tag{25}$$

3. Results and Discussion

3.1. Graphical Description of a Direct-Reduction Shaft Working Point in ArcelorMittal Contrecoeur

We studied the operating point of Midrex NG^{TM} Module 2 in the ArcelorMittal Contrecoeur plant obtained in April 2016 using the Rist operating diagram, following the methodology developed in the previous section.

To describe this operating point, we assumed that in situ methane reforming was localized in the metallization zone. This assumption is widely accepted, although not measured, because this reaction is favored both by the local temperatures and by the presence of iron, which acts as a catalyst [4,5]. Therefore, a constant methane rate was assumed throughout the reduction zone, equal to that of the top gas. The CH_4 content at points **W** and **M** was finally equal to that of the top gas, and their corresponding coefficient α can be calculated with Equation (25).

Figure 4 shows the operating line for this operating point, as well as the forbidden zone calculated for a gas-Wustite and gas-Magnetite equilibrium at a temperature of 800 °C, in agreement with in-situ measurements provided by Takenaka and Kimura [15].

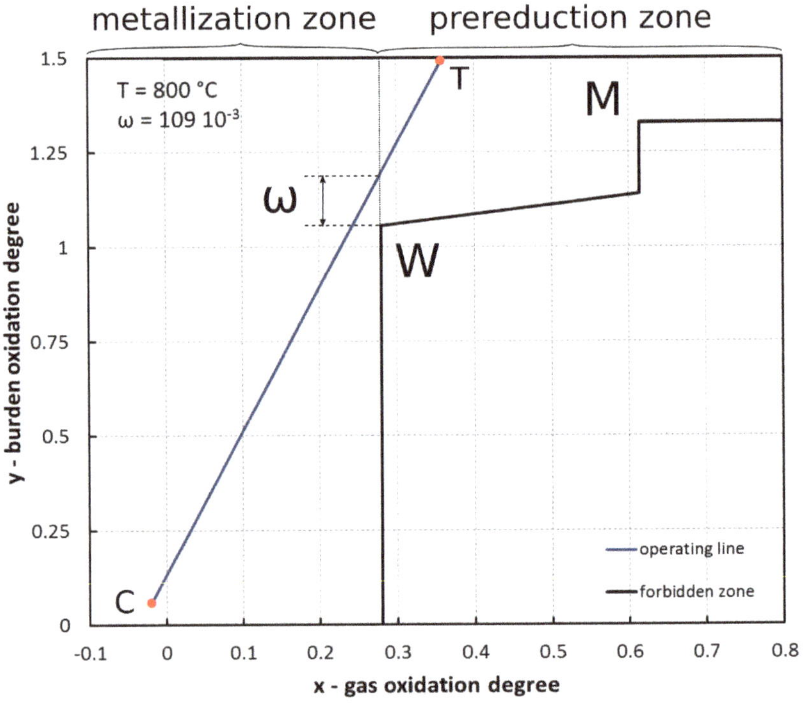

Figure 4. Operating diagram for Midrex NG^{TM} module 2 in ArcelorMittal Contrecoeur plant on April 2016, assuming 100% in situ reforming in the metallization zone.

Point **C** represents the bottom of the reduction zone, assumed to be localized above DRI carburization. Point **T** corresponds to the top gas zone. Point **W** defines the critical value of the gas oxidation degree x_M above which the Wustite cannot be reduced to ferrous metal. In this diagram, we considered, on both sides of point **W**, the metallization zone and the prereduction zone.

The figure shows a distance between the operating line and point **W**. Rist defines this distance as the deviation from ideality ω, here equal to 0.109 (Figure 4). It corresponds to the difference between the average oxidation state of the iron bearing material and that of the pure Wustite at the time of metallization start-up.

In this case, the average oxidation degree of iron oxides was 1.165 ($y_M + \omega$), when metallization was initiated. In 2015, in the framework of experimental laboratory tests on the reduction of pellets in a steady-state counter-current configuration, under conditions as close as possible to the industrial reactor, we showed that the reduction of Hematite to Magnetite was almost complete when Wustite appeared, at point **M** [16]. Therefore, in the present situation, it is reasonable to assume that there was no residual Hematite at the time of metallization start-up. The degree of oxidation of iron oxides therefore corresponds to a mixture of 40% Magnetite and 60% Wustite ($1.165 \approx 0.4 \times 1.33 + 0.6 \times 1.056$).

The reduction of iron oxides follows successive and staggered reactions, which is inherent to the counter-current configuration. We emphasize that this staggered effect is less pronounced here than in the blast furnace where metallization starts when sinter is mainly prereduced into Wustite.

3.2. *Influence of In Situ Reforming*

To show the impact of in situ reforming on the operating point of the shaft furnace, we now assume that methane cracking occurs in the prereduction zone and not in the metallization zone. This obviously contradicts what is commonly accepted, for the reasons detailed above. According to this hypothesis, the methane content taken into account for plotting the forbidden zone corresponds to that of the reducing gas injected in the lower part (mixture of the bustle gas with the natural gas injected in the transition zone and in the cooling zone).

A higher methane content changes the value of the α coefficient, calculated with relation (25), used to transpose the coordinates of the Chaudron diagram to the operating diagram. The increase in the coefficient α induces a decrease in the values x_W and x_M. As a result, the forbidden zone is shifted to the left. On the other hand, the operational line remains unchanged.

Figure 5 depicts this hypothetical operating point that would obey these assumptions. It shows that the deviation to ideality ω is negative this time, because the operating line crosses the forbidden zone. This situation, which is thermodynamically impossible, proves that in situ reforming cannot take place in the prereduction zone.

3.3. *Definition of the Ideal Working Point*

In the situation previously described in Figure 5, the most direct way to restore metallization is to increase specific consumption, either by increasing gas flow rates or simply by decreasing DRI production.

Figure 6 shows the change in the process working point required to generate metallization in the shaft. This is an optimized (or ideal) configuration where we rotated the operating line around the point **C** so that it passes through the point **W**.

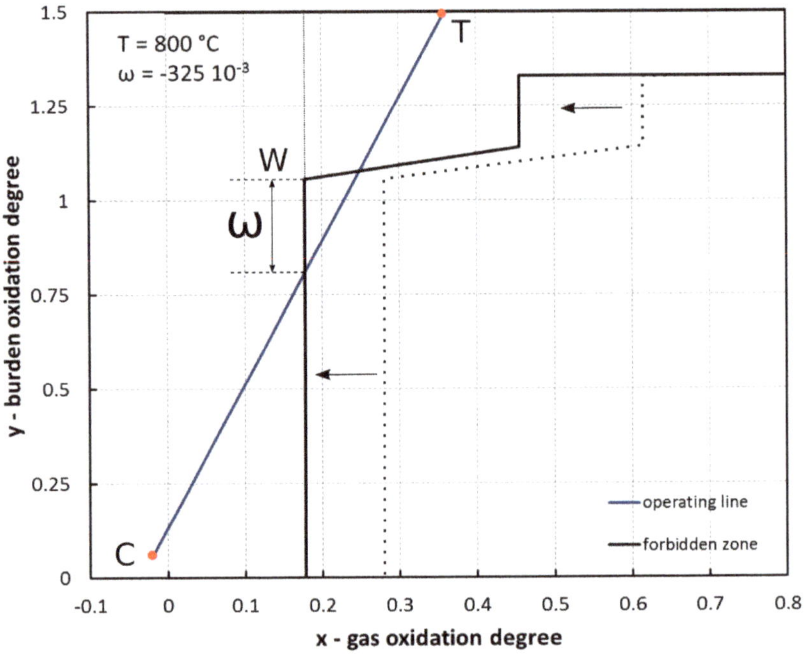

Figure 5. Operating diagram for Midrex NG^{TM} module 2 in ArcelorMittal Contrecoeur plant on April 2016, assuming 100% in situ reforming in the prereduction zone.

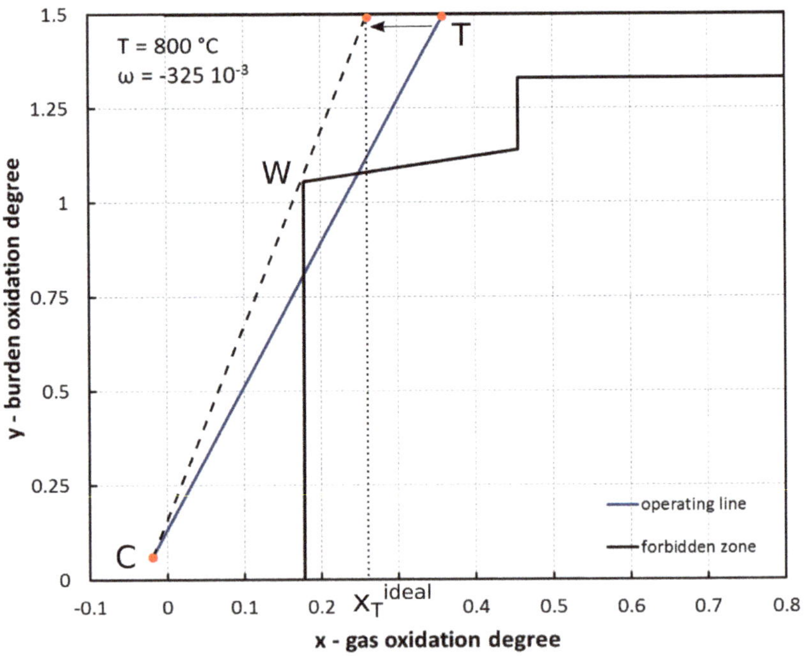

Figure 6. Operating diagram for an ideal configuration.

We thus defined a maximum oxidation degree for the top gas, denoted x_T^{ideal}, necessary for metallization to occur. This graphical description highlights a direct link between the efficiency of in situ reforming and the minimal required top gas quality (or degree of oxidation).

Finally, we must also point out that this ideal working point may not be achievable, as we did not consider the limits related to the heat balance in the metallization zone. In the situations described above, we assumed that metallization occurs at 800 °C. However, the energy required to heat the ferrous material and the metallization was provided by the sensible heat of the reducing gas, which has a maximum acceptable temperature of about 950 °C in the bustle zone to avoid DRI clustering. The only way to provide sufficient energy is therefore to inject a minimum flow of gas, which implies the existence of a minimum critical specific consumption. This condition will be particularly severe for higher levels of H_2 where the metallization is more endothermic than with CO.

4. Conclusions

We presented the adaptation of the operating diagram proposed by Rist, a graphical tool describing the operating point of the blast furnace, to direct reduction.

To achieve this, we generalized the initial approach by taking into account the presence of hydrocarbons in the reducing gas, the cracking phenomena in the shaft furnace, and the carburization of the DRI.

The diagram was thus redefined for direct reduction, also integrating the information provided by the Chaudron diagram concerning the thermodynamic equilibria between iron oxides and the reducing gas.

We used this diagram to describe the operating point of the Midrex NG^{TM} module n°2 of the ArcelorMittal Contrecoeur plant. We thus highlighted the staggered character of the reduction within the shaft furnace, distinguishing between a prereduction zone and a metallization zone. Similar to the blast furnace, the beginning of the metallization is the critical point of reduction from a thermodynamic point of view.

We studied the impact of in situ reforming on the process limits to ensure metallization, which allowed us to define the minimum top gas quality required to meet this condition. This graphical tool can be built with basic information, such as solid and gas chemical compositions and flows. It can allow us to estimate the plant performance by calculating the distance between the operating point and thermodynamic limits in order to minimize the gas/solid ratio. Such a minimal ratio should lead to a lower natural gas consumption.

This graphical tool can be used to optimize the operating point of the direct-reduction shaft furnace or to define prospective operating points (e.g., coupling with other processes—hydrogen reduction). It is also an interesting pedagogical tool to understand the reduction zone of the shaft as a counter-current gas–solid reactor.

Author Contributions: Conceptualization, T.Q.; methodology, T.Q.; software, T.Q.; validation, T.Q. and J.B.; formal analysis, T.Q.; resources, P.G. and G.L.; writing—original draft preparation, T.Q.; writing—review and editing, T.Q., J.B. and P.G. All authors have read and agreed to the published version of the manuscript.

Funding: This research received no external funding.

Institutional Review Board Statement: Not applicable.

Informed Consent Statement: Not applicable.

Data Availability Statement: Not applicable.

Conflicts of Interest: The authors declare no conflict of interest.

Abbreviations

The following abbreviations are used in this manuscript:

Greek letters	
μ	specific consumption
$\vec{\phi}$	molar flux ($mol \cdot m^{-3} \cdot s^{-1}$)
Latin letters	
a_j	volumetric fraction of the molecule j in a mix gas
c_i	local molar concentration of the atomic element i ($mol \cdot m^{-3}$)
n_i^j	number of atoms i in the molecule j
Q_m	mass flow rate ($kg \cdot s^{-1}$)
Q_v	volumetric flow rate ($m^3 \cdot s^{-1}$)
V_m	molar volume of the gas ($m^3 \cdot mol^{-1}$)
w	mass fraction
x^{gas}	gas oxidation degree (derived definition)
X^{gas}	gas oxidation degree (original definition)
y	burden oxidation degree
superscript	
gas	related to gaseous element
s	related to solid element (iron bearing material)
subscript	
i	atomic element (C, H, O, or Fe)

References

1. Usui, T.; Ohmi, M.; Yamamura, E. Analysis of Rate of Hydrogen Reduction of Porous Wustite Pellets Basing. *ISIJ Int.* **1990**, *30*, 347–355. [CrossRef]
2. Vannucci, M.; Colla, V.; Corbo, G.; Fera, S. A simplified approach to the simulation of direct reduction of iron ore. *Metall. Res. Technol.* **2010**, *107*, 195–204. [CrossRef]
3. Parisi, D.R.; Laborde, M.A. Modeling of counter current moving bed gas-solid reactor used in direct reduction of iron ore. *Chem. Eng. J.* **2004**, *104*, 35–43. [CrossRef]
4. Shams, A.; Moazeni, F. Modeling and Simulation of the MIDREX Shaft Furnace: Reduction, Transition and Cooling Zones. *JOM* **2015**, *67*, 2681–2689. [CrossRef]
5. Hamadeh, H.; Mirgaux, O.; Patisson, F. Detailed Modeling of the Direct Reduction of Iron Ore in a Shaft Furnace. *Materials* **2018**, *11*, 1865. [CrossRef]
6. Rist, A.; Bonnivard, G. Reduction d'un lit d'oxydes de fer par un gaz, première partie. *Rev. Métall.* **1963**, *60*, 23–37.
7. Rist, A.; Meysson, N. Étude graphique de la marche du haut fourneau avec vent humide et injections aux tuyères. *Rev. Métall.* **1965**, *62*, 995–1040. [CrossRef]
8. Rist, A.; Bonnivard, G. Réduction d'un lit d'oxydes de fer par un gaz, deuxième partie. *Rev. Métall.* **1966**, *63*, 296–312. [CrossRef]
9. Rist, A.; Dancoisne, P.; Jon, R. Connaissance du haut fourneau. *Rev. Métall.* **1967**, *64*, 97–118. [CrossRef]
10. Cordier, J.; Guerbois, J.; Hazard, P.; Malvaux, R.; Reveil, C. Les injections au haut fourneau. *Rev. Métall.* **1962**, *59*, 791–815. [CrossRef]
11. Rist, A.; Meysson, N. Recherche graphique de la mise au mille minimale du haut fourneau a faible température de vent. *Rev. Metall.* **1964**, *61*, 121–146. [CrossRef]
12. Meysson, N.; Maaref, A.; Rist, A. Emploi de boulettes prereduites au haut fourneau etude graphique. *Rev. Metall.* **1965**, *62*, 1161.
13. Nicolle, R.; Capelani, R.; Guillot, J.B.; Rist, A. La réduction dans le haut fourneau et la recherche des très faibles mises au mille de coke. *Rev. Métall.* **1980**, *77*, 769–779. [CrossRef]
14. Rist, A. Analogue Diagrams for Process Metallurgists. *ISIJ Int.* **1992**, *32*, 1034–1043. [CrossRef]
15. Takenaka, Y.; Kimura, Y.; Narita, K.; Kaneko, D. Mathematical model of direct reduction shaft furnace and its application to actual operations of a model plant. *Comput. Chem. Eng.* **1986**, *10*, 67–75. [CrossRef]
16. Quatravaux, T.; Barros, J.; Maurice, Y.; Pierret, H.; Chen, E.; Tsvik, G. Laboratory scale study of reduction kinetics of iron oxides on direct reduction conditions. In Proceedings of the METEC & 2nd Estad 2015, Düsseldorf, Germany, 15–19 June 2015.

metals

MDPI

Article

Experimental and Numerical Investigations on Charging Carbon Composite Briquettes in a Blast Furnace

Qiang Wang, Zhu Liu, Huiqing Tang * and Zhancheng Guo

State Key Laboratory of Advanced Metallurgy, University of Science and Technology Beijing, Beijing 100083, China; g20199322@xs.ustb.edu.cn (Q.W.); g20199317@xs.ustb.edu.cn (Z.L.); zcguo@ustb.edu.cn (Z.G.)

* Correspondence: hqtang@ustb.edu.cn; Tel.: +86-10-82375842

Abstract: In the present research, charging carbon composite briquettes (CCB) in a blast furnace (BF) was investigated. The CCB used contained 29.70 wt.% Fe_3O_4, 39.70 wt.%, FeO, 1.57 wt.% iron, 8.73 wt.% gangue, and 20.30 wt.% carbon. Its reaction kinetics in BF was examined by nonisothermal tests and modeled. Thereafter, the influence of replacing 10% ore with CCB on BF performance was studied by numerical simulations. Results showed that the CCB reaction behavior in BF could be modeled using the previously proposed model under a_{gs} = 1900 $m^2 \cdot m^{-3}$. Numerical simulations on a BF with a production of 6250 t hot metal per day (tHM/day) showed that replacing 10% ore with CCB efficiently improved the BF operation for coke saving. In the CCB charging operation, the CCB reached a full iron-oxide reduction above the cohesive zone (CZ) and a carbon conversion of 85%. By charging CCB, the thermal state in the BF upper part was significantly changed while it was not influenced in the BF lower part; the ore reduction was retarded before the temperature reached 1073 K and was prompted after and the local gas utilization tends to increase above the CZ. By the CCB reduction above the CZ, BF top gas temperature was decreased by 8 K, the BF top gas utilization was increased by 1.3%, the BF productivity was decreased by 17 tHM/day, the coke rate was decreased by 52.2 kg/tHM, and ore rate was decreased by 101 kg/tHM. Considering the energy consumption of sintering and coking, charging the CCB could have a significant energy-saving and CO_2-emission-reducing effect for BF iron making.

Keywords: carbon composite briquette; blast furnace ironmaking; reaction kinetics; numerical simulation; coke saving

Citation: Wang, Q.; Liu, Z.; Tang, H.; Guo, Z. Experimental and Numerical Investigations on Charging Carbon Composite Briquettes in a Blast Furnace. *Metals* **2021**, *11*, 1669. https://doi.org/10.3390/met11111669

Academic Editors: Thomas Echterhof, Ko-Ichiro Ohno and Ville-Valtteri Visuri

Received: 18 August 2021
Accepted: 14 October 2021
Published: 20 October 2021

1. Introduction

The development of the economy and society is increasing the demand and production of iron and steel. In 2019, crude steel production in China reached 996.3 million tons, representing 53.3% of global crude steel production [1,2]. The blast furnace ironmaking-basic oxygen furnace steelmaking (BF-BOF) route is the main route for the production, producing approximately 70% of the total crude steel [3]. Nowadays, the iron and steel industry is facing pressures of energy-saving and CO_2 emission reduction [4,5]. As the BF sector (including coking and sintering) is responsible for approximately 80% of the total energy consumption and generating most CO_2 emissions in the BF-BOF route [6], low-carbon technologies in BF ironmaking have attracted increasing attention from scholars worldwide [7,8].

Charging carbon composite briquettes (CCB) is considered to be a promising technology to improve BF efficiency [9,10]. The CCB refers to carbonaceous materials mixed with iron-bearing materials into agglomerates. Using CCB as a partial charge in BF offers the following benefits of (1) less coking and sintering [11–13], (2) utilization of low-grade iron ores or carbon materials [14–16], (3) the process is completed faster than that with pellets or sinter [17,18], and (4) energy consumption tends to decrease [19]. Several methods of preparation of CCB for BF have been proposed, including hot briquetting using the thermal plasticity of coal [20], and cold briquetting using cement as a binder [21,22], cold briquetting using coking tar as a binder [23]. The authors of the current study previously proposed to prepare CCB using cold briquetting followed by heat treatment [24,25]. By this method, various noncoking coals and iron-rich metallurgical dust could be used as raw materials, which could significantly reduce the CCB cost. The behavior of a single CCB prepared using this method under simulated BF conditions and in actual BF have been elucidated and the results showed that the CCB reaction in BF includes five stages: reduction by BF gas, partial self-reduction with reduction by BF gas, full self-reduction partial self-reduction with gasification by BF gas and gasification by BF gas [25]. However, to improve BF efficiency, it is more important to understand the influence of charging CCB on BF performance.

The BF ironmaking is a complex process with high temperature, high pressure, and hazardous environment, so conditions of lab-scale experiments could not fully simulate the actual BF in-furnace state and thus results may be unreliable. Nowadays, novel processes involved in BF ironmaking are usually investigated by numerical simulations. Using numerical simulations, investigators can gain very detailed information to examine the feasibility, understand the mechanisms, and optimize operation conditions towards the envisaged BF processes [26–29]. Simulations on BF operations with CCB charging have been conducted by Chu et al. [30] and Yu et al. [31]. However, in their studies, the reaction model of CCB was significantly simplified and could not reflect the real behavior in BF, which may lead to some misunderstanding in interpreting the influence of CCB charging on BF performance.

In this research, the reaction behavior of CCB under BF conditions was experimentally studied and modeled. Thereafter, BF operation with replacing 10% ore by CCB was investigated by numerical simulations.

2. Experimental

2.1. CCB Sample

The CCB sample used in the present research was the same as that in reference [25]. It was prepared by cold briquetting followed by heat treatment. The raw materials for preparing the CCB sample were hematite fines, quartz fines, and coal fines. The quartz fines were employed as an additive. The hematite fines and the quartz fines were the chemical reagents. The coal fines were provided by the BF PCI (pulverized coal injection) sector. The hematite fines, quartz fines, and coal fines were thoroughly mixed under a mass ratio of hematite:quartz:coal = 67:3:30. After the addition of 10.0% distilled water, 2.0% organic binder, the briquettes were made by pressing these moistened fines using a die under a pressure of 15 MPa. The briquettes were dried in the air followed by drying at 423 K and were then hardened by heat treatment. The heat treatment was carried out under an N_2 atmosphere. The thermal route was the following. The furnace was heated from room temperature to 1073 K at a rate of 5 K/min. After holding for 10 min, the furnace was cooled naturally. The prepared CCB is cylindrical with diameter and height of 14 mm. Its mass is 4.7 g. Its mineralogical composition is listed in Table 1.

Table 1. Mineralogical composition of CCB: (wt %).

Carbon	Fe_3O_4	FeO	Metallic Iron	Gangue
20.30	29.70	39.70	1.57	8.73

2.2. Non-Isothermal Reaction Tests

The experimental setup is detailed elsewhere [25]. The following is an outline. The setup mainly consists of a gas supply system, and a temperature-controlled furnace with an accuracy of ±2 K, and a computer for data acquisition. The furnace was heated using super-canthal ($MoSi_2$) elements, producing a 50-mm hot zone in the reaction tube (Diameter: 55 mm). The sample holder was made of a heat-resistant alloy (Fe-Cr-Al) wire. In each test, the furnace was heated up to 1073 K and stabilized for 30 min under N_2 atmosphere. A single CCB was loaded at a time. After being preheated for 5 min in the upper part of the tube, the sample was lowered into the constant- temperature zone. The furnace was then heated up under a predetermined heating rate and the gas flow was switched from N_2 to a CO-CO_2-N_2 mixture (CO:CO_2:N_2 = 4:1:5 (volume)). The mass loss of CCB was recorded via a computer. In the test, the total gas flow rate was maintained at 3000 $cm^3 \cdot min^{-1}$ (standard temperature and pressure). The test was completed after the temperature reached 1373 K. Pre-experimental results showed that, after heat hardening, volatiles and organic binder could be completely removed from the CCB, so its mass loss fraction at time t (f_m) was calculated by Equation (1).

$$f_{\rm m} = (m - m_{\rm b})/(m_{\rm C,0} + m_{\rm O,0}) \quad (1)$$

where, m is the mass of CCB at time t, (g); m_b is the initial mass of CCB, (g); $m_{C,0}$ and $m_{O,0}$ are the initial mass of carbon and iron-oxide oxygen in CCB, (g), and they are determined according to Table 1 and the initial mass of CCB.

3. Model Development

3.1. Description of BF Operation with CCB Charging

Size of the BF for numerical simulations is given by Tang et al. [32], and its normal operation data is given in Table 2. In the present investigation, two cases (case A and case B) were simulated and compared. Case A was the BF operation under normal conditions and considered as the base case. Case B was with CCB charging. In case B, 10% (mass) ore was replaced with CCB. In CCB charging of case B, the CCB is assumed to be fully mixed with the ore (sinter, pellet, and lump ore).

Table 2. BF operation data.

Variable	Value
Productivity ($tHM \cdot day^{-1}$)	6250
Blast temperature (K)	1523
Blast rate ($Nm^3 \cdot min^{-1}$)	4800
Oxygen enrichment (mol%)	4.0
Top absolute pressure (Pa)	2.8×10^5
PC injection rate ($kg \cdot tHM^{-1}$)	180
Ore rate ($kg \cdot tHM^{-1}$)	1680
Coke rate ($kg \cdot tHM^{-1}$)	335
Batch weight of ore (ton)	76
Batch weight of coke (ton)	15
Solid inlet temperature (K)	300
Ore particle property	Composition: TFe: 55.8 wt.%, FeO: 6.8 wt.%, CaO: 4.60, SiO_2: 4.97 wt.%, Al_2O_3: 2.19 wt.%, TiO_2: 2.0 wt.% Porosity: 0.35; Bulk density: 1750 kg/m^3; Average particle size: 20 mm.
Coke particle property	Composition: Fixed Carbon: 90 wt.%, and Ash: 10 wt.%; Porosity: 0.50; Bulk density: 500 kg/m^3; Average particle size: 40 mm.
PC property	Composition: C: 80.0 wt.%, H: 4.0 wt.%, O: 3.5 wt.%, N: 2.0 wt.%, and S: 0.32 wt.%; H_2O: 4.0 wt.%, and Ash: 7.0 wt.%.
Liquid phase (molten iron and slag) property	[%C]: 4.0 wt.%, Temperature: 1753 K, Average heat capacity: 1000 J/kg, and Slag rate: approximately 400 kg/tHM

tHM: ton hot metal, PC: pulverized coal.

3.2. BF Model

The present model is based on a total BF model developed by current authors [32]. The validity of the total BF model was confirmed by the comparison of the simulation results with the averaged industrial data. In the present model, the reaction kinetics of CCB is modeled and is incorporated into the total BF model. The model is two-dimensional, axisymmetric, and steady. In the model, the gas-phase behavior and solid-phase behavior in BF are represented by the conservation of mass, momentum, energy, and species. The computation grid is shown in Figure 1. It is a two-dimensional structure grid, including 780 cells. The computational zone is based on 12 degrees in the circumferential direction. The positions of cohesive zone (CZ), deadman, raceway (RW) are predefined. The porosity of CZ, dripping zone (DZ), and deadman are fixed at 0.15, 0.30, and 0.15, respectively. The Equations involved in the model are listed in Table 3. Equations (1)–(3) in Table 3 are CO gaseous reduction of ore, Equations (4) and (5) in Table 3 are coke solution-loss reaction and combustion. Equations (6)–(9) are CCB Equations. Equations (10)–(12) in Table 3 are the melting of the ore. Equations (12)–(15) in Table 3 are the melting of CCB. the melting CCB is assumed to be similar to that of the ore. Rates of Equations (6)–(9) and (16) in Table 3 are given in the following sections, Reaction heats of all Equations and rates of Equations (1)–(5), (10)–(15) in Table 3 are given elsewhere [32].

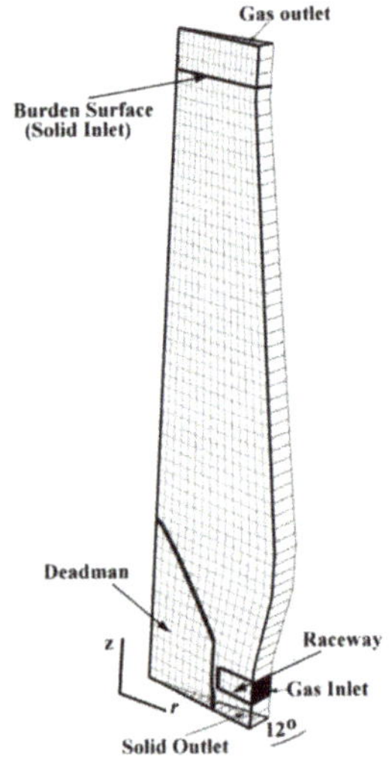

Figure 1. Mesh placement for numerical simulation.

Table 3. Equations involved in BF model.

	Reaction	Reaction Rate (kmol·m^{-3}s^{-1})	Explanation
1	$3Fe_2O_3$ (ore, s) + CO(g) = $2Fe_3O_4$(ore, s) + CO_2(g)	R_1	stepwise reduction of ore (sinter, pellet and lump ore) by CO
2	Fe_3O_4 (ore, s) + CO(g)=3 FeO(ore, s) + CO_2(g)	R_2	
3	FeO (ore, s) + CO(g) = Fe(ore, s) + CO_2(g)	R_3	
4	C (coke) + CO_2 (g) = 2CO (g)	R_4	coke solution-loss reaction
5	C (coke) + $1/2O_2$ (g) = CO (g)	R_5	coke combustion
6	$3Fe_2O_3$ (CCB, s) + CO (g) = $2Fe_3O_4$ (CCB, s) + CO_2 (g)	R_6	CCB reactions
7	Fe_3O_4(CCB, s) + CO(g) = 3FeO(CCB, s) + CO_2(g)	R_7	
8	FeO (CCB, s) + CO (g) = Fe (CCB, s) + CO_2 (g)	R_8	
9	C (CCB) + CO_2 (g) = 2CO (g)	R_9	
10	Fe (ore,s) = Fe (l)	R_{10}	melting reactions of ore
11	FeO (ore,s) = FeO (l)	R_{11}	
12	Gangue (ore,s) = Slag (l)	R_{12}	
13	Fe (CCB,s) = Fe (l)	R_{13}	melting reactions of CCB
14	FeO (CCB,s) = FeO (l)	R_{14}	
15	Gangue (CCB,s) = Slag (l)	R_{15}	
16	FeO (l) + C (s) = Fe (l) + CO (g)	R_{16}	direct reduction of molten FeO

Behaviors of other phases (molten iron, molten slag, and PC fines) are treated using simplified methods. The PC particles are gasified in the raceway zone reaching a burnout rate of more than 90% within 20 ms. Therefore, the combustion products of the blast and the PC through Equation (5) form the inlet condition for the gas phase in the model. The liquid phase includes the molten iron and the molten slag. Droplets of the molten iron and the molten slag are generated in the cohesive zone with an initial temperature equivalent to the local solid temperature. After generation, they flow down through the dripping zone, acquiring heated by the gas phase and the coke bed and reaching the final tapping temperature in the hearth. On their flowing path, these droplets undergo coalescing, splitting, or flying a short distance with the strong bosh gas, so it is difficult to give precise mathematical descriptions upon the gas-liquid and solid-liquid heat exchanges. In the BF bosh, heat is mainly generated by the combustion of oxygen with coke and PC in the raceway and the gas phase has the highest temperature. As the gas flows upward, the heat is transferred from the gas to the coke bed, and to the liquid droplets; simultaneously, the heat is also transferred from the coke bed to the liquid droplets. This analysis shows that the required heat for the liquid phase could be simplified as an energy source of the gas phase. The temperature of the hearth is considered to be 1753 K, therefore, the overall heat loss rate (Q_l) from the gas-solid system to the liquid phase is Equation (2), Assuming that the heat loss rate is uniformly distributed in the DZ, an enthalpy source Equation (3) is added to energy Equation of the gas phase in the DZ.

$$Q_l = \sum_{i}^{CZ}(M_{Fe}(R_{10} + R_{13}) + M_{FeO}(R_{11} + R_{14}) + M_{Gangue}(R_{12} + R_{15}))Cp_l V_{cell}(1753 - T_S) \quad (2)$$

$$E_{gl} = Q_l / \sum_{i}^{DZ} V_{cell} \quad (3)$$

Molten FeO in the slag droplets is reduced fast in the DZ through Equation (9). In the view of the mass balance of molten FeO in CZ and DZ, the rate of Equation (16) in Table 3 is described using Equation (4), in which, Equation (16) is assumed to uniformly proceed in the DZ.

$$R_{16} = \sum_{i}^{CZ}((R_7 + R_{14})V_{cell}) / \sum_{i}^{DZ} V_{cell} \quad (4)$$

The above method of treating the behavior of the liquid phase was demonstrated to be helpful for the model to reach a high convergence of the model.

The gas flow is considered to be the flow through the porous bed. The gas phase consists of CO, CO_2, O_2, and N_2, and is considered to be an ideal gas. The general governing Equation of the gas phase is Equation (5), in which, the superficial gas velocity is adopted. Terms to represent ϕ, Γ_ϕ and S_ϕ in Equation (5) are listed in Table 4.

$$\mathrm{div}(\rho_g \vec{\mathbf{U}}_g \phi) = \mathrm{div}(\Gamma_\phi \mathrm{grad}\phi) + S_\phi \quad (5)$$

Table 4. Dependent variables and sources in Equation (5).

Equation	ϕ	S_ϕ
Mass	1	$M_O \sum_{i=1}^{3} R_i + M_O \sum_{i=6}^{8} R_i + M_C(R_4 + R_9) + M_C R_5 + M_{CO} R_{16}$
Momentum	$\vec{\mathbf{U}}_g$	$-\nabla P_g - \vec{\mathbf{F}}_{gs}$
Energy	H_g	$0.5 \sum_{i=1}^{9} R_{i,}(-\Delta H_i) - E_{gs} - E_{gl} + E_{add}$
Species	y_{O_2}	$M_{O_2}(-0.5R_5)$
	y_{CO}	$M_{CO}(-R_1 - R_2 - R_3 + 2R_4 + R_5 - R_6 - R_7 - R_8 + 2R_9 + R_{16})$
	y_{CO_2}	$M_{CO_2}(R_1 + R_2 + R_3 - R_4 + R_6 + R_7 + R_8 - R_9)$
	y_{N_2}	0

A non-slip wall condition for the gas velocity and an impermeable condition for the gas species are defined on the BF wall. The heat loss of gas phase on the BF wall is calculated by 5.0 (T_g-353) [33]. The PC particles are assumed to be gasified completely in the raceway. Therefore, the combustion products of the blast with the PC through Equation (5) in Table 3 form the inlet conditions for the gas phase. At the gas outlet, a fully-developed gas flow is assumed.

The solid flow is treated as a viscous flow. The CCB is treated as one component of the solid phase. As a consequence, the solid phase consists of coke, ore, and CCB. Each component has its physical properties. Above the CZ, the overall physical properties of the solid phase are calculated by averaging the physical properties of the components based on their volume fractions. Regarding the chemical species, even the same species in CCB and in ore/coke are treated separately because it undergoes different reaction schemes. The general governing Equation of the solid phase is Equation (6), in which, the solid bulk density and the solid physical velocity are adopted. Terms to represent φ, Γ_φ, and S_φ in Equation (6) are listed in Table 5. In actual BF, the iron-bearing burden is transformed to molten iron and slag, and the coke is completely consumed by combustion, carbon-solution loss reaction, carburization, and other equations. However, the present BF model is developed based on the gas-solid two-phase flow. Therefore, the present BF model needs a solid outlet. For ensuring a stable solid flow, the consumption of coke is not included in Table 5.

$$\mathrm{div}(\rho_s \vec{\mathbf{V}}_S \varphi) = \mathrm{div}(\Gamma_\varphi \mathrm{grad}\varphi) + S_\varphi \quad (6)$$

Table 5. Dependent variables and sources in Equation (6).

Equation	φ	Γ_φ	S_φ
Continuity	1	0	$-\left(M_O\sum_{i=1}^{3}R_i + M_O\sum_{i=6}^{8}R_i\right) - (M_C R_9) - M_{Fe}(R_{10}+R_{13}) - M_{FeO}(R_{11}+R_{14})$ $-M_{gangue}(R_{12}+R_{15})$
Momentum	$\vec{\mathbf{V}}_s$	$\mu_{s,eff}$	$-\nabla P_s$
Energy	H_s	$\lambda_{s,eff}/Cp_s$	$0.5\sum_{i=1}^{9}(R_{i,}(-\Delta H_i)) + \sum_{i=10}^{16}(R_{i,}(-\Delta H_i)) + E_{gs}$
Species	$y_{Coke,C}$	0	0
	y_{ore,Fe_2O_3}	0	$M_{Fe_2O_3}(-3R_1)$
	y_{ore,Fe_3O_4}	0	$M_{Fe_3O_4}(2R_1 - R_2)$
	$y_{ore,FeO}$	0	$M_{FeO}(R_2 - R_3 - R_{11})$
	$y_{ore,Fe}$	0	$M_{Fe}(R_3 - R_{10})$
	$y_{CCB,C}$	0	$M_C(-R_9)$
	y_{CCB,Fe_2O_3}	0	$M_{Fe_3O_4}(-3R_6)$
	y_{CCB,Fe_3O_4}	0	$M_{Fe_2O_3}(2R_6 - R_7)$
	$y_{CCB,FeO}$	0	$M_{FeO}(R_7 - R_8 - R_{14})$
	$y_{CCB,Fe}$	0	$M_{Fe}(R_8 - R_{13})$
	$y_{CCB,gangue}$	0	$M_{gangue}(-R_{15})$

A fluid-slip boundary is applied for the solid velocity on the BF wall. Heat loss of the solid phase on the BF wall is not considered. Inlet conditions of the solid phase are established according to BF operation conditions. At the solid outlet, the solid phase reaches a fully-developed flow. As, in actual BF operation, coke is completely consumed in BF, the enthalpy loss owing to the solid flow at the solid outlet is compensated by adding a source (E_{add}) on the gas enthalpy Equation in RW, which is expressed by Equation (7).

$$E_{add} = (m_{C,coke} T_{s,out} Cp_s) / \sum_{i}^{RW} V_{cell} \quad (7)$$

3.3. CCB Model

The CCB model developed by Tang et al. [34] was used in this research. The following is a brief introduction of the CCB model. The shape of CCB is nearly spherical, so the model is one-dimensional in a radial direction. The model is developed based on mass conservation of gas species, mass conservation of solid species, and mass transfer between CCB and environment. The gas species include CO, CO_2, and N_2, and the solid species are the components in the CCB. The Equations in the CCB model are listed in Table 6. Equations (1)–(3) in Table 6 are the gaseous reductions of iron particles, and Equation (4) in Table 6 is the carbon solution-loss reaction of carbon particles.

Table 6. Equations in CCB model.

No	Reaction	Reaction Rate/($mol \cdot m^{-3} \cdot s^{-1}$)
1	$3Fe_2O_3$(fine, s) + CO(g) = $2Fe_3O_4$(fine, s) + CO_2(g)	$r_i = \frac{(P_{CO}-P_{CO_2}/K_i)/(8.314T)}{(K_i/(k_i(1+K_i))}(1-f_i)^{2/3}\Big)a_{gs}$, (i = 1,2,3),
2	Fe_3O_4(fine, s) + CO(g) = 3FeO(fine, s) + CO_2(g)	$k_1 = \exp(-1.445 - 6038/T)$, $K_1 = \exp(7.255 + 3720/T)$
3	FeO(fine, s) + CO(g) = Fe(fine, s) + CO_2(g)	$k_2 = 1.70\exp(2.515 - 4811/T)$, $K_2 = \exp(5.289 - 4711/T)$ $k_3 = \exp(0.805 - 7385/T)$, $K_3 = \exp(-2.946 + 2744.63/T)$
4	C(fine,s) + CO_2(g) = 2CO(g)	$r_4 = \rho_{C,0}k_4(1-f_4)^{2/3}(P_{CO_2}/1.01 \times 10^5)/M_C$, $k_4 = 1400\exp(-138000/RT)$

The mass conservation of the gas species in the CCB gives Equations (8) and (9).

$$\frac{\partial(\alpha P_{co_2})}{\partial t} = \frac{1}{r^2}\frac{\partial}{\partial r}\left(r^2 D_{CO_2-N_2,eff}\frac{\partial P_{co_2}}{\partial r}\right) + RT(R_1 + R_2 + R_3 - R_4) \quad (8)$$

$$\frac{\partial(\alpha P_{co})}{\partial t} = \frac{1}{r^2}\frac{\partial}{\partial r}\left(r^2 D_{CO-N_2,eff}\frac{\partial P_{co}}{\partial r}\right) + RT(2R_4 - R_1 - R_2 - R_3) \quad (9)$$

where, $\alpha = 0.5$, $D_{eff,CO-N_2} = D_{CO-N_2}\alpha^2/\sqrt{3}$, and $D_{eff,CO_2-N_2} = D_{CO_2-N_2}\alpha^2/\sqrt{3}$.

For Equations (8) and (9), the boundary conditions are Equations (10)–(12).

$$r = 0: \frac{\partial P_{CO}}{\partial r} = 0, \frac{\partial P_{CO2}}{\partial r} = 0 \quad (10)$$

$$r = d/2: D_{eff,CO-N_2}\frac{\partial P_{CO}}{\partial r} = \left(D_{CO-N_2}\left(2.0 + 0.6Re^{1/2}Sc^{1/3}_{CO-N_2}\right)/d\right)(P_{CO} - P_{CO,e}) \quad (11)$$

$$r = d/2: D_{eff, CO_2-N_2}\frac{\partial P_{CO_2}}{\partial r} = \left(D_{CO_2-N_2}\left(2.0 + 0.6Re^{1/2}Sc^{1/3}_{CO_2-N_2}\right)/d\right)(P_{CO_2} - P_{CO_2,e}) \quad (12)$$

where, $Re = u_{g,e}\rho_{g,e}d/\mu_{g,e}$, $Sc_{CO-N_2} = \mu_{g,e}/(\rho_{g,e}D_{CO-N_2})$, and $Sc_{CO_2-N_2} = \mu_{g,e}/(\rho_{g,e}D_{CO_2-N_2})$.

For Equations (8)–(9), the initial conditions are provided by Equation (13).

$$t = 0, r \in (0, d/2): P_{CO} = P_{CO,e}, P_{CO_2} = P_{CO_2,e} \quad (13)$$

The mass conservation of the solid species in the CCB gives Equation (14).

$$\partial\rho_j/\partial t = S_j \quad (14)$$

where j = Fe_2O_3, Fe_3O_4, FeO, Fe, and C; $S_{Fe_2O_3} = 0.003M_{Fe_2O_3}(-r_1)$, $S_{Fe_3O_4} = 0.001M_{Fe_3O_4}(2r_1 - r_2)$, $S_{FeO} = 0.001M_{FeO}(3r_2 - r_3)$, $S_{Fe} = 0.001M_{Fe}R_3$, and $S_C = -0.001M_Cr_4$.

The initial condition for Equation (14) is provided by Equation (15).

$$t = 0, r \in (0, d/2): \rho_j = \rho_{j,0} \quad (15)$$

3.4. Solution Strategy

The simulations were performed using the following strategy. Firstly, the rates of the CCB Equations (6)–(9) in Table 3 were initialized. The BF model Equations (5) and (6) were solved numerically solved using PHOENICS [35] and an in-house developed code. After the BF model reached a primary convergence, the Lagrangian method was used to adjust these reaction rates in all cells. These adjustments continued until the BF model reached the final convergence.

The Lagrangian method to adjust the rates of Equations (6)–(9) in Table 3 is illustrated in Figure 2 and detailed elsewhere [34]. For a given cell, the solid-phase streamline through the cell is determined after the BF model reaches a primary convergence Figure 2a. The CCB reaction behavior along the streamline is calculated using the above CCB model. The descending time of the CCB to reach the cell center is calculated by $\int_0^s \left(1/\left|\vec{\mathbf{V}}_s\right|\right)ds$, where, s is the distance from the cell center to the burden surface, (m). The boundary conditions of the CCB model are determined by the corresponding BF variables along the streamline. Thus, the radial distributions of r_1, r_2, r_3, and r_4 of Equations (1)–(4) Table 6 in CCB reaching the cell center are obtained. It is considered that the CCB reaching the cell center is representative of all briquettes in the cell Figure 2b. Therefore, the rates of Equations (6)–(9) in Table 3 in the cell are Equation (16).

$$R_j = 0.001\alpha_{\text{CCB}}N_{\text{CCB}}4\pi\int_0^{d/2} r_i r^2 dr \tag{16}$$

where, i = 1, 2, 3, and 4 of the Equations in Table 5 for j = 6, 7, 8, and 9 of the Equations in Table 3, respectively; α_{CCB} is the volume fraction of CCB in the solid burden, (-); and N_{CCB} is the number density of CCB, $(1/\text{m}^3)$.

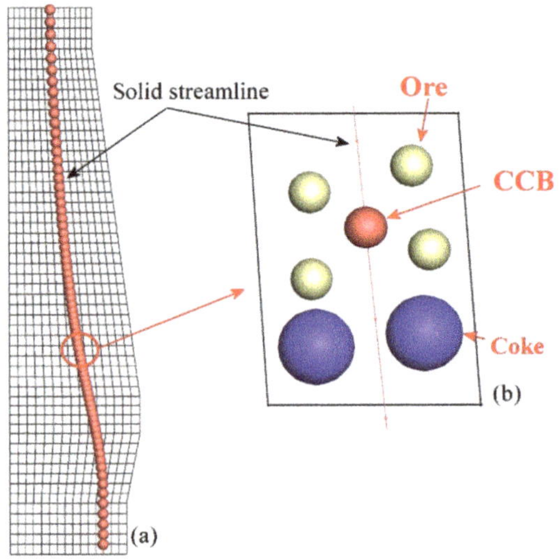

Figure 2. Illustration of Lagrangian method in model: (**a**) solid-flow stream line in BF (blast furnace), and (**b**) illustration of CCB in solid burden.

In addition to the examination of the convergence of gas and solid flow fields, the mass balance of the removable element O and of the element Fe are examined and the convergence criteria are Equations (17) and (18).

$$\left|m_{\text{O}} - \sum M_{\text{O}}(R_1 + R_2 + R_3 + R_6 + R_7 + R_8)V_{\text{cell}}\right|/m_{\text{O}} < 0.01 \tag{17}$$

$$\left|m_{\rm Fe} - \sum M_{\rm Fe}(R_{10} + R_{11} + R_{13} + R_{14} + R_{16})V_{\rm Cell}\right| / m_{\rm Fe} < 0.01 \quad (18)$$

where, m_O is the mass supply rate of the element O in the solid phase at the solid inlet, and m_{Fe} is the mass supply rate of element Fe in the solid phase at the solid inlet.

Detailed gas and solid inlet conditions of cases A and B are shown in Table 7. The results of case A are kept as reference values for case B. In case B, the coke supply rate is determined by trial and error, and the convergence criteria for the mass balance of element C in the coke is Equation (19).

$$\left|m_{\rm C,coke} - \sum M_{\rm C}(R_4 + R_5 + R_9 + R_{16})V_{\rm cell} - m_{\rm Fe}[\%{\rm C}]/(1.0 - [\%{\rm C}]) - m_{\rm C,other}\right| / m_{\rm C,coke} < 0.01 \quad (19)$$

where, $m_{C,coke}$ and m_{Fe} are the mass supply rates of element C in the coke, and element Fe in the solid burden, respectively; [%C] is the carbon content in molten iron; and $m_{C,other}$ is the rate of carbon consumed by other Equations (e.g., silica and manganese oxide reductions), which is determined by case A.

Table 7. Gas and solid inlet conditions of cases A and B (1/30 BF volume).

Condition	Variable	Case A	Case B
Solid inlet conditions	Ore supply rate (kg/s)	4.05	3.64
	CCB supply rate (kg/s)	0	0.41
	Coke supply rate (kg/s)	0.81	-
	Solid temperature (K)	300	
Gas inlet conditions	Gas supply rate (kg/s)	3.88	
	Gas composition (mass fraction, -)	CO: 20, O_2:13, N_2:67	
	Gas temperature (K)	2350	

4. Results and Discussion

4.1. Determination of Parameter in CCB Model

In this case, a_{gs} in the reaction rates of Equations (1)–(3) in Table 6 was difficult to determine owing to the sintering of iron-oxide particles [25]. Different from CCB reduction in some direct reduction processes (e.g., rotary hearth furnace), CCB in BF undergoes slow heating. During the heating, its self-reduction and Equations with BF gas proceed. Under the BF environment, changes in CCB volume and porosity become obvious with the increase of temperature [34,36]. Therefore, compared to the isothermal tests, the nonisothermal tests under simulated BF environment are more suitable to determine the CCB model parameters. The value of a_{gs} was determined by trial and error. The experimental data points for fitting are shown in Figure 3a. They were selected at time intervals of 300 s on the curve of 2 K·min^{-1}, 120 s on the curve of 5 K·min^{-1}, and 60 s on the curve of 10 K·min^{-1}. Five values of a_{gs} (1000 m^2·m^{-3}, 1300 m^2·m^{-3}, 1600 m^2·m^{-3}, 1900 m^2·m^{-3}, and 2200 m^2·m^{-3}) were examined. the fitness of each value was evaluated by *MSE*, which is expressed by Equation (20).

$$MSE = \left(\sum_i^{N_P} (v_{\rm sim} - v_{\rm exp})^2\right) / N_{\rm P} \quad (20)$$

where, v_{sim} is model-predicted value, v_{exp} is experimental value, and N_P is the total number of data points in Figure 3a.

The fitting results showed that, under a_{gs} = 1000 m^2·m^{-3}, *MSE* = 0.064; under a_{gs} = 1300 m^2·m^{-3}, *MSE* = 0.058; under a_{gs} = 1600 m^2·m^{-3}, *MSE* = 0.055; under a_{gs} = 1900 m^2·m^{-3}, *MSE* = 0.054, and under a_{gs} = 2200 m^2·m^{-3}, *MSE* = 0.055. Thus, it was considered that a_{gs} = 1900 m^2·m^{-3} was optimal. Measured mass-loss curves and optimal model-predicted ones are plotted in Figure 3b. It could be seen in Figure 3b that the model predictions agree well with the experimental measurements.

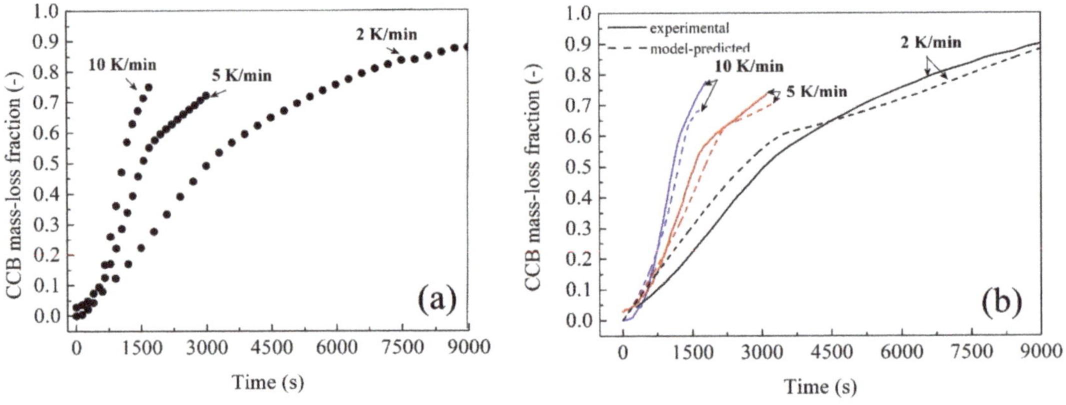

Figure 3. Determination of CCB model parameter of a_{gs}: (**a**) experimental data points for fitting, and (**b**) comparison between model prediction curve and experimental curve under different heating rates.

4.2. CCB Behavior in BF

The simulation results of CCB behavior in case B are shown in Figure 4. Figure 4a shows that the CCB iron oxide starts reduction at approximately 673 K, and it reaches a full reduction at approximately 1123 K, reflecting that the CCB reducibility is high. Figure 4b shows that the CCB carbon starts gasification at approximately 923 K. Above the CZ, its overall conversion is 85%, indicating that 15% of the CCB carbon would enter the BF lower part. In the present investigation, the influence of the ungasified CCB carbon in the BF lower part was not considered as its behavior has not been distinctly disclosed so far [37].

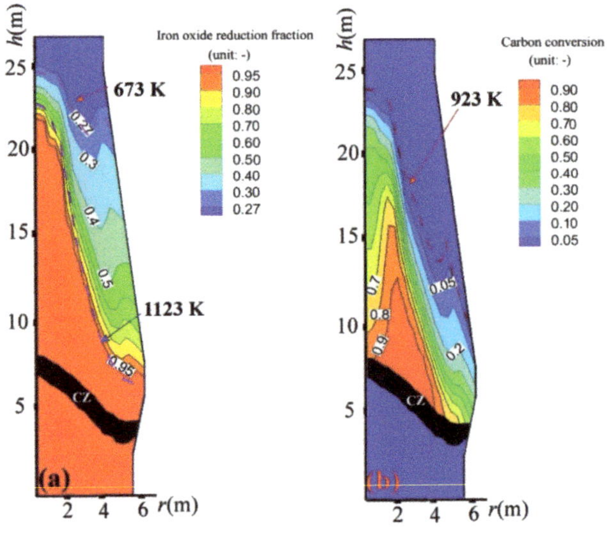

Figure 4. Reaction behavior of CCB in case B: (**a**). profile of CCB iron-oxide reduction fraction, (**b**). profile of CCB carbon conversion.

Figure 5 shows the behavior of a single CCB along a solid flowing path in the BF. The path is near the BF mid-radius Figure 5a. From Figure 5b, it can be seen that, in the temperature range from 923 K to 1123 K, self-reduction occurs in the CCB. However, the

CO potential ($P_{CO}/(P_{CO}+P_{CO_2})$) in CCB becomes higher than that in BF gas after the temperature reaches 1083 K.

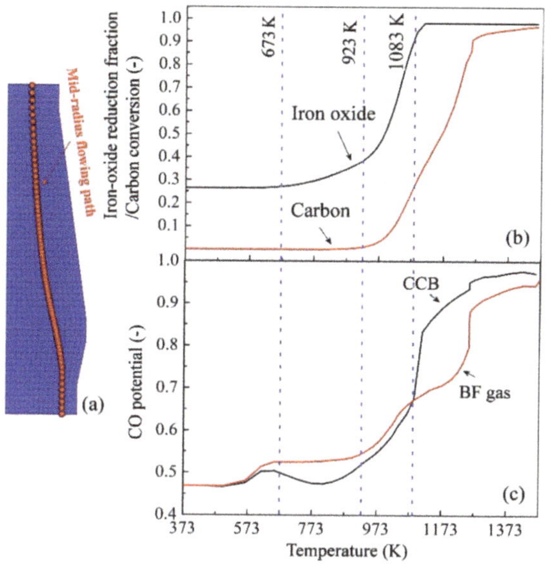

Figure 5. Reaction behavior of a single CCB: (**a**) illustration of the mid-radius solid flowing path, (**b**) changes of iron-oxide reduction fraction and carbon conversion in CCB along the path, and (**c**) changes of CO potential in CCB along the path.

4.3. Influence on BF in-Furnace State

The influence of CCB charging on the BF thermal state is shown in Figures 6 and 7. In Figure 6, compared to case A, lines 873 K and 1073 K move downward in case B. These tendencies are also displayed in Figure 7. In CCB charging operation, the CCB self-reduction is a strongly endothermic reaction and needs more heat than the ore gaseous reduction or the coke gasification. The CCB self-reduction occurs in the temperature range from 923 K to 1123 K Figure 5a, so the gas-solid heat transfer is enhanced there, and the gas-solid heat transfer is weakened above, resulting in a considerable change of thermal state in the BF upper part. After 1273 K, the CCB Equations are finished, so the influence on the BF thermal state becomes negligible. It is observed that, in Figure 6, locations of line 1273 K in both cases are nearly the same, and the heights of CZ in both cases don't show a significant difference. Furthermore, in Figure 7, gas temperature profiles in both cases exhibit similar patterns in the BF lower part.

The influence of charging CCB on gas and solid reaction behaviors is shown in Figures 8–11. By charging CCB, the ore reduction is retarded in the BF upper part (e.g., in Figure 8, in comparison to case A, the distance between lines 0.1 and 0.2 in case B increases). This is mainly attributed to the delay of solid temperature increase. However, after the solid temperature reaches 1073 K, the ore reduction is prompted (e.g., in Figure 8, comparing cases A and B, lines of 0.2 and 0.9 are closer in case B, and, in the mid radius zone, the distance between line 0.9 and CZ decreases in case B). This is attributed to the increase of CO potential in BF gas by CCB (Figure 5a). As a result, charging CCB has a positive effect on the overall ore reduction above CZ. Before the temperature reaches 1073 K, the CCB reduces the CO potential in BF gas (Figure 5b), leading to a decrease of CO volume fraction. After the temperature reaches 1073 K, the ore reduction is intensified by CCB, thus, CO volume fraction above CZ tends to decrease by charging CCB (e.g., in Figure 9, compared to case A, lines of 0.25 and 0.40 move downward in case B), and CO_2 volume fraction above CZ tends to increase (Figure 10). Accordingly, local gas utilization ($P_{CO_2}/(P_{CO}+P_{CO_2})$)

increases above CZ (e.g., in Figure 11, compared to case A, lines of 0.4 and 0.5 in case B move downward).

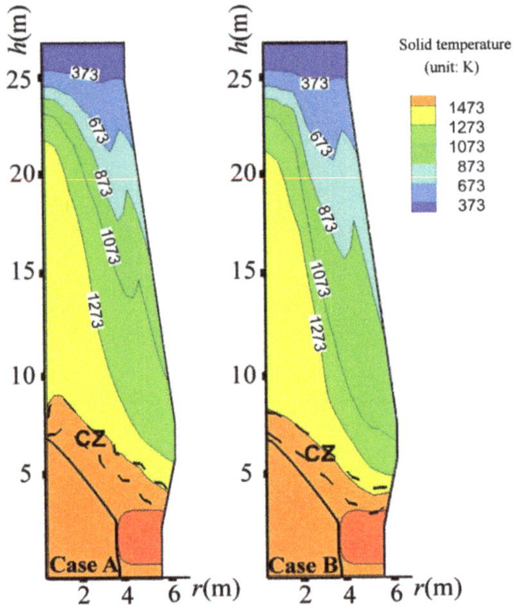

Figure 6. Profiles of solid temperature in cases A and B.

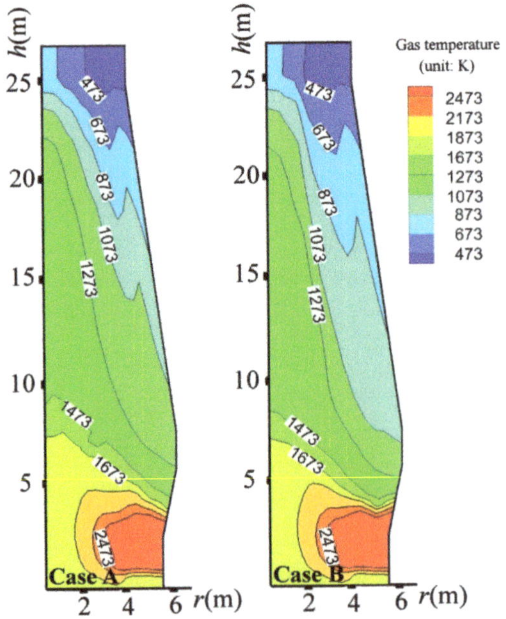

Figure 7. Profiles of gas temperature in cases A and B.

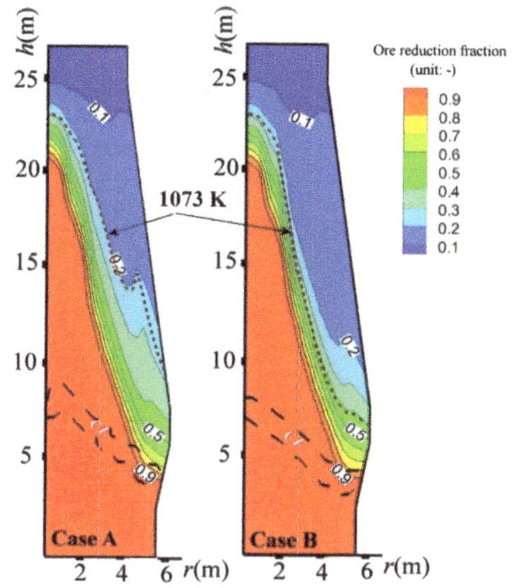

Figure 8. Profiles of ore reduction fraction in cases A and B.

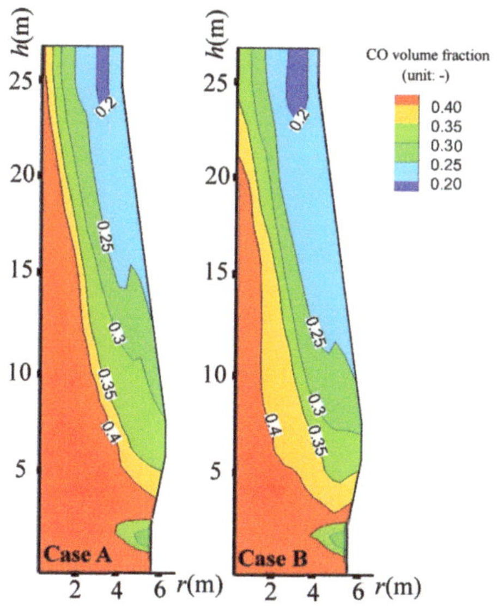

Figure 9. Profiles of CO volume fraction in cases A and B.

4.4. Coke-Saving Analysis

Table 8 lists some operation indices for cases A and B. By replacing 10% ore with CCB, the BF top gas temperature is decreased by 8 K, the BF top gas utilization is increased by 1.3%, BF productivity is decreased by 17 tHM/day, and the coke rate is decreased by 52.2 kg/tHM (this decrease is only with the consideration of gasification of CCB carbon above CZ).

The BF ironmaking process includes sintering, coking, and ironmaking. Replacing 10% ore with CCB, the CCB rate is 179 kg/tHM and the ore rate is decreased to 1579 kg/tHM, Therefore, in the CCB charging operation, 101 kg sinter and 52.2 kg could be saved to produce one-ton hot metal. Considering the energy consumption of sintering and coking, the CCB charging operation could have a significant energy-saving and CO_2-emission-reducing effect for BF ironmaking.

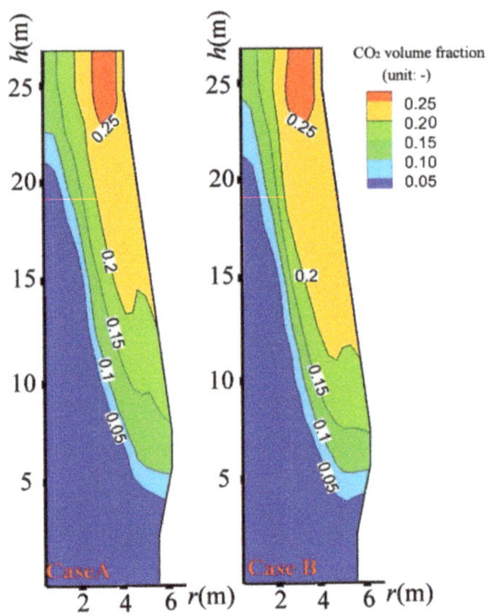

Figure 10. Profiles of CO_2 volume fraction in cases A and B.

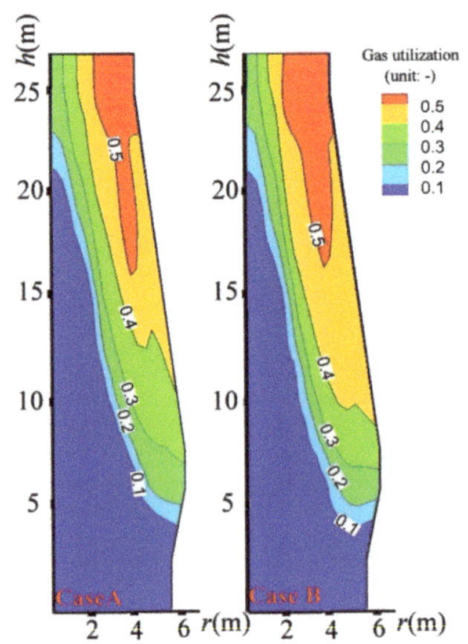

Figure 11. Profiles of local gas utilization in cases A and B.

Distributions of coke consumption in cases A and B are listed in Table 9. From Table 9, it is known that in the total coke-rate reduction by charging CCB, 14.3 kg/tHM is from the carbon solution loss reaction of coke, 38.4 kg/tHM from the direct reduction of molten FeO. These findings indicate that charging CCB can suppress the coke solution-loss reaction and significantly reduce the coke consumption in the direct reduction of molten FeO.

Table 8. Simulation results of some BF indices.

Index	Case A	Case B
Productivity ($tHM \cdot day^{-1}$)	6250	6233
Top gas temperature (K)	463	455
Top gas utilization (%)	51.3	52.6
Fuel rate ($kg \cdot tHM^{-1}$)	PC: 180 Coke:335, CCB carbon: 0	PC: 180.5, Coke:282.8, CCB carbon gasified above CZ: 30 above CZ:30.0 PC:180.5, and CCB carbon gasified above CZ:30.0

Table 9. Distributions of coke consumption in cases A and B (kg/tHM).

Item	Case A	Case B
Combustion	175.0	175.5
Carbon solution loss in upper BF	56.8	42.5
Direct reduction of molten FeO in lower BF	49.2	10.8
Carburization of molten iron	45.0	45.0
Other reactions	9.0	9.0
Total	335.0	282.8

5. Conclusions

In this study, the CCB for BF application was prepared using cold briquetting followed by heat treatment. The CCB contained 29.70 wt.% Fe_3O_4, 39.70 wt.% FeO, 1.57 wt.% Fe, 8.73 wt.% gangue, and 20.30 wt.% carbon. its reduction kinetics in BF conditions were examined by nonisothermal tests and modeled. The BF operation with replacing 10% ore with CCB was investigated by numerical simulations. Some conclusions were obtained.

1. The CCB reaction behavior in BF could be modeled using the previously proposed model by the current authors. Under a_{gs} = 1900 $m^2 \cdot m^{-3}$, the agreement between experimental measurements and model predictions was satisfying.
2. In the CCB charging operation, the CCB reached a full iron-oxide reduction and a carbon conversion of 85% above CZ.
3. By charging CCB, the thermal state in the BF upper part was significantly changed; however, the BF thermal state in the BF lower part was inconsiderably influenced.
4. By charging CCB, the ore reduction was retarded before the temperature reached 1073 K and was prompted after; and the local gas utilization tended to increase above the CZ.
5. By the CCB reduction above the CZ, the BF top gas temperature was decreased by 8 K, the BF top gas utilization was increased by 1.3%, the BF productivity was decreased by 17 tHM/day, 101 kg sinter and 52.8 kg could be saved to produce one-ton hot metal. Considering the energy consumption of sintering and coking, charging the CCB could have a significant energy-saving and CO_2-emission-reducing effect for BF iron making.

Author Contributions: Conceptualization, H.T.; methodology, H.T.; software, H.T.; validation, Q.W. and Z.L.; formal analysis, Q.W. and Z.L.; investigation, Q.W.; writing—original draft preparation, Q.W.; writing—review and editing, H.T.; visualization, Z.L.; supervision, Z.G. All authors have read and agreed to the published version of the manuscript.

Funding: This research was funded by the National Natural Science Foundation of China, grant number U1960205.

Institutional Review Board Statement: Not applicable.

Informed Consent Statement: Not applicable.

Data Availability Statement: Not applicable.

Acknowledgments: The authors thank the National Natural Science Foundation of China and the State Key Laboratory of Advanced Metallurgy USTB for the financial support for this work.

Conflicts of Interest: The authors declare no conflict of interest.

Nomenclature

a_{gs}	specific surface of iron-oxide particles in CCB, ($m^2 \cdot m^{-3}$)
Cp	heat capacity, ($J \cdot kg^{-1} \cdot K^{-1}$)
d	diameter of CCB, (m)
D, D_{eff}	gas diffusivity, effective gas diffusivity, ($m^2 \cdot s^{-1}$)
E	enthalpy source, ($J \cdot m^{-3} \cdot s^{-1}$))
f_i	reduction fraction or carbon conversion of reaction i in Table 5, (-)
H	total enthalpy ($J \cdot kg^{-1}$)
ΔH_i	reaction heat of reaction i in Table 2, ($J \cdot kmol^{-1}$)
k_i	reaction rate constant of reaction i in Table 5, ($m \cdot s^{-1}$, $kg \cdot s^{-1} atm^{-1}$)
K_i	equilibrium constant of reaction i in Table 5, (-)
m	mass supply/consumption rate of the given element, ($kg \cdot s^{-1}$)
M	molar weight, ($kg \cdot kmol^{-1}$)
P	pressure, (pa)
Pr	Prandtl number, (-)
R	gas constant, (8.314 $J \cdot mol^{-1} \cdot K^{-1}$)
Re	Reynolds number, (-)
R_i	chemical reaction rate of reaction i in Table 2, ($kmol \cdot m^{-3} \cdot s^{-1}$)
r	radial direction, (m)
r_i	chemical reaction rate of reaction i in Table 5, ($mol \cdot m^{-3} \cdot s^{-1}$)
S	source, units vary
Sc	Schmidt number (-)
T	temperature, (K)
t	time, (s)
V_{cell}	cell volume, (m^3)
y	mass fraction, (-)
Greek letter	
α	CCB porosity, (-)
ϕ, φ	general dependent variable
Γ	general difusion coeffcient
ρ	density, ($kg \cdot m^{-3}$)
λ	thermal conductivity, ($W \cdot m^{-1} \cdot K^{-1}$)
μ	fluid viscosity, ($kg \cdot m^{-1} \cdot s^{-1}$)
Vector	
$\vec{\mathbf{U}}_g$	superficial gas velocity, ($m \cdot s^{-1}$)
$\vec{\mathbf{V}}_s$	solid physical velocity, ($m \cdot s^{-1}$)
$\vec{\mathbf{F}}_{gs}$	gas flow resistance, ($N \cdot m^{-3}$)
Subscript	
0	initial
CCB	CCB variable
coke	coke variable
ore	ore variable
g	gas variable
l	liquid variable
s	solid variable
e	environment variable
species or element name	variable of assigned species or element

References

1. Wang, Y.; Zuo, H.; Zhao, J. Recent progress and development of ironmaking in China as of 2019: An overview. *Ironmak. Steelmak.* **2020**, *47*, 640–649. [CrossRef]
2. Wang, Z.; Zhang, J.; Liu, Z.; Wang, G.; Yang, T. Status, technological progress, and development directions of the ironmaking industry in China. *Ironmak. Steelmak.* **2019**, *46*, 937–941. [CrossRef]
3. Naito, M.a.K.; Takeda, K.; Matsui, Y. Ironmaking technology for the last 100 years: Deployment to advanced technologies from introduction of technological know-how, and Evolution to Next-generation Process. *ISIJ Int.* **2015**, *55*, 7–35. [CrossRef]
4. An, R.; Yu, B.; Ru, L.; Wei, Y.M. Potential of energy savings and CO_2 emission reduction in China's iron and steel industry. *Appl. Energy* **2018**, *226*, 862–880. [CrossRef]
5. Tan, X.; Li, H.; Guo, J.; Gu, B.; Zeng, Y. Energy-saving and emission-reduction technology selection and CO_2 emission reduction potential of China's iron and steel industry under energy substitution policy. *J. Clean. Prod.* **2019**, *222*, 823–834. [CrossRef]
6. Ariyama, T.; Murai, R.; Ishii, J.; Sato, M. reduction of CO_2 emissions from integrated steel works and its subjects for a future study. *ISIJ Int.* **2005**, *45*, 1371–1378. [CrossRef]
7. Chowdhury, S.; Chakraborti, N.; Sen, P.K. Energy optimization studies for integrated steel plant employing diverse steel-making route: Models and evolutionary algorithm-based approach. *Miner. Process. Extr. Metall. Rev.* **2020**, *42*, 1–12. [CrossRef]
8. Liu, Z.J.; Zhang, J.L.; Yang, T.J. Low Carbon Operation of Super-Large Blast Furnaces in China. *ISIJ Int.* **2015**, *55*, 1146–1156. [CrossRef]
9. Ahmed, H. New trends in the application of carbon-bearing materials in blast furnace iron-making. *Minerals* **2018**, *8*, 561. [CrossRef]
10. Sharma, T. Composite pre-reduced pellet (CPR)—A new way for fines utilisation. *Miner. Process. Extr. Metall. Rev.* **1993**, *12*, 115–124. [CrossRef]
11. Ahmed, H.M.; Viswanathan, N.; Bjorkman, B. Composite pellets—A potential raw material for iron-making. *Steel Res. Int.* **2014**, *85*, 293–306. [CrossRef]
12. Kasai, A.; Toyota, H.; Nozawa, K.; Kitayama, S. Reduction of reducing agent rate in blast furnace operation by carbon composite iron ore hot briquette. *ISIJ Int.* **2011**, *51*, 1333–1335. [CrossRef]
13. Yokoyama, H.; Higuchi, K.; Ito, T.; Oshio, A. Decrease in carbon consumption of a commercial blast furnace by using carbon composite iron ore. *Tetsu Hagane* **2012**, *100*, 601–609. [CrossRef]
14. Andersson, A.; Gullberg, A.; Kullerstedt, A.; Wedholm, A.; Wikstrm, J.; Ahmed, H.; Kvist, L.S. Recycling of blast furnace sludge to the blast furnace via cold-bonded briquettes: Evaluation of feasibility and influence on operation. *ISIJ Int.* **2019**, *59*, 1786–1795. [CrossRef]
15. Nikai, I.; Garbers-Craig, A.M. Use of iron ore fines in cold-bonded self-reducing composite pellets. *Miner. Process. Extr. Metall. Rev.* **2016**, *37*, 42–48. [CrossRef]
16. Ueda, S.; Yanagiya, K.; Watanabe, K.; Murakami, T.; Inoue, R.; Ariyama, T. Reaction model and reduction behavior of carbon iron ore composite in blast furnace. *ISIJ Int.* **2009**, *49*, 827–836.
17. Mourao, M.B.; Cyro, T. Self-reducing pellets for ironmaking: Reaction rate and processing. *Miner. Process. Extr. Metall. Rev.* **2003**, *24*, 183–202. [CrossRef]
18. Watanabe, K.; Ueda, S.; Inoue, R.; Ariyama, T. Enhancement of reactivity of carbon iron ore composite using redox reaction of iron oxide powder. *ISIJ Int.* **2010**, *50*, 524–530. [CrossRef]
19. Kasai, A.; Matsui, Y. Lowering of thermal reserve zone temperature in blast furnace by adjoining carbonaceous material and iron ore. *ISIJ Int.* **2004**, *44*, 2073–2078. [CrossRef]
20. Kawanari, M.; Matsumoto, A.; Ashida, R.; Miura, K. Enhancement of reduction rate of iron ore by utilizing iron ore/carbon composite consisting of fine iron ore particles and highly thermoplastic carbon material. *ISIJ Int.* **2011**, *51*, 1227–1233.
21. Kemppainen, A.; Iljana, M.; Heikkinen, E.P.; Paananen, T.; Mattila, O.; Fabritius, T. Reduction behavior of cold-bonded briquettes under simulated blast furnace conditions. *ISIJ Int.* **2014**, *54*, 1539–1545. [CrossRef]
22. Singh, M.; Björkman, B. Testing of cement-bonded briquettes under laboratory and blast furnace conditions Part 1—Effect of processing parameters. *Ironmak. Steelmak.* **2017**, *34*, 41–53. [CrossRef]
23. Mochizuki, Y.; Nishio, M.; Tsubouchi, N.; Akiyama, T. Reduction behavior and crushing strength of carbon-containing pellet prepared from COG tar. *Fuel Process. Technol.* **2016**, *142*, 287–295. [CrossRef]
24. Tang, H.; Liu, S.; Tao, R. Preparation of High-Carbon Metallic Briquette for Blast Furnace Application. *ISIJ Int.* **2019**, *59*, 12–20. [CrossRef]
25. Tang, H.; Sun, Y.; Rong, T. Experimental and numerical investigation of reaction behavior of carbon composite briquette in blast furnace. *Metals* **2020**, *10*, 49. [CrossRef]
26. Kuang, S.; Li, Z.; Yu, A. Review on modeling and simulation of blast furnace. *Steel Res. Int.* **2018**, *89*, 1700071. [CrossRef]
27. Okosun, T.; Silaen, A.K.; Zhou, C.Q. Review on computational modeling and visualization of the ironmaking blast furnace at purdue university northwest. *Steel Res. Int.* **2019**, *90*, 1900046. [CrossRef]
28. Sahai, Y. Computer/mathematical modeling in materials processing operations. *Miner. Process. Extr. Metall. Rev.* **1992**, *9*, 1–26. [CrossRef]
29. Ueda, S.; Natsui, S.; Nogami, H.; Yagi, J.I.; Ariyama, T. Recent progress and future perspective on mathematical modeling of blast furnace. *ISIJ Int.* **2010**, *50*, 914–923. [CrossRef]

30. Chu, M.; Nogami, H.; Yagi, J.I. Numerical analysis on charging carbon composite agglomerates into blast furnace. *ISIJ Int.* **2004**, *44*, 510–517. [CrossRef]
31. Yu, X.; Shen, Y. Computational fluid dynamics study of the flow and thermochemical behaviors of a carbon composite briquette in an ironmaking blast furnace. *Energy Fuels* **2019**, *33*, 11603–11616. [CrossRef]
32. Tankg, H.; Rong, T.; Fan, K. Numerical investigation of applying high-carbon metallic briquette in blast furnace ironmaking. *ISIJ Int.* **2019**, *59*, 810–819. [CrossRef]
33. Kuang, S.B.; Li, Z.Y.; Yan, D.L.; Qi, Y.H.; Yu, A.B. Numerical study of hot charge operation in ironmaking blast furnace. *Miner. Eng.* **2014**, *63*, 45–56. [CrossRef]
34. Tang, H.; Sun, Y.; Rong, T.; Guo, Z. Reaction model and reaction behavior of carbon composite briquette in blast furnace. *Powder Technol.* **2021**, *377*, 832–842. [CrossRef]
35. Cham. *PHOENICS User Document*; CHAM Ltd.: London, UK, 2000.
36. Tanaka, Y.; Ueno, T.; Okumura, K.; Hayashi, S. Reaction behavior of coal rich composite iron ore hot briquettes under load at high temperatures until 1400 °C. *ISIJ Int.* **2011**, *51*, 1240–1246. [CrossRef]
37. Gavel, D.J. A review on nut coke utilization in the ironmaking blast furnaces. *Mater. Sci. Technol.* **2017**, *33*, 381–387. [CrossRef]

metals

MDPI

Article

Numerical Analysis on Characteristics of Reduction Process within a Pre-Reduction Rotary Kiln

Chenghong Liu [1], Xueyong Ding [1,*], Hegong Liu [1], Xinlin Yan [2], Chao Dong [3] and Jia Wang [4]

[1] School of Metallurgy, Northeastern University, Shenyang 110819, China; hcl123love@126.com (C.L.); liu306827@163.com (H.L.)
[2] Institute of Solid State Physics, Vienna University of Technology, Wiedner Hauptstr. 8-10, 1040 Vienna, Austria; yan@ifp.tuwien.ac.at
[3] Tangsteel New District Ironmaking Business Department, Hebei Iorn & Steel Group, Tangshan 063000, China; dongchaoaizww@163.com
[4] Hebei Yanxing Machinery Co., Ltd., Zhangjiakou 075000, China; guayujibuzailai@sina.com
* Correspondence: xyding@mail.neu.edu.cn; Tel.: +86-024-8367-2068

Abstract: The reduction process inside the ore pre-reduction rotary kiln involves a series of physico-chemical reactions, and in-depth understanding of the reduction behavior is helpful to improve the product quality and productivity. This paper reports a three-dimensional steady state mathematical model based on computational fluid dynamics, which considers heat transfer, mass transfer and chemical reactions inside the rotary kiln. A user-defined functions (UDFs) program in C language is developed to define physical parameters and chemical reactions, and calculate the heat and mass transfer between freeboard and bed regions. The model is validated by measurement data and is then used to investigate the detailed information inside the rotary kiln. The results show that there is a temperature gradient in the bed, which is maximal near the kiln tail and decreases gradually as the reduction process progresses. The result also confirms that the reduction of FeO to Fe is the limiting step of the whole reduction process because this reaction requires a higher reduction potential. Furthermore, the influence of C/O mole ratio and fill degree are analyzed by comparing the average bed temperature, reduction potential and metallization ratio.

Keywords: rotary kiln; reduction process; numerical simulation; pre-reduction

Citation: Liu, C.; Ding, X.; Liu, H.; Yan, X.; Dong, C.; Wang, J. Numerical Analysis on Characteristics of Reduction Process within a Pre-Reduction Rotary Kiln. *Metals* **2021**, *11*, 1180. https://doi.org/10.3390/met11081180

Academic Editor: Thomas Echterhof

Received: 21 June 2021
Accepted: 22 July 2021
Published: 25 July 2021

1. Introduction

As the most efficient technique, blast furnace ironmaking is currently the most widely used ironmaking process. However, it has inherent defects, such as long process, high dependence on coke, environmental pollution and so on. Additionally, with the depletion of coking coal and the strict emission standards of various countries, developing an alternative ironmaking process has always been a hot topic [1,2]. In recent decades, many research institutions and steel companies have carried out research on smelting reduction technology [3–6]. Among such research, a new smelting reduction process for iron ore processing has been presented by China Iron and Steel Research Institute Group (CISRI) [7,8]. The core of the process consists of two reactors: one is the pre-reduction rotary kiln, and the other one is the smelting reduction furnace. The main purpose of this process using a reliable rotary kiln as a pre-reduction reactor is to use the thermal energy and chemical energy of gas from the final reduction reactor to lighten the load of iron oxide reduction in the smelting reduction process, further improving the energy utilization of the whole system. Undoubtedly, the stable pre-reduction process is also the necessary guarantee for the smelting reduction process. Therefore, it is necessary to study the characteristics of a rotary kiln-based process on the reduction behavior of iron oxide.

Due to the enclosed process and the limitation of measuring equipment, it is difficult to directly obtain the comprehensive information by measurement. To solve this problem, many researchers have suggested employing mathematical methods. For example, Kingsley et al. [9] established a one-dimensional model of mass and energy transfer based on the FORTRAN program, and the obtained result provides a powerful approach for kiln operational optimization. However, due to the one-dimensional nature, the mass and energy balance equations are solved for the solid and gas assuming only axial temperature variation, that is, the solid and gas phases are fully mixed in cross-section, which is obviously far from the actual production. With the development of computational fluid dynamics (CFD), the numerical simulation technique has been widely used to solve such problems. A two-dimensional numerical simulation has been frequently used. Mujumdar et al. [10,11] developed a comprehensive framework based on a two-dimensional model. They divided the computational domain into two regions: freeboard region and bed region, and then connected heat and mass transfer through a common interface. The model was able to predict qualitative and quantitative information about the key parameters. Furthermore, a similar approach was used by Gaurav et al. [12,13] to investigate the influence of input parameters on the metallization ratio and temperature profile. Unfortunately, the two-dimensional model only considers one axial section of the rotary kiln, which makes it difficult to reflect the overall behavior of the rotary kiln. In order to be more relevant to practical production, it is necessary to develop a more practical three-dimensional computational fluid dynamics (CFD) model. Manju et al. [14] developed a three-dimensional numerical model of the pneumatic coal injection and combustion process using the Eulerian–Lagrangian approach. The influence of operating parameters such as particle size distribution and wall temperature on the distribution of injected particles was investigated. However, the reduction process in the bed region was not included. In fact, the flame stability, combustion characteristics and temperature distribution directly affect the reduction process and product quality, and the release of gas from the bed region also affects the distribution of gas and temperature in the freeboard region.

As mentioned above, this work proposed a strategy of a three-dimensional steady state mathematical model with the Eulerian method to investigate the characteristics inside the rotary kiln. This model involved mass, momentum and heat transfers, as well as combustion and reduction reactions both in the freeboard and bed regions. The simulation results show detailed information from inside the rotary kiln, such as the temperature field, variation of gas and solid phase species. Additionally, the effects of key operational parameters on the average bed temperature, reduction potential and metallization ratio are discussed. These findings are helpful to control and optimize the reduction process in the pre-reduction rotary kiln.

2. Model Description

2.1. Description of the Process and Assumption

The schematic diagram of the pre-reduction rotary kiln is shown in Figure 1. The kiln is 48 m long, with a 2.3 m internal diameter and is inclined by 2 degrees. In order to obtain the appropriate temperature profile, the air supply was staged, and five blowers were placed around the kiln at equal spacing to inject air. The raw materials were charged from the kiln tail in a predetermined ratio and heated by the hot flue gas flowing in reverse direction. The iron oxide was gradually reduced to metallic iron, and then the products of kiln were hot-charged into the smelting reduction furnace.

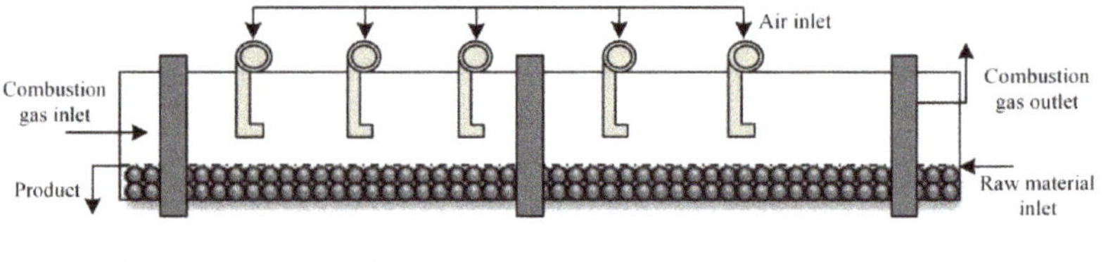

Figure 1. Schematic diagram of the pre-reduction rotary kiln.

The reduction process in the rotary kiln is complicated, including mass transfer, heat transfer, various chemical reactions and so on. Hence, it is necessary to make some simplifications for developing a mathematical model. In this study, the rotary kiln is divided into two separate regions: freeboard region and bed region, as shown in Figure 2. It is assumed that only gas phase exists in the freeboard region, and both gas and solid phases exist in the bed region. The two regions are coupled through a common boundary, and mass transfer and heat exchange occur between freeboard region and bed region via this boundary.

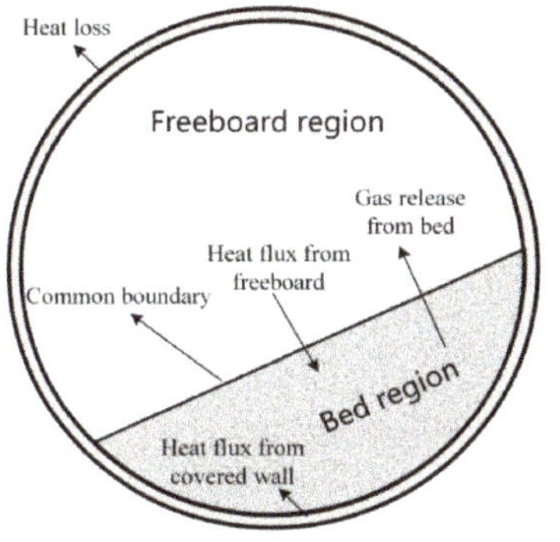

Figure 2. Heat and mass transfer between freeboard region and bed region.

2.2. Governing Equations

In this work, the rotary kiln model is a 3D steady-state model. Both gas and solid phases are treated as interpenetrating continua phases using the Eulerian method. The governing equations of mass, momentum, energy and species transport for the gas phase and solid phase can be written as follows:

$$\nabla(\varepsilon_p \cdot \rho_p \cdot \phi \cdot \vec{v}_p) = \nabla(\varepsilon_p \cdot \Gamma_\phi \cdot \nabla(\phi)) + S_\phi \quad (1)$$

where the diffusive coefficient (Γ_ϕ), variable (ϕ) and source term (S_ϕ) change with different kinds of governing equations, as summarized in Table 1.

Table 1. Variables in Equation (1).

Phase	Eq.	ϕ	Γ_ϕ	S_ϕ
Gas	Mass	1	0	$S_{\phi,g}$
	Momentum	$\vec{v}_g$	0	$\varepsilon_g(-\nabla P + \rho_g \cdot \vec{g}) + \vec{F}_{gs}$
	Energy	H_g	$\lambda_g / C_{P,g}$	$h_{gs}A_s(T_g - T_s) + M_o \cdot \sum_{n=1}^{n}(R_n \cdot \Delta H_n^T)$
	Species i	$Y_{i,g}$	$\rho_g D_i$	$\sum_k R_{i,k} M_i$
Solid	Mass	1	0	$S_{\phi,s}$
	Momentum	$\vec{v}_s$	0	$\varepsilon_s(-\nabla P + \rho_s \cdot \vec{g}) - \vec{F}_{gs}$
	Energy	H_s	$\lambda_{s,eff} / C_{P,s}$	$h_{gs}A_s(T_s - T_g) - M_o \cdot \sum_{n=1}^{n}(R_n \cdot \Delta H_n^T)$
	Species j	$Y_{j,s}$	$\rho_s D_j$	$\sum_k R_{j,k} M_j$

2.3. Reaction Kinetics Model

2.3.1. Coal Devolatilization

In the coal devolatilization process, it is considered that coal forms volatile matter and char in the gas phase and solid phase, respectively [15]. This process consists of two steps and can be described as follows:

$$\text{Coal} = \psi_1\text{Char} + \text{Volatile matter} \tag{2}$$

$$\text{Volatile matter} = \psi_2 CO + \psi_3 CO_2 + \psi_4 H_2 + \psi_5 H_2O + \psi_6 CH_4 + \psi_7 N_2 \tag{3}$$

The ψ is the mass fraction of pyrolysis products, which is determined by the coal proximate and elemental analysis [16]. Moreover, the release of individual gas species from volatile matter is assumed to be a self-degradation [17], and the first-order reaction model developed by Donskoi was used [18]:

$$R_{vi} = k_{vi,0}\exp\left(-\frac{E_{va,i}}{RT}\right)\frac{W_{vi}}{M_{vi}} \tag{4}$$

2.3.2. Homogeneous Reaction

In this model, the solid–solid reactions are neglected because they only take place at the contact surface of particles. For gas–gas reactions, gases' combustion and water–gas shift reactions are mainly considered. In the complex turbulent reacting flow, the reaction rate is usually controlled by chemical reaction kinetics and turbulent mixing rate. Therefore, the finite rate/eddy-dissipation model was adopted to describe the influence of turbulence–chemistry interaction on the reaction process [19]. This model computes both the Arrhenius rate and the mixing rate, and the smaller is favored. The kinetic parameters of homogenous reactions are shown in Table 2. The reaction rates can be calculated by the following equations:

$$\hat{R}_{i,r} = (v''_{i,r} - v'_{i,r})\left(k_{f,r}\prod_{j=1}^{N_r}[C_{j,r}]^{\eta'_{j,r}} - k_{b,r}\prod_{j=1}^{N_r}[C_{j,r}]^{\eta''_{j,r}}\right) \tag{5}$$

$$R_{i,r} = \min\left\{v'_{i,r}M_{w,i}A\rho\frac{\varepsilon}{k}\min_R\left(\frac{Y_R}{v'_{R,r}M_{w,R}}\right), v'_{i,r}M_{w,i}AB\frac{\varepsilon}{k}\frac{\sum_p Y_p}{\sum_j^N v^n_{j,r}M_{w,j}}\right\} \tag{6}$$

Table 2. Kinetic parameters of homogenous reactions [20,21].

Reaction	A_r (s^{-1})	E_a (J/kmol)	β_r	Reaction Order
$2CO + O_2 = 2CO_2$	2.2×10^{12}	1.7×10^8	0	$[CO][O_2]$
$2H_2 + O_2 = 2H_2O$	6.8×10^{15}	1.67×10^8	−1	$[H_2]^{0.25}[O_2]^{1.5}$
$CO + H_2O = CO_2 + H_2$	2.75×10^9	8.4×10^7	0	$[CO][H_2O]$

2.3.3. Heterogeneous Reaction

For gas–solid reactions, the main reactions are the carbon gasification (by CO_2 and H_2O) and the reduction of iron oxides (by CO and H_2). Since the pellet is made of very fine particles, the time scale for the reactions at high temperature is much shorter than the time scale for diffusion in the pores. Therefore, it can reasonably be assumed that pores will not contribute significantly to the reaction rate, that is, the gas–solid reactions are limited to the reaction interface. The shrinking core model developed by Sun and Lu was used to describe the heterogeneous reaction process [22,23]. This model takes into account the effects of heat and mass transfer on the reaction rate. The kinetic parameters of these heterogeneous reactions are shown in Table 3. The reaction rates can be calculated by the following equations:

$$R_i^{CO} = S_{f,i}n_i 4\pi r_i^2 (1-f_i)^{\frac{2}{3}} k_{i,o} \exp(-\frac{E_{a,i}}{RT})(C_{co} - \frac{C_{co_2}}{K_{e,i}}) \tag{7}$$

$$R_i^{H_2} = S_{f,i}n_i 4\pi r_i^2 (1-f_i)^{\frac{2}{3}} k_{i,0} \exp(-\frac{E_{a,i}}{RT})(C_{H_2} - \frac{C_{H_2O}}{K_{e,i}}) \tag{8}$$

$$R_c^{CO_2} = S_{f,c}n_c 4\pi r_c^2 (1-f_c)^{\frac{2}{3}} k_{c,0} \exp(-\frac{E_{a,c}}{RT})(C_{co_2} - \frac{C_{co}^2}{K_{e,c}}) \tag{9}$$

$$R_c^{H_2O} = S_{f,c}n_c 4\pi r_c^2 (1-f_c)^{\frac{2}{3}} k_{c,0} \exp(-\frac{E_{a,c}}{RT})(C_{H_2O} - \frac{C_{H_2} \cdot C_{CO}}{K_{e,c}}) \tag{10}$$

Table 3. Kinetic parameters of heterogeneous reactions [24].

Reactions	$k_{i,0}$ (m·s^{-1})	E_a (J/mol)	K_e	Rate Expressions
$C + CO_2 = 2CO$	1.87×10^8	221,800	exp (−20,765.92/T + 32.8)	Equation (9)
$C + H_2O = CO + H_2$	6.05×10^5	172,700	exp (−16,142.19/T + 28.16)	Equation (10)
$3Fe_2O_3 + CO = 2Fe_3O_4 + CO_2$	2700	113,859	exp (5815.5/T + 5.5076)	Equation (7)
$3Fe_2O_3 + H_2 = 2Fe_3O_4 + H_2O$	160	92,000	exp (2065/T + 8.102)	Equation (8)
$Fe_3O_4 + CO = 3FeO + CO_2$	23	71,100	exp (−4685.22/T + 5.19)	Equation (7)
$Fe_3O_4 + H_2 = 3FeO + H_2O$	30	63,600	exp (−1857.51 + 1.01)	Equation (8)
$FeO + CO = Fe + CO_2$	17	69,454	exp (2376.46/T−2.82)	Equation (7)
$FeO + H_2 = Fe + H_2O$	30	63,600	exp (−1857.51/T + 1.01)	Equation (8)

2.4. Turbulence Model

The standard k-ε model was adopted to simulate the gas phase turbulence in the freeboard region. For the bed region, the details of the movement of solids in the radial direction and variation of bed height are ignored, and solids are regarded as pseudo-homogeneous fluids. The flow velocity of solid particles in the axial direction is very small, and the turbulence has little effect on the reduction process. Hence, the bed region is assumed to be a laminar zone. The standard k-ε model is a two-equation turbulence model based on the transport equation of turbulence kinetic energy (k) and its dissipation rate (ε). The expression is as follows [25,26]:

$$\frac{\partial(\rho k)}{\partial t} + \nabla(\rho k v) = \nabla\left[\left(\mu + \frac{\mu_t}{\sigma_k}\right) \cdot \nabla k\right] + C_k - \rho\varepsilon \tag{11}$$

$$\frac{\partial(\rho \varepsilon)}{\partial t} + \nabla(\rho \varepsilon v) = \nabla\left[\left(\mu + \frac{\mu_t}{\sigma_\varepsilon}\right) \cdot \nabla \varepsilon\right] + \frac{\varepsilon}{k}(C_{\varepsilon 1}C_K - C_{\varepsilon 2}\rho\varepsilon) \tag{12}$$

where $C_{\varepsilon 1} = 1.44$, $C_{\varepsilon 2} = 1.92$, $C_\mu = 0.09$, $\sigma_k = 1.0$, $\sigma_\varepsilon = 1.3$.

2.5. Radiation Model

In this work, the discrete ordinates (DO) radiation model was used to calculate the radiation flux. The DO model is suitable for solving radiation problem in the entire range of optical thicknesses, especially for the problem of local heat source. The model solves the radiative transfer equation (RTE) for a finite number of discrete solid angles, each associated with a vector direction fixed in the global Cartesian system [27], and the equation can be written as follows:

$$\nabla(I(\vec{r},\vec{s})\vec{s}) + (a+\sigma_s)I(\vec{r},\vec{s}) = an^2\frac{\sigma T^4}{\pi} + \frac{\sigma_s}{4\pi}\int_0^{4\pi} I(\vec{r},\vec{s}')\Phi(\vec{s},\vec{s}')d\Omega \tag{13}$$

where a is the absorption coefficient of gas phase. Here, the weighted-sum-of-gray-gases model (WSGGM) is used to specify the variable absorption coefficient [28].

2.6. Heat Transfer in the Bed Region

As shown in Figure 2, the heat transfer mechanisms of the top and bottom surfaces of the bed are different. The top surface of the bed is heated mainly by convection and radiation from the gas and radiation from the inner exposed wall. The heat flux of top surface can be defined as [29,30]:

$$q_{top} = h_{gb}(T_g - T_b) + e_g X_{bg}(E_g - J_b) + (1 - e_g) X_{bw}(J_{ew} - J_b) \tag{14}$$

$$h_{gb} = 0.46\lambda_g \mathrm{Re}^{0.535} \mathrm{Re}_{\omega}^{0.104} \eta_s^{-0.341} / D_e \tag{15}$$

The bottom surface of the bed receives heat from conduction, convection and radiation from the covered wall. Tscheng and Watkinson [30] have reported the combined heat transmission coefficient between bottom surface and covered wall, and the heat flux is given by:

$$q_{bottom} = h_{wb}(T_w - T_b) \tag{16}$$

$$h_{wb} = \frac{11.6\lambda_s}{R\theta}\left(\frac{\omega R^2 \theta}{\alpha_s}\right)^{0.3} \tag{17}$$

Moreover, the rotation speed of the pre-reduction rotary kiln is very low at 0.6 rpm, and the active layer makes up a very small portion of the bed region. Hence, it is reasonable to treat the heat transfer process of the bed in the rotary kiln approximately according to the heat transfer process of the plug flow region, which is a common method and has been widely used [31–33]. The expression of effective thermal conductivity in the bed is as follows [33]:

$$\lambda_{s,eff} = \frac{\varepsilon_s}{\frac{1}{\lambda_s} + \frac{1}{4\sigma e d_p T^3}} + 4\varepsilon_g \sigma e d_p T^3 \tag{18}$$

2.7. Numerical Method and Boundary Condition

Figure 3 shows the geometric model of the rotary kiln. To simplify the model and improve the grid quality, the elbows are simplified with cylinders as air inlets, which has no effect on the simulation results. Figure 4 presents the computational grid of the rotary kiln created using ICEM CFD, containing 4,118,224 hexahedral cells in all regions. The CFD software FLUENT was used for the simulation process. The segregated independent and 3D steady solver was adopted. The governing equations were discretized with second order upwind scheme using the control volume method and solved by SIMPLE arithmetic, and the simulation process was considered as convergence when the residuals are less than 1×10^{-5}. Furthermore, the mass transfer and heat exchange on the interface were defined as mass source and energy sink, and the UDF program in C language was developed and linked to FLUENT to calculate the energy absorbed by bed surface and gas escaping from the bed surface.

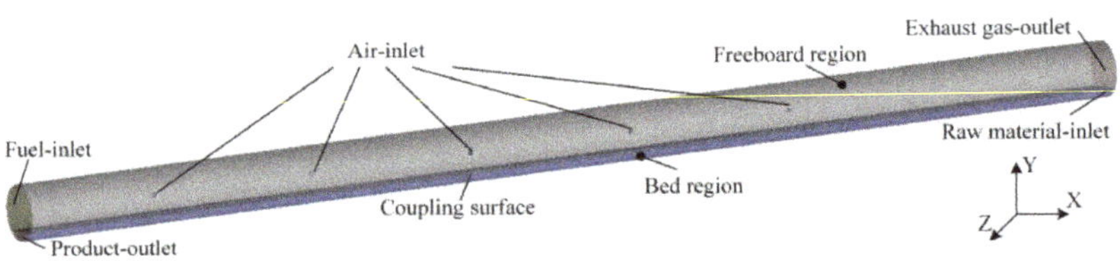

Figure 3. Simplified geometric model of the rotary kiln.

Figure 4. Computation mesh.

The boundary condition of the rotary kiln for numerical simulation is shown in Figure 3. The boundary conditions of fuel-inlet, air-inlet and raw material-inlet are "mass-flow-inlet". The boundary conditions of gas-outlet and product-outlet are "pressure-outlet". The mass flow rate and temperature of fuel gas are 0.468 kg/s and 1473 K, respectively. The mass flow rate and temperature of raw material are 2.833 kg/s and 300 K, respectively. The mass flow rate and temperature of combustion air are 0.443 kg/s and 300 K, respectively. The particle size was set to 13 mm. The chemical composition of fuel gas and raw material are listed in Tables 4 and 5. In addition, the shear condition at the kiln wall is the no-slip condition, and heat transfer coefficient for convection and radiation from kiln wall to environment was calculated by [34]:

$$h_{we}^{c} = \frac{0.11k_a}{D_{sh}}\left((0.5\mathrm{Re}_{\omega}^{2} + Gr)\mathrm{Pr}_a\right)^{0.35} \tag{19}$$

$$h_{we}^{r} = \sigma e \frac{T_{sh}^{4} - T_a^{4}}{T_{sh} - T_a} \tag{20}$$

Table 4. Chemical composition of fuel gas.

Species	CO	CO_2	H_2	H_2O	N_2
Mole fraction	56.7	20.0	14.3	4.8	4.2

Table 5. Chemical composition of raw material.

Species	Fe_2O_3	FeO	Al_2O_3	SiO_2	CaO	MgO	TiO_2	C	Vol	Ash
Mass fraction	38.6	24.6	2.6	3.5	1.1	2.7	5.0	15.3	4.7	1.9

3. Results and Discussion

3.1. Model Validation

The comparison between the computational and measured results is listed in Table 6, mainly composed of gas composition and temperature of exhaust gas at the kiln tail, metallization rate (MR) of the reduced product outlet and relative error. The metallization rate (MR) and relative error are given by Equations (19) and (20). As shown in Table 6, the simulation results are in good agreement with the measured value. The maximum relative error between measured and calculated results is 10.8%. Generally speaking, the model in this work is reasonably reliable for predicting the inner characteristics of the rotary kiln.

$$MR = \frac{w_{\mathrm{Fe}}}{w_{\mathrm{Fe}} + \frac{56}{72} \times w_{\mathrm{FeO}} + \frac{112}{160} \times w_{\mathrm{Fe_2O_3}} + \frac{168}{232} \times w_{\mathrm{Fe_3O_4}}} \tag{21}$$

$$\text{Relative Error} = \frac{|\text{Measured value} - \text{Simulated value}|}{\text{Measured value}} \tag{22}$$

Table 6. Comparison between measured and simulated results.

Parameters	Measured	Simulated	Relative Error
CO(%)	19.4	20.2	4.3%
CO_2(%)	24.2	22.9	5.1%
H_2(%)	8.1	9.0	10.8%
H_2O(%)	15.2	14.9	1.6%
T(K)	1023	1002	2.1%
Metal(%)	70.0	72.4	3.4%

3.2. *Basic Analysis*

3.2.1. Temperature Distribution

Figure 5 displays the temperature distribution of the gas phase along the axial section of the pre-reduction rotary kiln. The high temperature zones near the flames due to combustion reactions of H_2 and CO are visible, and the maximum temperature at the edge of flames is about 2200 K. Moreover, the combustion-supporting air enters the rotary kiln through five different nozzles to form five combustion regions, which is beneficial to reasonably controlling the axial temperature distribution and preventing local high temperature zones near the bed surface.

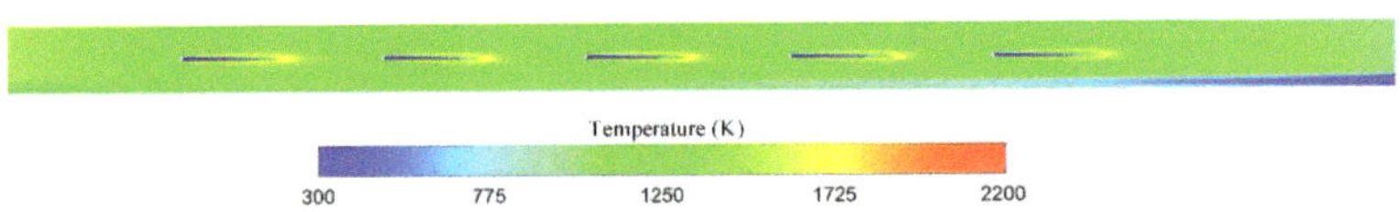

Figure 5. Temperature distribution of the gas phase along the axial section.

Figure 6 shows the temperature distribution of the solid phase within the bed cross-section at different positions. "L" represents the distance from the cross-section to the kiln tail. As expected, the temperature of the cross-section within the bed is not uniform, and there is a temperature gradient in the radial direction. The temperature of the bed surface and bottom is higher than that of the core. To better describe the evolution of temperature, the temperature of the cross-section along the axis of the rotary kiln is illustrated in Figure 7. T_{ave}, T_{max}, T_{min} and ΔT represent the average temperature, the maximum temperature, the minimum temperature, and the difference between maximum temperature and minimum temperature of the cross-section, respectively. After the raw materials are charged into the kiln at ambient temperature, the raw materials near the outer region are heated rapidly, while the heating rate of raw materials near the core region is relatively slow, and the temperature gradient at the kiln tail is larger than others. As the raw materials move toward the kiln head, the temperature gradient gradually decreases and changes gently after moving to the middle part of the kiln. This is mainly because the heating rate is restrained by carbon gasification reactions (by CO_2 and H_2O), which are strongly endothermic reactions.

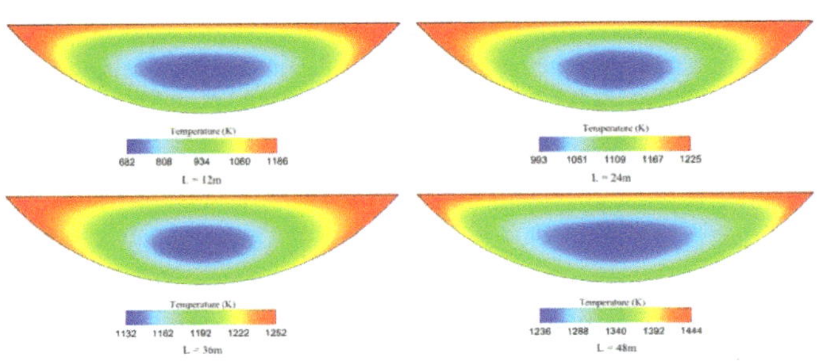

Figure 6. Temperature distribution of the solid phase within the bed cross-section.

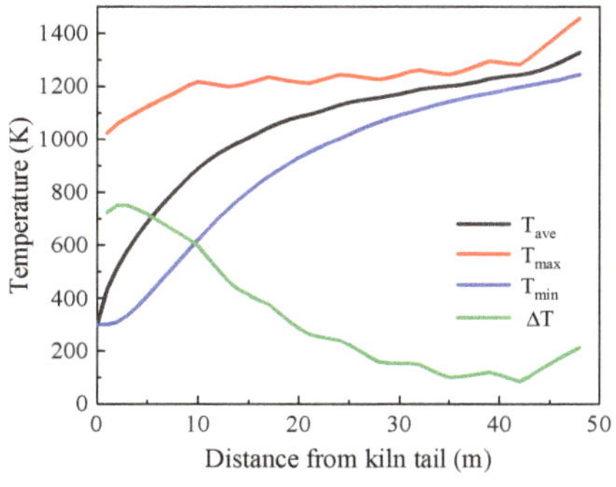

Figure 7. Temperature distribution of the cross-section along the axis of the rotary kiln.

3.2.2. Distribution of Gas Composition

The detailed information of the main gas species' distribution (O_2, CO, H_2, CO_2, H_2O) in the rotary kiln is obtained by calculation. Figure 8 shows the contour of the gas phase in mass fraction along the axial section. To maintain the reducing atmosphere in the kiln, the air supply is insufficient. It can be seen that there is almost no O_2 in the kiln except near the nozzles, which is beneficial to reducing the risk of re-oxidization of the metalized Fe. At the combustion reaction zones, a portion of CO and H_2 are consumed rapidly. The concentrations of CO and H_2 gradually decrease in the direction of gas flow; on the contrary, the concentrations of H_2O and CO_2 gradually increase.

Figure 9 shows the average mole fraction of gas species (CO, H_2, CO_2, H_2O) in the bed region as a function of distance from the kiln tail, which are the intermediate gases for the reduction of iron oxides by carbon. It is observed that the concentrations of gas species (CO, H_2, CO_2, H_2O) increase linearly near the inlet of raw materials. This is mainly attributed to the rapid release of volatile matter. After that, the volatile mater is gradually exhausted and the concentrations of H_2 and H_2O gradually decrease. When the materials move to the middle of the rotary kiln, there is almost no H_2 and H_2O in the system. On the contrary, with the increase of material temperature, the rate of carbon gasification is accelerated. When the generation rate of CO exceeds its consumption rate, the concentration of CO keeps increasing. Moreover, the variation in concentration of CO_2 is different from that of CO. The increase of CO_2 concentration in the early stage of reduction process may be attributed to the generation of volatile matter and reduction of iron oxide by CO, while the decrease of CO_2 concentration in the later stage of the reduction process is attributed to the rapid consumption of carbon gasification.

3.2.3. Variation of Iron Oxides

Figure 10 presents the average mass fraction of solid species (Fe, FeO, Fe_2O_3, Fe_3O_4) in the bed region as a function of distance from the rotary kiln tail. When the material temperature is higher than 570 °C, the reduction process of Fe_2O_3 to Fe includes three steps: $Fe_2O_3 \rightarrow Fe_3O_4 \rightarrow FeO \rightarrow Fe$. The reduction of Fe_2O_3 to Fe_3O_4 occurs rapidly, which is due to the low reduction potential required for this reaction. It can be seen that this reaction has been completed when the material reaches the middle of kiln. Since the reduction rate of Fe_2O_3 to Fe_3O_4 is faster than that of Fe_3O_4 to FeO, the concentration of Fe_3O_4 increases first and then decreases. Once Fe_3O_4 is completely reduced to FeO, only the reduction of FeO to Fe remains, and then the concentration of FeO decreases gradually. Obviously, the reduction of FeO to Fe is the rate limiting step of the reduction process because this reaction requires a higher reduction potential. In addition, there is almost no H2 in the system after the materials reach the middle of the rotary kiln (Figure 9), and the reduction reaction of iron mainly occurs in the second half of the rotary kiln. Therefore, it can be considered that H_2 makes little contribution to the reduction of FeO to Fe, and CO is the major reducing agent for this reduction reaction.

3.3. Influences of C/O Ratio

The C/O mole ratio is an important parameter in industrial production. To investigate the effect of C/O mole ratio on reduction process, the C/O mole ratio is set to 0.8, 1.0, 1.2 and 1.4, respectively, and the fill degree of the rotary kiln is kept at 11.6%. Figure 11 shows the variation of average bed temperature along the axial section with different C/O mole ratios. It is clear that the trends of average bed temperature for different C/O mole ratios are similar, and the heating rate is first fast and then slow. The fast heating rate in the early stage is mainly due to the large temperature gradient between the bed and surrounding environment. Moreover, the reductions of Fe_2O_3 and Fe_3O_4 are the main reactions in this stage, thus the amount of heat consumption by reactions is small. Besides, the average bed temperatures change very little between different C/O mole ratios, which indicates that the effect of the C/O mole ratio on the average bed temperature is negligible.

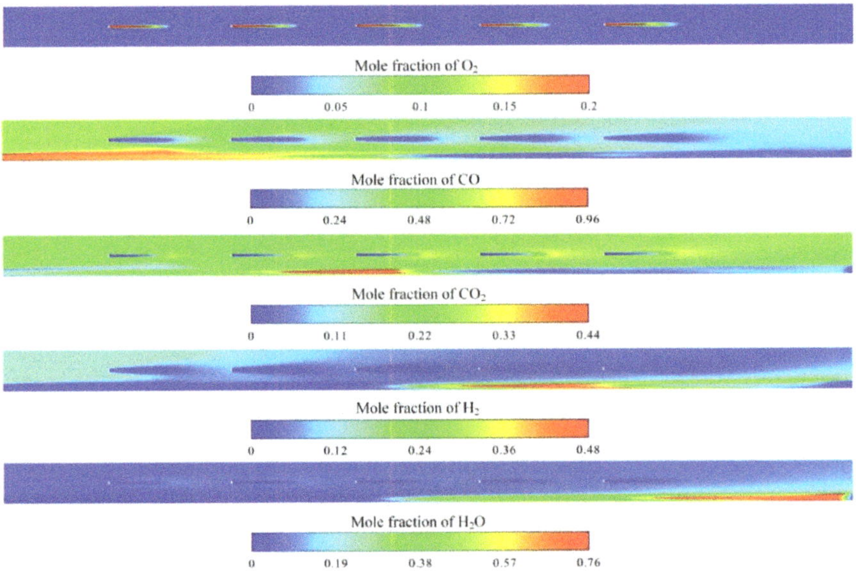

Figure 8. Composition distributions of the gas phase along the axial section.

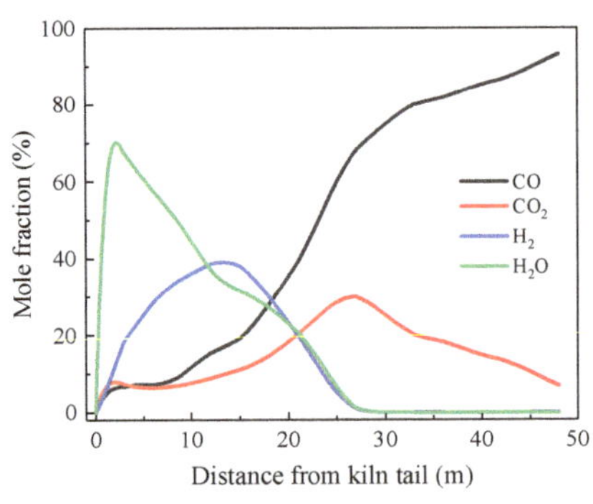

Figure 9. Average mole fraction of gas composition in the bed region along the axial section.

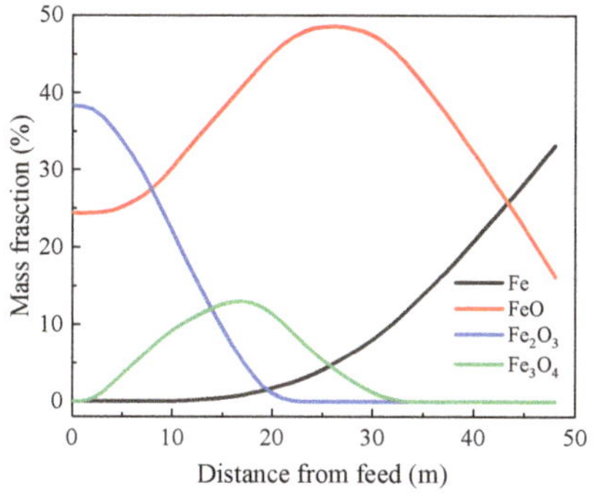

Figure 10. Average mass fraction of solid composition in the bed region along the axial section.

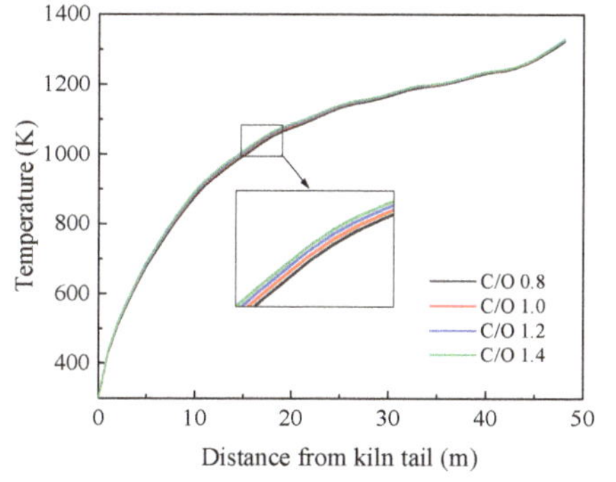

Figure 11. Average bed temperature along the axial section for different C/O mole ratios.

The reduction potential is defined by Equation (23), which is an important index with which to evaluate the reduction ability of intermediate reducing gas. Figure 12 shows the variation of reduction potential along the axial section with different C/O mole ratios. As shown in Figure 12, comparing the four simulation cases at the same position, the reduction potential increases significantly with the increase of C/O mole ratio. In general, the reduction potential is determined by the gas composition, and the higher reduction potential represents the higher concentration of reducing gas. Clearly, the carbon concentration in the bed increases with the increase of C/O mole ratio. This is helpful to accelerate the carbon gasification reactions, leading to an increase in the concentration of reducing gas.

$$\text{Reduction Potential} = \frac{\varphi C_O + \varphi_{H_2}}{\varphi_{CO} + \varphi_{H_2} + \varphi_{CO_2} + \varphi_{H_2O}} \tag{23}$$

Figure 13 shows the variation of metallization rate along the axial section with different C/O mole ratios. It can be observed that the decrease of C/O mole ratio is unfavorable to the production of metallic iron. This can be explained by the obtained results from Figures 11 and 12. Both the average bed temperature and reduction potential are very important for the reduction process. Although the C/O mole ratio has little effect on the average bed temperature, it has a strong effect on the reduction potential. At the same bed temperature, the higher reduction potential of reducing gas leads to a stronger reduction ability, which promotes the reduction of FeO to Fe.

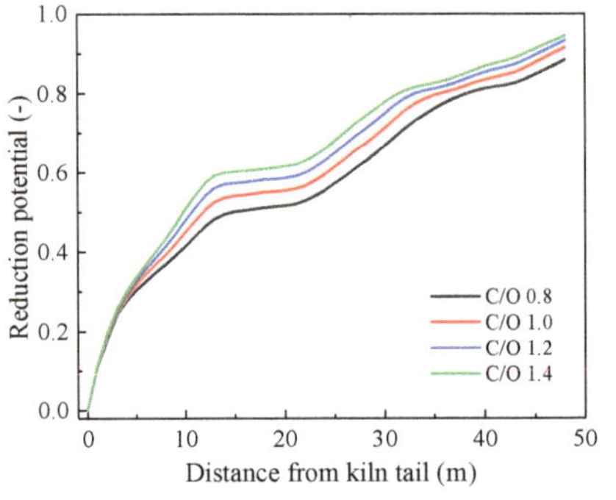

Figure 12. Reduction potential along the axial section for different C/O mole ratios.

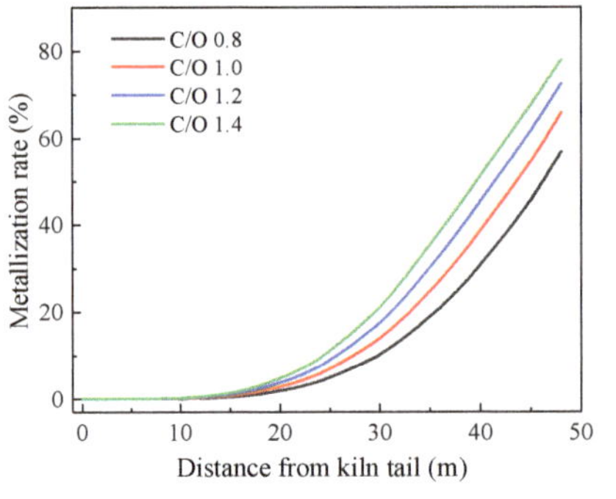

Figure 13. Metallization rate along the axial section for different C/O mole ratio.

3.4. Influences of Fill Degree

In the previous discussion, the fill degree of the rotary kiln is fixed. This section examines the influence of fill degree on the reduction process in the rotary kiln, and the fill degree is set to 11.6%, 16.6%, 21.6% and 26.6%, respectively. Figure 14 shows the variation of average bed temperature along the axial section for different fill degrees. It appears that the fill degree has a strong effect on the average bed temperature. As expected, the higher fill degree lengthens the path of heat transfer throughout the bed, which leads to lower bed temperature.

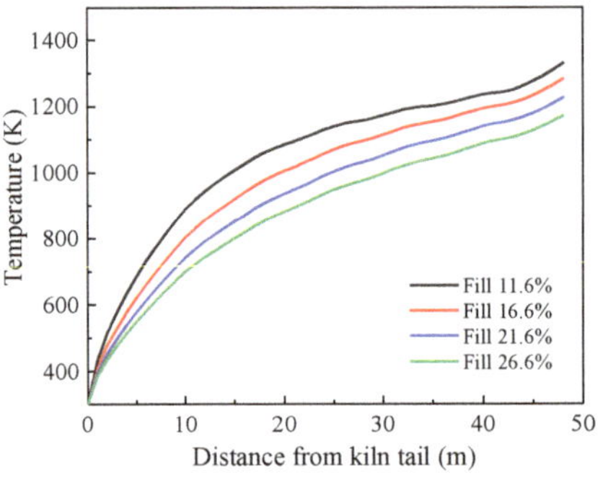

Figure 14. Average bed temperature along the axial section for different fill degrees.

Figure 15 shows the variation of reduction potential along the axial section for different fill degrees. It is found that the reduction potential decreases with the increase of fill degree. The curves are mainly composed of three parts. In the first part, the reduction potential rapidly increases due to the release of volatile matter. With the increase of fill degree, the bed temperature decreases correspondingly, which weakens the release rate of volatile matter and prolongs the residence time in the first part. The second part is a period with a very slow rising rate, and the reduction potential remains nearly constant in this part. This implies that the supply of reducing gas just meets the requirement of reduction reactions. Similarly, the increase of the fill degree also extends the residence time of the second part. With the increases of bed temperature, the carbon gasification reactions become significant, and the reduction potential rises again, which marks the beginning of the third part and lasts until the product is discharged from the kiln head.

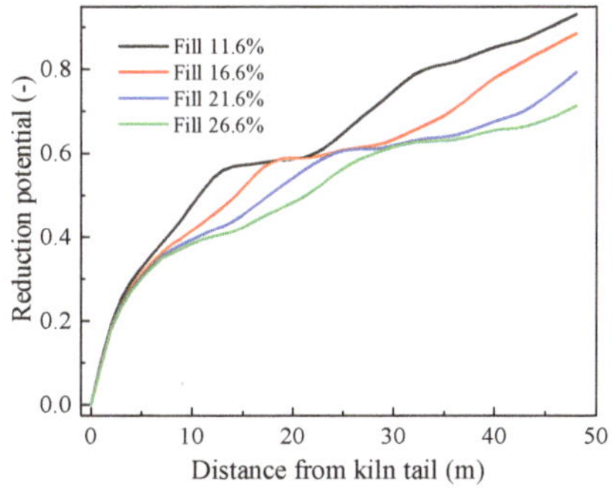

Figure 15. Reduction potential along the axial section for different fill degree.

Figure 16 shows the variation of metallization rate along the axial section for different fill degrees. It can be seen that there is almost no metallic iron in the bed at the early stage, and then the metallization rate gradually increases as the reduction reaction progresses. A lower metallization rate is found in the case of higher fill degree at the same position. When the fill degree increases to 26.6%, the final metallization rate is only 26.3%. This is because higher fill degree leads to lower bed temperature and lower reduction potential, which is unfavorable to the reduction reactions. Therefore, in order to ensure product quality, the fill degree should not be too high in practical production.

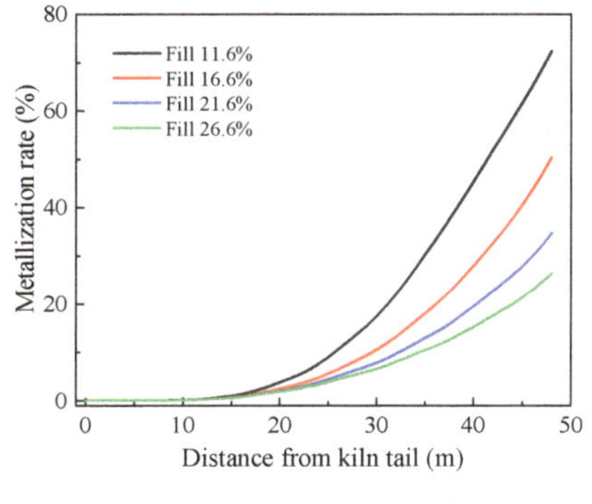

Figure 16. Metallization rate along the axial section for different fill degrees.

4. Conclusions

In the present work, the reduction process in a pre-reduction rotary kiln is studied by the computational fluid dynamic approach. Based on the simulation results, the following conclusions can be drawn:

(1) A temperature gradient (ΔT) was found in the bed, which is maximal near the kiln tail and decreases gradually as the reduction process progresses.

(2) The reductions of Fe_2O_3 to Fe_3O_4 and Fe_3O_4 to FeO are always fast, and the reduction of FeO to Fe is the limiting step of the whole reduction process because this reaction requires a higher reduction potential. H_2 makes little contribution to the reduction of FeO to Fe, and CO is the major reducing agent for this reduction reaction.

(3) The reduction potential curves can be divided into three parts. The curves increase rapidly in the first part and then remain nearly constant in the second part. In the third part, the curves increase again until the products are discharged from the kiln head.

(4) The C/O mole ratio has little effect on the average bed temperature but has a strong effect on the reduction potential. The higher reduction potential promoted the reduction process.

(5) With the increase of fill degree, both average bed temperature and reduction potential decrease. When the fill degree increases to 26.6%, the final metallization rate is only 26.3%.

Author Contributions: Conceptualization, C.L. and X.D.; methodology, X.Y.; software, C.L.; validation, H.L., C.D. and J.W.; formal analysis, C.L.; investigation, J.W.; resources, C.D.; data curation, H.L.; writing—original draft preparation, C.L.; writing—review and editing, X.D.; visualization, X.Y.; supervision, X.D.; project administration, X.Y.; funding acquisition, X.D. All authors have read and agreed to the published version of the manuscript.

Funding: This research was funded the National Key R&D Program of China (2017YFB0603801).

Institutional Review Board Statement: Not applicable.

Informed Consent Statement: Not applicable.

Data Availability Statement: Not applicable.

Conflicts of Interest: The authors declare no conflict of interest.

Nomenclature

A_s	Area (m^2)
C_p	Specific heat capacity (J/kg·K)
D_i	Mass diffusion coefficient (m^2/s)
D_{sh}	Diameter of kiln (m)
$E_{a,i}$	Activation energy of reaction i (J/kg·mol)
F_{gs}	Drag force (N/m^2)
Gr	Grashof number (-)
H_p	Enthalpy of phase p (J/mol)
ΔH_i	Enthalpy of reaction i (J/mol)
h	Heat transfer coefficient (W/m^2·K)
J	Radiation intensity (W/m^2)
k	Turbulence kinetic energy (J)
$K_{e,i}$	Equilibrium constant of reaction i (-)
M_i	Molecular weight of species i (kg/mol)
P	Pressure (Pa)
Pr	Prandtl number (-)
q	Heat flux (J/m^2·s)
R	Molar gas constant (J/mol·K)
R_i	Rate of reaction i (mol/m^3·s)
Re	Reynolds number (-)
$S_{\phi,p}$	Source term for phase p (kg/m^3·s)
T	Temperature (K)
v_p	Velocity of phase p (m/s)
Y_i	Mole fraction of species i (-)
Greek Symbols	
α_s	Thermal diffusivity (m^2/s)
ε_p	Volume fraction of phase p (-)
θ	Central angel (rad)
λ_p	Thermal conductivity (W/m·K)
ρ_p	Density of phase p (kg/m^3)
σ	Stefan-Boltzmann constant (W/m^2·K^4)
Subscripts	
g	Gas phase
s	Solid phase

References

1. Quader, M.A.; Ahmed, S.; Dawal, S.Z.; Nukman, Y. Present needs, recent progress and future trends of energy-efficient Ultra-Low Carbon Dioxide (CO_2) Steelmaking (ULCOS) program. *Renew. Sustain. Energy Rev.* **2016**, *55*, 537–549. [CrossRef]

2. Xu, W.; Cao, W.; Zhu, T.; Li, Y.; Wan, B. Material Flow Analysis of CO_2 Emissions from Blast Furnace and Basic Oxygen Furnace Steelmaking Systems in China. *Steel Res. Int.* **2015**, *86*, 1063. [CrossRef]
3. Qu, Y.; Yang, Y.; Zou, Y.; Zeilstra, C.; Meijer, K.; Boom, R. Thermal decomposition behavior of fine ore particle. *ISIJ Int.* **2014**, *53*, 2196–2205. [CrossRef]
4. Srivastava, B.; Roy, S.K.; Sen, P.K. An Approach for Simulation of COREX Process Smelter Gasifier for Prediction of Coal Rate and Silicon in Hot Metal. *Metall. Mater. Trans. B* **2010**, *41*, 935–939. [CrossRef]
5. Wu, S.; Xu, J.; Yang, S.; Zhou, Q.; Zhang, L. Basic Characteristics of the Shaft Furnace of COREX(R) Smelting Reduction Process Based on Iron Oxides Reduction Simulation. *ISIJ Int.* **2010**, *50*, 1032–1039. [CrossRef]
6. Liu, X.; Pan, G.; Wang, G.; Wen, Z. Mathematical model of lump coal falling in the freeboard zone of the COREX melter gasifier. *Energy Fuels* **2011**, *25*, 5729–5735. [CrossRef]
7. Zhang, X.H.; Zhao, K.; Bai, G.C.; Zhang, Y.Z.; Gao, J.J.; Zhang, Y.Y. Thermal analysis of rotary kiln oxygen combustion melting furnace under multi-constraint conditions. *Iron Steel* **2020**, *55*, 9–15.
8. Gao, J.J.; Wan, X.Y.; Qi, Y.H.; Wang, F. Technical analysis on ironmaking process of rotary kiln pre-reduction and smelting by coal and oxygen. *J. Iron Steel Res.* **2018**, *30*, 91–96.
9. Kingsley, T.C.; Kritzinger, H.P. Modelling and optimization of a rotary kiln direct reduction process. *J. South. Afr. Inst. Min. Metall.* **2015**, *115*, 51–60.
10. Mujumdar, K.S. Modeling of Rotary Cement Kilns: Applications to Reduction in Energy Consumption. *Ind. Eng. Chem. Res.* **2016**, *45*, 2315–2330. [CrossRef]
11. Mujumdar, K.S.; Ranade, V.V. CFD modeling of rotary cement kilns. *Asia-Pac. J. Chem. Eng.* **2008**, *3*, 106–118. [CrossRef]
12. Gaurav, G.K.; Khanam, S. Computational fluid dynamics analysis of sponge iron rotary kiln. *Case Stud. Therm. Eng.* **2016**, *9*, 14–27. [CrossRef]
13. Gaurav, G.K.; Khanam, S. Analysis of temperature profile and % metallization in rotary kiln of sponge iron process through CFD. *J. Taiwan Inst. Chem. Eng.* **2016**, *63*, 473–481. [CrossRef]
14. Manju, M.S.; Savithri, S. Three dimensional CFD simulation of pneumatic coal injection in a direct reduction rotary kiln. *Fuel* **2012**, *102*, 54–64. [CrossRef]
15. Vicente, W.; Ochoa, S.; Aguillon, J.; Barrios, E. An Eulerian model for the simulation of an entrained flow coal gasifier. *Appl. Therm. Eng.* **2003**, *23*, 1993–2008. [CrossRef]
16. Merrick, D. Mathematical models of the thermal decomposition of coal. *Fuel* **1983**, *62*, 540–546. [CrossRef]
17. Kuwauchi, Y.; Barati, M. A Mathematical Model for Carbothermic Reduction of Dust-Carbon Composite Agglomerates. *ISIJ Int.* **2013**, *53*, 1097–1105. [CrossRef]
18. Donskoi, E.; Mcelwain, D.L.S.; Wibberley, L.J. Estimation and modeling of parameters for direct reduction in iron ore/coal composites: Part II. Kinetic parameters. *Metall. Mater. Trans. B* **2003**, *34*, 255–266. [CrossRef]
19. Magnussen, B.F.; Hjertager, B.H. On mathematical modeling of turbulent combustion with special emphasis on soot formation and combustion. *Symp. Combust.* **1977**, *16*, 719–729. [CrossRef]
20. Silaen, A.; Wang, T. Effect of turbulence and devolatilization models on coal gasification simulation in an entrained-flow gasifier. *Int. J. Heat Mass Transf.* **2010**, *53*, 2074–2091. [CrossRef]
21. Mei, Z.; Li, P.; Wang, F.; Zhang, J.; Mi, J. Influences of Reactant Injection Velocities on Moderate or Intense Low-Oxygen Dilution Coal Combustion. *Energy Fuels* **2013**, *28*, 369–384. [CrossRef]
22. Sun, S.; Lu, W.K. Mathematical Modeling of Reactions in Iron Ore/Coal Composites. *ISIJ Int.* **2007**, *33*, 1062–1069. [CrossRef]
23. Sun, S.; Lu, W.K. Building of a Mathematical Model for the Reduction of Iron Ore in Ore/Coal Composites. *ISIJ Int.* **2007**, *39*, 130–138. [CrossRef]
24. Sun, K.; Lu, W.K. Mathematical Modeling of the Kinetics of Carbothermic Reduction of Iron Oxides in Ore-Coal Composite Pellets. *Metall. Mater. Trans. B* **2009**, *40*, 91–103. [CrossRef]
25. Jones, W.P.; Launder, B.E. The prediction of laminarization with a two-equation model of turbulence. *Int. J. Heat Mass Transf.* **1972**, *15*, 301–314. [CrossRef]
26. Jaw, Y.; Chen, C. Fundamentals of turbulence modeling. *J. Fluid Mech.* **1998**, *371*, 379–381.
27. Chui, E.H.; Raithby, G.D. Computation of radiant heat transfer on a nonorthogonal mesh using the finite-volume method. *Numer. Heat Transf. Part B Fundam.* **1993**, *23*, 269–288. [CrossRef]
28. Smith, T.F.; Shen, Z.F.; Friedman, J.N. Evaluation of coefficients for the weighted sum of gray gases model. *ASME J. Heat Transf.* **1982**, *104*, 602–608. [CrossRef]
29. Gorog, J.P.; Adams, T.N.; Brimacombe, J.K. Regenerative heat transfer in rotary kilns. *Metall. Trans. B* **1982**, *13*, 153–163. [CrossRef]
30. Tscheng, S.H.; Watkinson, A.P. Convective heat transfer in a rotary kiln. *Can. J. Chem. Eng.* **1979**, *57*, 433–443. [CrossRef]
31. Marias, F.; Roustan, H.; Pichat, A. Modelling of a rotary kiln for the pyrolysis of aluminium waste. *Chem. Eng. Sci.* **2005**, *60*, 4609–4622. [CrossRef]
32. Boateng, A.A. On flow-induced kinetic diffusion and rotary kiln bed burden heat transport. *Chem. Eng. Commun.* **1998**, *170*, 51–66. [CrossRef]
33. Boateng, A.A.; Barr, P.V. A thermal model for the rotary kiln including heat transfer within the bed. *Int. J. Heat Mass Transf.* **1996**, *39*, 2131–2143. [CrossRef]

34. Agrawal, A.; Ghoshdastidar, P.S. Numerical simulation of heat transfer during production of rutile titanium dioxide in a rotary kiln. *Int. J. Heat Mass Transf.* **2017**, *106*, 263–279. [CrossRef]

metals

Article

Adaptive First Principles Model for the CAS-OB Process for Real-Time Applications

Kasper Linnestad [1,*], Seppo Ollila [2], Stein O. Wasbø [1], Agne Bogdanoff [2] and Torstein Rotevatn [3]

1 Cybernetica AS, Leirfossvegen 27, 7038 Trondheim, Norway; stein.wasbo@cybernetica.no
2 SSAB Europe OY, Rautaruukintie 155, 92100 Raahe, Finland; seppo.ollila@ssab.com (S.O.); agne.bogdanoff@ssab.com (A.B.)
3 Adigo AS, Berghagan 3, 1405 Langhus, Norway; torstein.rotevatn@adigo.no
* Correspondence: kasper.linnestad@cybernetica.no

Abstract: A model-based system for real-time monitoring and operational support has been developed for the Composition Adjustment by Sealed argon-bubbling with Oxygen Blowing (CAS-OB) process. The model of the system is based on a previously developed dynamic model using first principles, i.e., mass and energy balances, reaction kinetics, and thermodynamics. Adaptive estimation of state variables has been implemented using a Kalman filter to ensure that the model is able to correct for deviations between measured and calculated temperatures in real-time operation. The estimation technique reduces the standard deviation of the predicted end temperature from 19.5 °C to 5.5 °C in a data series with more than 1000 heats. The system also includes an endpoint optimisation, which calculates the amount of scrap or oxygen to be added to achieve the target temperature at the end of the heat. The model has been implemented in the Cybernetica® CENIT™ framework. The overall model can be regarded as a hybrid digital twin, where a first principles model is adapted in real time using process measurements. The system also includes user interfaces for operators where process predictions can be followed, and suggested optimised inputs are presented. The system has been deployed at two refining stations at SSAB Europe OY in Raahe, Finland. The optimized suggestions for oxygen and scrap are presented to the operators in the graphical user interface. A projected temperature profile is calculated into the near future using the planned operational procedure as well as the projected temperature profile using optimised inputs. Both profiles are displayed in the user interface. Based on these trajectories, the operator can decide on whether to follow the nominal trajectory, or the recommendation from the optimisation This will help the operators make better decisions, which in turn reduces the number of rejected heats in the CAS-OB process.

Keywords: real-time model; estimation; model predictive control; steel refining

Citation: Linnestad, K.; Ollila, S.; Wasbø, S.O.; Bogdanoff, A.; Rotevatn, T. Adaptive First Principles Model for the CAS-OB Process-for Real-Time Applications. *Metals* **2021**, *11*, 1554. https://doi.org/10.3390/met11101554

Academic Editors: Thomas Echterhof, Ko-Ichiro Ohno and Ville-Valtteri Visuri

Received: 24 August 2021
Accepted: 24 September 2021
Published: 29 September 2021

Publisher's Note: MDPI stays neutral with regard to jurisdictional claims in published maps and institutional affiliations.

1. Introduction

Composition Adjustment by Sealed argon-bubbling with Oxygen Blowing (CAS-OB) is a secondary steelmaking process step used for alloying and temperature control. The CAS-OB-process was developed by Nippon Steel Corporation to improve on existing refining and alloying stations. The CAS-OB process provides a buffer station in the secondary steelmaking area where chemical composition and steel temperature can be adjusted as well as homogenized in a controlled manner.

The main component of the CAS-OB design is a refractory lined ladle with a porous plug for gas bubbling in the bottom. The operation of the process is accomplished through the use of three items: (1) a refractory bell on top of the ladle, which can be used to contain an inert atmosphere of argon or the chemical reaction between oxygen and aluminium, (2) supply of argon gas, and (3) a water-cooled oxygen lance. These functional items form the heart of the process (Figure 1).

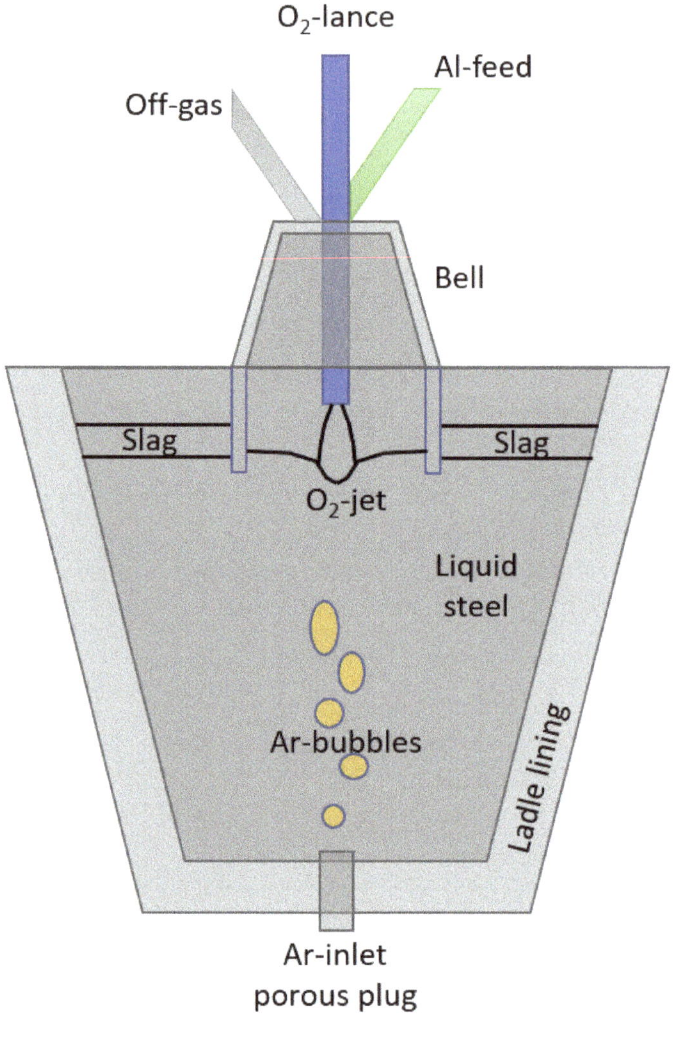

Figure 1. Illustration of the CAS-OB process.

Molten steel from a primary steelmaking process (Basic Oxygen Furnace, BOF) is initially poured into a ladle. The ladle consists of a steel casing with a refractory brick layer on the inside. The refractory layer provides insulation and is resistant to the corrosive environment of the steel bath, having a high temperature of around 1600 °C. During treatment, steel components will react with added oxygen to form a slag phase. Additions in form of oxides will also dissolve into the slag phase. Typical slag components are FeO, SiO_2, MgO, Al_2O_3, and CaO. In the CAS-OB-process, the slag phase takes part in the reactions and protects the molten steel from the atmosphere and works as an insulating layer.

The CAS-OB process is designed to create an inert atmosphere above the steel to allow the addition of alloys without contact with atmospheric oxygen or an oxide slag. This is accomplished by first creating a slag free area (known as an eye) at the surface of the molten steel by the introduction of argon into the steel through a porous plug at the bottom of the ladle. Argon bubbles reaching the surface of the steel push aside the slag layer on top of the ladle, creating a slag free area. The amount of argon flow required to produce the required size of eye on top of the ladle varies with the condition of the porous plug, the depth of slag on top of the ladle, and the fluidity of the slag. Once the eye is created, the refractory bell can be lowered into the slag free area.

The bell provides a protected environment for adding alloying materials, where the steel surface is open, but still protected from contact with the surrounding atmosphere. This also ensures that the amount of absorbed nitrogen can be kept at a low level. The argon bubbling also provides stirring for homogenizing temperature and composition of the steel [1–4].

2. Materials and Methods

The numerical model used for describing the dynamic behaviour of the CAS-OB process is based on previous work by Rotevatn et al. [5] and Järvinen [2]. In recent years, CAS-OB model updates have have been reported by Sulasalmi et al. [6], Visuri et al. [7], and Kärnä et al. [8]. The latter have done a comprehensive work on utilizing computational fluid dynamics (CFD) calculations to develop an improved model of the CAS-OB. Industrial trials using a pyrometer were also made and measurements applied for validation of temperature development in the model. The model developed in this work consists of a set of ordinary differential equations describing the mass and energy balance of the process, including component balances, main reactions, and thermodynamics. The approach in this work is to provide a model that is as simple as possible to describe the main process dynamics, focusing primarily on temperature development, in order to obtain fast enough real-time calculation for optimization of the industrial process. The model used in this work is the same as the one presented in Rotevatn et al. [5], but with the addition of a stirring-dependent heat transfer coefficient for the convective heat transfer between the metal and the ladle wall. In addition, considerable work has been done in preparing the model for online signals, process measurements, and material and product analyses.

2.1. Control Volumes and State-Space Variables

The model is formulated as a continuous state space model, which also contains an integrator providing a time discrete state vector. The model has control volumes where the masses of each species and the temperature are calculated for each time step. These control volumes include liquid steel in the bulk of the vessel, free surface on the top of the metal inside the bell, and the slag phase at the top of the metal. The model also includes states describing the temperature distribution through the refractories in the ladle and the bell.

2.1.1. Liquid Steel Phases

The liquid steel is divided into two control volumes, (1) the main bulk volume of the treatment ladle and (2) the free surface. The modeled metal components are the same as for the two control volumes: Al, Si, Mn, Fe, C, P, S, Nb, V, Ni, Ti, B, Ca, N, Cr, Cu, and Mo. For each of the metal components, a time dynamic component balance is calculated, based on material flows and reaction rates. The main control volume in the model comprises the main bulk of the metal in the ladle, typically about 90% of the total mass, whereas the rest, typically 10%, is assumed to be part of the free surface control volume.

There is an exchange of mass between the main bulk control volume and the free surface. The amount of exchange is varying with degree of bubbling in the process. Higher volumes of stirring gas will increase the flow of material exchange between the two control volumes.

Material additions as well as the oxygen blowing are assumed to enter the free surface control volume. Added materials are heated up from ambient temperature to the free surface temperature and are assumed to melt instantly into the free surface. The energy required is taken from the free surface control volume. For numerical stability and accuracy, this assumption means that the additions should not be too large compared to the mass and available energy in the free surface control volume. If larger and energy demanding material additions are applied, the model dimensions, calculation intervals, or equation solver should be adjusted accordingly. Both the free surface and the main bulk are assumed to be liquid at all times during a treatment batch.

2.1.2. Slag Phase

The slag phase is floating on top of the hot metal. The slag is assumed to be in contact with the hot metal and chemical components in the slag will interact with the metal through interface reactions between hot metal and slag. In the model, it is assumed here that there is no contact between the free surface control volume and slag. The following chemical components are assumed present in the slag phase: Al_2O_3, SiO_2, MnO, FeO, MgO, TiO_2, P_2O_5, CaO. The slag is assumed to be liquid at all times during a batch treatment.

2.1.3. Gas Phase

In the model, it is assumed that there is no gas phase inside the liquid hot metal. This is a simplification that has been done to save computation time. The aim of this model is to be able to track the main dynamics of the process during a treatment. The gas dynamics of bubbles inside the hot metal is in this respect considered to be much faster than the component balances and the temperature. The bubbling gas is argon which is inert to reactions. In the cases where oxygen heating is applied, the oxygen is assumed to react instantly with metal components forming oxides that are subsequently reacting further, where the slag–metal equilibrium will decide the reaction potentials.

2.1.4. Ladle

The ladle refractory is modeled as a number of equally thick shells from the inside in contact with the hot metal to the outside in contact with the steel casing, which in turn is in contact with the surrounding air. The number of refractory shell is configurable but typically a value of five shells is applied. This means that there are in total six control volumes describing the temperature profile of the ladle refractory and steel shell. The ladle only has equations describing the heat evolution. No chemical reactions or mass balance equations are included. Refractory wear can be modeled by adjusting refractory thickness between heats. Consumed refractory is not assumed to influence slag composition in single heats.

2.1.5. Bell

Similarly, as for the ladle, the bell is divided into refractory shells and steel casing. Typically, six control volumes are applied. On the inside of the bell, the refractory is in contact with the gas phase and also exposed to radiation from the free surface. Through the shells and to the steel casing, heat is transferred by conduction. From the steel casing, heat is lost to the surrounding air by convection, as well as radiation.

2.2. Chemical Reactions

2.2.1. Dissolution of Added Material

Added metallic material, such as aluminium, is assumed to melt instantaneously into the free surface control volume:

$$Al(s) \longrightarrow Al(f) \tag{R1a}$$

Dissolution of other metallic additions is modeled similarly.

Added oxides are assumed to melt instantaneously into the slag phase:

$$CaO(s) \longrightarrow (CaO) \tag{R1b}$$

Dissolution of other non-metallic additions are modeled similarly.

Instantaneous dissolution of additions is a simplification and approximation. The additions are assumed to be small in sizing and amounts for the dissolution to be fast, such that the introduced temporary error in mass and energy balance will be negligible with respect to the required accuracy for the industrial application of the model.

Solid phase from added materials is denoted (*s*), chemical components in the free surface volume, a liquid phase, is denoted (*f*), gas phase is denoted (*g*), while liquid hot metal is denoted by underline, and liquid slag phase by parentheses.

2.2.2. Free Surface Reactions

Furthermore, the metallic components including dissolved carbon are assumed to react with added oxygen. All free surface reactions involving oxygen are assumed to be irreversible:

$$2Al(f) + 1.5O_2(g) \longrightarrow (Al_2O_3) \tag{R2}$$

$$Si(f) + O_2(g) \longrightarrow (SiO_2) \tag{R3}$$

$$2Mn(f) + O_2(g) \longrightarrow 2(MnO) \tag{R4}$$

$$2Fe(f) + O_2(g) \longrightarrow 2(FeO) \tag{R5}$$

$$2C(f) + O_2(g) \longrightarrow 2CO(g) \tag{R6}$$

All of these reactions are assumed to be irreversible reactions, driven far to the right. The reaction rates are calculated based on the activities of the reacting species and calculated thermochemical equilibrium. The available oxygen is distributed according to an oxygen yield factor and individual reaction rates. The oxygen available to take part in reactions are given by $\theta_{O_2,yield}F_{O_2,in}$, where $F_{O_2,in}$ is the molar flow of oxygen from the lance. The oxygen yield factor, $\theta_{O_2,yield}$, should be a number between 0 and 1.

As an example, the nominal reaction rate for (R2) is given by:

$$R_2^{'} = k_r(a_{Al}^2 a_{O_2}^{1.5} - \frac{a_{Al_2O_3}}{K_2}) \tag{1}$$

where K_2 is the thermochemical equilibrium constant.

The total reaction rate of (R2)–(R6) is limited by the available oxygen:

$$\theta_{O_2,yield}F_{O_2,in} = \sum_{j=2}^{6} R_j \tag{2}$$

which leads to the following actual reaction rates:

$$R_i = \theta_{O_2,yield}F_{O_2,in}\frac{R_i^{'}}{\sum_{j=2}^{6} R_j^{'}}, i = 2 - 6 \tag{3}$$

With these reaction rates, the reactions will compete for the oxygen based on thermochemical terms, yet only consuming the available oxygen.

2.2.3. Slag-Metal Interface Reactions

On the interface between slag and free surface equilibrium reactions are assumed to take place:

$$1.5(SiO_2) + 2Al(f) \longrightarrow Al_2O_3 + 1.5Si(f) \tag{R7}$$

$$3(MnO) + 2Al(f) \longrightarrow (Al_2O_3) + 3Mn(f) \tag{R8}$$

$$3(FeO) + 2Al(f) \longrightarrow (Al_2O_3) + 3Fe(f) \tag{R9}$$

$$2(FeO) + Si(f) \longrightarrow (SiO_2) + 2Fe(f) \tag{R10}$$

$$(FeO) + Mn(f) \longrightarrow (MnO) + Fe(f) \tag{R11}$$

$$2(MnO) + Si(f) \longrightarrow (SiO_2) + 2Mn(f) \tag{R12}$$

2.3. Heat Transfer

The refractory layer of the bell typically has a radial temperature gradient, where heat conduction transports heat from the hot inside to the cooler outside of the bell. Heat is transferred by heat convection from the gas phase inside the bell to the inside of the bell. Radiation takes place between the hot free surface area and the inside of the bell. From the steel cap to the surrounding air, heat is transferred by convection, as well as a term for radiation losses to the surroundings.

Similarly, in the ladle refractory, heat transfer is assumed to take place by convection from the hot metal to the inner refractory layer. Heat is further transferred by conduction through the refractories to the steel cap. From the steel cap, heat is lost by radiation to the surroundings, as well as through heat convection to the air.

The varying heat convection from the hot metal to the inner refractory layer due to flow conditions in the hot metal and free surface control volumes inside the ladle has a substantial impact on the energy balance. The internal exchange of mass between the free surface and the bulk hot metal is influenced strongly by stirring gas and oxygen blowing.

Heat is also transferred by convection between slag and hot metal, and between free surface and hot metal. The free surface and hot metal are also mixed by the stirring, which provides a substantial exchange of heat and mass between the two control volumes.

2.4. Mass and Energy Balances

A list of symbols used in the mathematical description and equations is provided in Abbreviation.

2.4.1. Free Surface

The added aluminium is assumed to melt as soon as it enters the free surface control volume. Reactions are consuming aluminium in this control volume, and stirring will mix the materials between hot metal and the free surface:

$$\frac{dm_{Al_f}}{dt} = -M_{Al}(R_2 + R_7 + R_8 + R_9) + w_{Al,L-f} + w_{Al,in} \tag{4}$$

where $w_{Al,in}$ is the sum of the representative aluminium part of metallic additions. $w_{Al,L-f}$ is the net mass of aluminium being exchanged between the liquid hot metal and the free surface control volumes due to mixing.

$$\frac{dm_{Si_f}}{dt} = M_{Si}(-R_3 + R_7 - R_{10} - R_{12}) + w_{Si,L-f} + w_{Si,in} \tag{5}$$

where $w_{Si,in}$ is the sum of the representative silicon part of metallic additions. $w_{Si,L-f}$ is the net mass of aluminium being exchanged between the liquid hot metal and the free surface control volumes due to mixing.

In general, for all components in the free surface, we can write:

$$\frac{dm_{i_f}}{dt} = M_i(\sum_j s_i R_j) + w_{i,L-f} + w_{i,in} \tag{6}$$

where $\sum_j s_i R_j$ is the net sum of reactions where component i takes part, weighted by the stoichiometric factor for the component.

The mixing is assumed to preserve the total mass, so the amount of liquid going into the free surface from the hot metal is the same as the amount going from the free surface into the hot metal. Therefore, the net flow for each component will be given by the chemical composition difference between the two control volumes and the composition in each control volume:

$$w_{i,L-f} = (c_{i,L} - c_{i,f})w_{L-f} \tag{7}$$

where $i = Al, Si, Mn, Fe, C, P, S, Nb, V, Ni, Ti, B, Ca, N, Cr, Cu, Mo$ are the components in the free surface and hot metal.

2.4.2. Liquid Steel

The energy balance for the liquid main bulk results in the following differential equation describing the temperature:

$$\begin{aligned} m_L Cp_L \frac{dT_L}{dt} =& w_{Ar,in} Cp_{Ar}(T_{in,Ar} - T_L) + Q_{L-f} - Q_{f-L} \\ & - \alpha_L S_f (T_L - T_f) - \alpha_{Li} S_{Li}(T_L - T_{Li}) - \alpha_L S_{Ls}(T_L - T_s) \end{aligned} \tag{8}$$

where

$$Q_{f-L} = Cp_f w_{f-L} \tag{9}$$

$$Q_{L-f} = Cp_L w_{L-f} \tag{10}$$

The terms $\alpha_L S_f(T_L - T_f)$, $\alpha_{Li} S_{Li}(T_L - T_{Li})$, and $\alpha_L S_{Ls}(T_L - T_s)$ are describing heat convection between the phases main bulk, free surface, inner ladle and slag, determined by the temperature difference between the neighbouring phases, the interface areas S_j and the heat transfer coefficients α_j. The flows $w_{f-L} = w_{L-f}$ are assumed equal.

2.4.3. Gas Phase

The gas phase is defined primarily to account for CO-gas produced in the process. The component balance is given as:

$$\frac{dm_{CO}}{dt} = m_{CO,produced} - m_{CO,exiting} \tag{11}$$

$$m_{CO,produced} = M_{CO} R_6 \tag{12}$$

where M_{CO} is the molar mass of CO(g) and R_6 is the molar rate of reaction.

$$m_{CO,exiting} = \frac{m_{CO}}{\tau_G} \tag{13}$$

where τ_G is a time constant for the gas dynamics.

The gas temperature is calculated as an algebraic relation:

$$T_G = \frac{T_f(w_{Ar} Cp_{Ar} + R_6 M_{CO} Cp_{CO}) + w_{O2,out} Cp_{O_2} T_{env} + \alpha_{Bi} S_{Bi} T_{Bi}}{w_{Ar,in} Cp_{Ar} + w_{O_2,out} Cp_{O_2} + R_6 M_{CO} Cp_{CO} + \alpha_{Bi} S_{Bi} + \alpha_G S} \tag{14}$$

where $w_{Ar,in}$ is the mass flow of argon being purged from the bottom of the ladle.

The reason for simplifying the gas temperature calculation is to avoid unnecessary complex equations that would cause stiffness to system and slow down the calculation speed.

2.4.4. Model Summary

The model consists of 61 ordinary differential equations describing the chemistry and temperature of the CAS-OB process. A total of 222 parameters and constants describe physical properties of materials, geometrical data of the system, etc. In the industrial implementation, there are 50 input variables, providing information on added materials, gas bubbling and oxygen blowing. Forty-three measurements are implemented, including slag and metal analysis and temperatures. Many of the measurements are not available during online operation, but data are often available after end of treatment. The logged data can be used for offline validation and improvements of the process model.

2.5. Recursive Parameter and State Estimation

In the CAS-OB process, there are considerable uncertainties related to the initial conditions before heat. The hot metal arriving from the BOF process has an end of BOF treatment temperature measurement. Occasionally, additional materials are added to adjust composition after BOF treatment and before CAS-OB treatment. Waiting time between the processes may vary. Refractory wear and net heat loss may vary between vessels. The initial ladle temperature is also unknown. These factors introduce uncertainties to the initial temperature of the CAS-OB process. To mitigate this issue, recursive estimation of ladle state in general and specifically ladle temperature has been introduced.

Estimation is implemented using an augmented and extended Kalman filter [9,10]. The metal temperature is typically measured three times during the treatment: once in the beginning of the heat, once midway through the heat, and once towards the end of the heat.

2.6. Real-Time Optimization

Real-time optimization is implemented using the nonlinear model predictive control schema similar to the method described by Maciejowski [11]. Two applications are implemented: (1) a monitoring application calculating the predicted result of the heat based on current and planned inputs. (2) an optimizing application where the desired end-point temperature and the delivery time to the subsequent continuous caster is specified. Based on this specification, the application optimizes the inputs of scrap metal addition and oxygen blowing such that the targets are met. The suggested optimized inputs are presented to the operator through the user interface.

The real-time optimization is repeated every five seconds, and new predictions and optimized inputs are calculated.

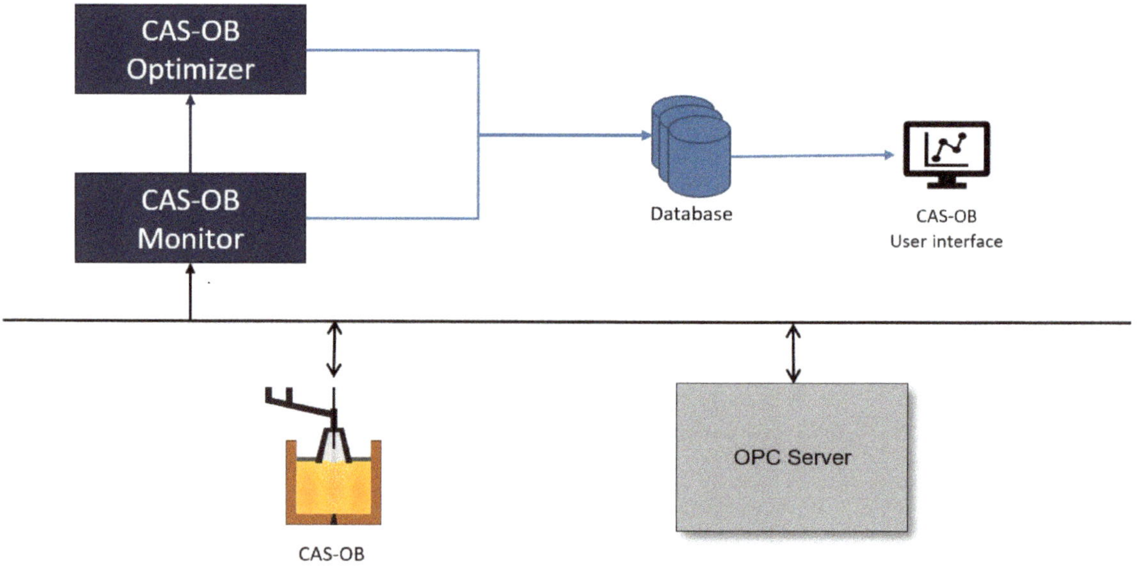

Figure 2. Real-time application configuration at SSAB.

3. Results

The monitoring and control application has been installed on two CAS-OB refining stations at SSAB Europe OY's Raahe plant with a configuration as illustrated in Figure 2. The application has been running in real-time operation since 2019 and validation of the results taken place over time by process engineers and recently also by process operators. Process data have been logged over the last two years, and the model agreement with process data studied based on a large set of data from this period.

3.1. Model Agreement with Process Data

An early version of the model was presented by Rotevatn [5] where model validation was also a part of the work. The model was in reasonable agreement with most of the chemical elements in the metal and slag. Not surprisingly, when studying the validation results, the slag components in general had the largest deviation. This is natural as the mass of slag is smaller and even moderate variations in metal components to be refined may have a larger impact on the resulting slag composition. Industrially, getting a representative sample of slag for analysis can also be more challenging.

The main use of this application has been in temperature control. The application optimizes the use scrap metal additions and the use of additional oxygen for reheating. Therefore, the main focus in this work has been on validation of the metal temperature. In Figure 3, a histogram of the prediction error of the end temperature is shown for a total of 1031 heats collected over a period of one month. The wider distribution curve in dark blue color shows the histogram for the prediction error for the end temperature using ballistic simulation. This corresponds to the nominal model accuracy. The standard deviation is 19.5 °C and a bias of 3.1 °C. The required specification for the application to give added value to the operation is approximately 10 °C.

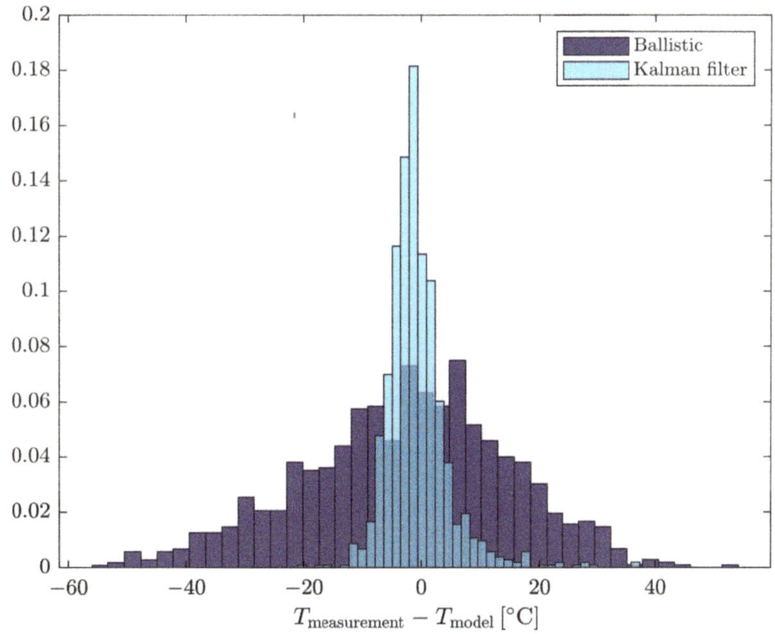

Figure 3. Histogram showing improvement of prediction errors by estimation on the final temperature estimates in the heat.

3.2. Recursive Estimation

In Figure 3, the effect of estimation on the final temperature is also illustrated. When applying the implemented Kalman-filtering, the standard deviation of prediction errors is reduced to 5.5 °C with a bias of 2.2 °C. The result is now well within the required specifications for the process improvements.

Figure 4 shows how the Kalman-filtering is adjusting the temperature estimates. In particular, the first measurement is very important since the initial temperature of the CAS-OB ladle is uncertain. After being adjusted by the first temperature measurement, the model is able to follow the remaining course of the process with high accuracy. This includes the following process events such as material additions and reheating by oxygen.

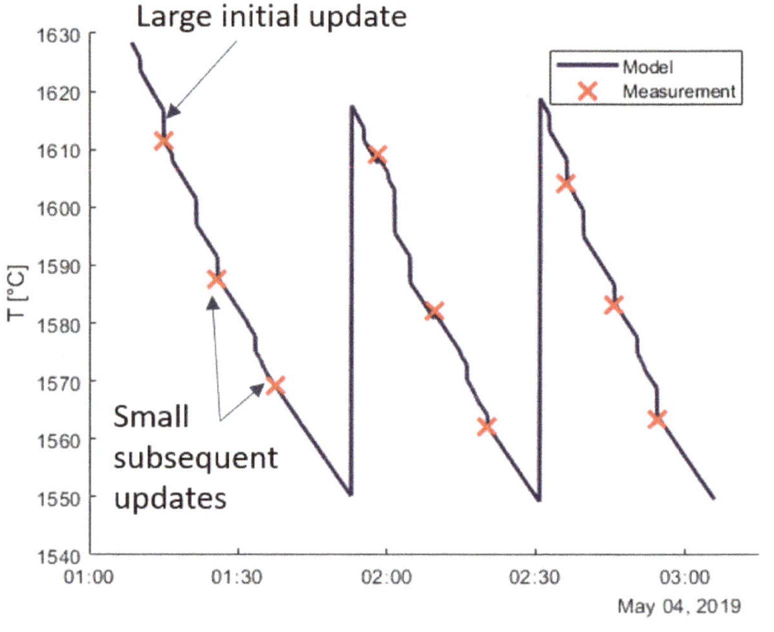

Figure 4. Temperature evolution and estimation improvement during the course of the heat.

3.3. Industrial Use and Application

The user interface has been developed in cooperation between SSAB Europe OY's engineers and operators, providing a suggestion for operators to reheat or add cooling scrap to meet the accurate target window of time and temperature for the so-called hook time. The hook time is the time when the refining ladle should be "hooked up" for transport to the casting station. Typically, the end of heat temperature from the CAS-OB station should be a few degrees higher than the specified temperature for casting to obtain optimal steel quality.

In the user interface (Figure 5), the vertical line represents the present time, and data to the left are historical values, while future model predictions are shown to the right. There are two predicted temperature curves; the dark line represents the prediction if the heat continues with the current plan for the heat. The light green curve shows the predicted temperature if the recommended optimized input is applied. The orange line shows that the target temperature has been specified for the hook time. In the historical data to the left, temperature measurements can also be seen as light blue crosses, while the target point for the previous heat is shown as an orange cross. The recommended amount and timing of oxygen reheating, aluminium and scrap metal additions are presented for the operators.

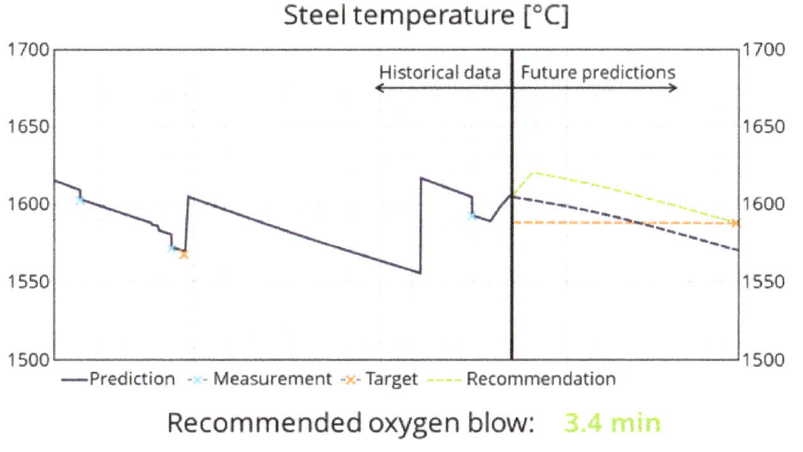

Figure 5. User interface for temperature control.

4. Conclusions

A monitoring and control application for temperature control of the CAS-OB secondary steelmaking process has been developed and tested industrially during the Morse project. The application is based on a first-principles model of the process, which provides a good problem structure. Since the model, provided with available input data, is not accurate enough for online use, an estimator is provided in the form of a Kalman filter. The estimator utilizes the available measurements to adjust the model in real time. The overall model may be regarded as a hybrid digital twin, utilizing the best of two worlds; the structure of first principles and the flexibility of a data driven approach. This hybrid twin is combined with a nonlinear model predictive control algorithm which is applied to optimize amount and timing of process inputs. The optimized inputs are presented to process operators along with the currently used heating profile, thus providing an operator support system for process control.

Author Contributions: Conceptualization, S.W., T.R. and K.L.; methodology, S.W., K.L. and T.R. ; software, K.L. and T.R.; validation, S.O., A.B. and K.L.; formal analysis, K.L., T.R. and S.O.; investigation, S.O. and K.L.; resources, A.B. and S.W.; data curation, S.O.; writing—original draft preparation, S.W.; writing—review and editing, S.W.; visualization, K.L.; supervision, A.B. and S.W.; project administration, A.B. and S.W.; funding acquisition, A.B. and S.W. All authors have read and agreed to the published version of the manuscript.

Funding: This research was funded by the European Commission, Directorate-General for Research and Innovation, through the project "Model-based optimisation for efficient use of resources and energy" (MORSE), Grant No. 768652. https://www.spire2030.eu/morse, accessed on 27 September 2021.

Institutional Review Board Statement: Not applicable.

Informed Consent Statement: Not applicable.

Acknowledgments: The authors acknowledge the assistance by Timo Veijola during the early validation phases of the online system.

Conflicts of Interest: The authors declare no conflict of interest.

Abbreviations

The following abbreviations are used in this manuscript:

BOF	Basic Oxygen Furnace
CAS-OB	Composition Adjustment by Sealed argon-bubbling with Oxygen Blowing
CFD	Computational Fluid Dynamics
MPC	Model Predictive Control
NMPC	Nonlinear Model Predictive Control

Symbol	Description	Unit
m_i	Mass of component i	kg
c_i	Concentration of component i	mass fraction
M_i	Specific molar mass of component i	kg/kmol
w_i	Mass flow of component i	kg/s
R_j	Reaction rate of reaction j	kmol/s
T_k	Temperature of control volume k	K
Q_{k-l}	Heat transfer between control volumes k and l	W
Cp_i	Heat capacity of component or control volume	J/(kg K)
S_k	Surface area of k	m^2
α_k	Heat transfer coefficient for k	$W/(m^2\ K)$

Subscript notation	
f	Free surface control volume
L	Liquid control volume
Li	Inner section of Ladle control volume
Bi	Inner section of Bell control volume
s	Slag control volume
f–L	From Free surface to Liquid control volumes
L–f	From Liquid to Free surface control volumes
in	Material addition

References

1. Ollila, S.; Syrjänen, A.; Robey, R. Commissioning and Initial Results from New Twin CAS-OB Plant at Ruukki, Raahe Steel Works. In Proceedings of the 6th European Oxygen Steelmaking Conference (EOSC), Stockholm, Sweden, 7–9 September 2011; pp. 170–183.
2. Järvinen, M.; Kärnä, A.; Ollila, S.; Fabritius, T. A novel approach for numerical modeling of the CAS-OB process. In Proceedings of the Scanmet IV, Luleå, Sweden, 10–13 June 2012; pp. 79–88.
3. Fruehan, R.J. (Ed.) *The Making, Shaping and Treating of Steel*; The AISE Steel Foundation: Pittsburgh, PA, USA, 1998; ISBN 0-930767-02-0.
4. Ghosh, A. *Secondary Steelmaking, Principles and Applications*; CRC Press: Boca Raton, FL, USA, 2001; ISBN 978-0-849-30264-0.
5. Rotevatn, T.; Wasbø, S.; Järvinen, M.; Fabritius, T.; Ollila, S.; Hammervold, A.; Blasio, C.D. A Model of the CAS-OB Process for Online Applications. In Proceedings of the Preprints of the IFAC 4th Workshop on Mining, Minerals and Metal Processing, Oulu, Finland, 2015.
6. Sulasalmi, P.; Visuri, V.V.; Kärnä, A.; Järvinen, M.; Ollila, S.; Fabritius, T. A Mathematical Model for the Reduction Stage of the CAS-OB Process. *Metall. Trans. B* **2016**, *47*, 3544–3556. [CrossRef]
7. Visuri, V.V.; Järvinen, M.; Pääskylä, K.; Kärnä, A.; Sulasalmi, P.; DeBlasio, C.; Ollila, S.; Fabritius, T. Preliminary Validation of a Numerical Model for the CAS-OB Process. In Proceedings of the 7th European Oxygen Steelmaking Conference (EOSC), Třinec, Czech Republic, 9–11 September 2014.
8. Kärnä, A.; Järvinen, M.; Sulasalmi, P.; Visuri, V.V.; Ollila, S.; Fabritius, T. An Improved Model for the Heat-Up Stage of the CAS-OB Process: Development and Validation. *Steel Res. Int.* **2018**, *89*, 1800141-16. [CrossRef]
9. Schei, T.S. On-line estimation for process control and optimization applications. *J. Process. Control* **2008**, *18*, 821–828. [CrossRef]
10. Simon, D. *Optimal State Estimation: Kalman, H Infinity, and Nonlinear Approaches*; Wiley: Hoboken, NJ, USA, 2006; ISBN 978-0-471-70858-2.
11. Maciejowski, J.M. *Predictive Control With Constraints*; Prentice Hall: Boston, MA, USA, 2002; ISBN 978-0-201-39823-6.

metals MDPI

Article

Evaluation of Mass Transfer Coefficient during Scrap Melting

Ming Gao, Jin-Tao Gao and Yan-Ling Zhang *

State Key Laboratory of Advanced Metallurgy, University of Science and Technology Beijing, Beijing 100083, China; gaoming0309@163.com (M.G.); gjt070@163.com (J.-T.G.)
* Correspondence: zhangyanling@metall.ustb.edu.cn

Abstract: Mass transfer is a critical scrap melting step. Herein, mass transfer coefficients (k) during scrap melting were calculated using laboratory-scale experiments. Correlation analysis and the entropy weight method were used to determine the effect of variables on k. The evaluation model under natural and forced convection was established. It was consistent with the experimental results. Under forced convection, at 1573 and 1673 K, when the rotation speed was increased from 141 to 423 r/min, k increased from 7.50×10^{-5} to 1.54×10^{-4} m/s and from 8.42×10^{-5} to 1.72×10^{-4} m/s, respectively. Furthermore, as the bath temperature was increased from 1573 to 1723 K, the k value of a stationary specimen increased from 3.14×10^{-5} to 5.31×10^{-5} m/s, respectively. Correlation analysis and the entropy weight method indicated that the effects of variables on k decreased as follows: molten pool stirring rate > bath temperature > scrap type. Moreover, the explicit functional relationships between k and the factors affecting k under natural and forced convection conditions were established, and the results were consistent with the experimental data. Our results can be used to determine the quantitative relationships between k and the factors affecting k.

Keywords: evaluation model; quantitative relationship; scrap melting; mass transfer coefficient; steelmaking process

Citation: Gao, M.; Gao, J.-T.; Zhang, Y.-L. Evaluation of Mass Transfer Coefficient during Scrap Melting. *Metals* **2021**, *11*, 1368. https://doi.org/10.3390/met11091368

Academic Editors: Thomas Echterhof, Ko-Ichiro Ohno and Ville-Valtteri Visuri

Received: 15 July 2021
Accepted: 26 August 2021
Published: 30 August 2021

1. Introduction

Scrap melting during the steelmaking process is the result of coupling heat and mass transfer. The evaluation of mass transfer, a critical scrap melting step, under different conditions is conducive to analyzing the melting process and optimizing the melting parameters [1,2].

Many recent studies have analyzed the changes in mass transfer coefficient (k) during scrap melting. Yang et al. [3] melted Φ20–50 mm steel bars in a 250 kg induction furnace in a bath temperature (T) range of 1573–1873 K and reported that the k values at 1573 and 1673 K were 6.3×10^{-5} and 6.4×10^{-5} m/s, respectively. Kosaka and Minowa [4] reported that the k value of a Φ10 mm steel bar subjected to a rotation speed of 200 r/min increased from 1.06×10^{-5} to 1.53×10^{-5} m/s as T was increased from 1573 to 1673 K, respectively. Isobe et al. [5] performed a scrap melting experiment using a 5 t converter. Upon increasing the stirring energy density of the molten pool from 3500 and 12,000 W/t, k increased from 8.3×10^{-5} to 19.4×10^{-5} m/s. Wei et al. [6] performed scrap melting experiments at T = 1673 K. The carbon content of the hot metal and the bottom-blowing flow rate ranged between 2.1 and 4.03 wt% and 3 and 7 L/min, respectively, and the calculated k values ranged between 8.0×10^{-5} and 10.0×10^{-5} m/s. In addition, the empirical formulas derived using the analogy method were used to analyze mass transfer [7,8]. Dimensionless correlations have been proposed for scrap melting under natural and forced convection conditions [4,9–12]. Wright [9], Kim and Pehlke [10], and Mineo and Susumu [11] calculated the dimensionless correlations during scrap melting under natural convection conditions. Conversely, Li [7], Don [8], Kim and Pehlke [10], and Mineo and Susumu [11] obtained the dimensionless correlations of the Colburn J-factor and Reynolds number (Re) under forced convection conditions for Re > 100. However, the aforementioned correlations were empirical and presented limitations and constraints. To date, the changes in k values during scrap melting have not been systematically analyzed. Therefore, an evaluation model should be established.

To evaluate the quantitative effects of several factors on k, the k values under different scrap melting conditions were calculated using laboratory-scale experiments. Correlation analysis and the entropy weight method were used to evaluate k during scrap melting. Furthermore, we established a theoretical model for determining k under natural and forced convection conditions. We anticipate that the model will serve as a theoretical foundation for online predicting and evaluating the scrap melting step of the steelmaking process.

2. Experimental Determination of Mass Transfer Coefficient

2.1. Experimental Set-Up and Description

A vertical tube furnace (BCMT-1973K, China) was used for the experiments. Approximately 1 kg of pig iron for steelmaking was melted in a 65 mm × 70 mm × 100 mm (I.D. × O.D. × H) alumina crucible to ensure that the bath depth reached 50 mm (Figure 1). For the tests involving rotating specimens, the rotational speeds were 141, 282, and 423 r/min. The cylinder was rotated after immersion into the hot metal. The material composition is summarized in Table 1.

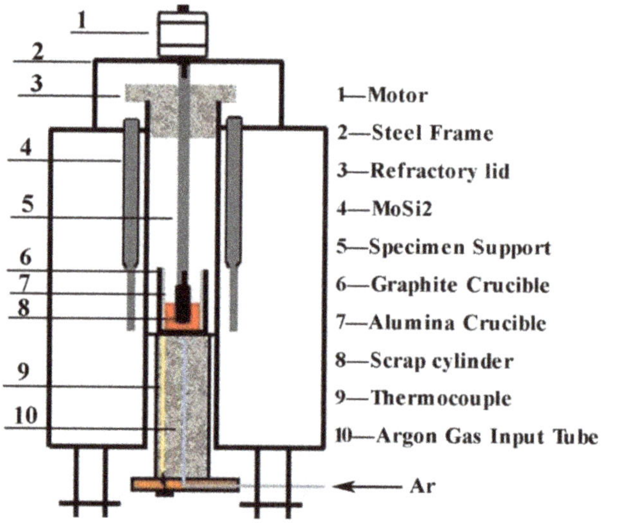

Figure 1. Schematic of experimental apparatus.

Before each experiment, the scrap cylinders were polished to ensure that their weights and diameters were identical. Argon gas was injected into the furnace to prevent oxidation of the hot metal. When the bath temperature was stabilized at the predetermined value, the scrap cylinders were immersed in the hot metal to a depth of approximately 30 mm without preheating and were held in the metal bath for different durations.

After removing them from the hot metal bath, the scrap cylinders were quenched in water. The quenched cylinders, especially the lower sections, contained frozen nodules, which transferred from the hot metal bath to the cylinders. Each specimen was cleaned to obtain a flat surface and subsequently weighed, and the final total length and diameters at the top, center, and bottom of the scrap cylinders were measured. Additionally, melt samples were collected before and after each test to determine their carbon concentrations using a carbon-sulfur analyzer (EMIA-920V2, Horiba, kyoto, Japan).

Table 1. Material composition (mass%).

	C	Si	P	Mn	S
Q235 low-carbon steel rod	0.168	0.110	0.041	0.391	0.0227
Q235 low-carbon steel plate	0.182	0.17~0.37	≤0.030	0.35–0.65	0.0011
45# steel rod	0.491	0.27	0.035	0.65	0.035
QT500-7 Ductile iron rod	3.69	1.653	0.0077	0.0825	0.0125
Iron/carbon bath	4.61	0.42	0.17	0.33	0.04

2.2. Calculation of Mass Transfer Coefficient

During scrap melting, the melting point of scrap is higher than bath temperature (T) owing to the low carbon content of scrap. The large carbon concentration gradient between the low-carbon scrap and the high-carbon molten pool promotes carbon mass transfer from the molten pool to the scrap surface. As the surface carbon content of scrap increases, the melting point of the scrap decreases. Therefore, carbon mass transfer is a critical step during scrap melting. Determining the k values under different conditions can provide a theoretical basis for improving melting efficiency. In previous studies, k (m/s) has been calculated using the melting rate (v; m/s), as follows [4,10–12]:

$$v = -(\frac{dr}{dt}) = kln(1+\frac{C_b - C_L}{C_L - C_s}) \tag{1}$$

where C_b and C_s are the carbon contents of the hot metal and scrap, respectively (wt%), and C_L is the carbon concentration in the liquid pool at the solid–liquid interface (wt%).

In previous studies, the melting point of scrap was assumed to be the liquidus temperature. The equilibrium temperature was calculated using the heat balance between the high-temperature molten pool and solid scrap. The C_L value was determined via linear approximation using the Fe–Fe_3C phase diagram [4,11–14]. The C_L values corresponding to the T values of 1573, 1623, 1673, and 1723 K were 2.972, 2.448, 1.87, and 1.293 wt%, respectively.

The v values under different conditions were substituted in Equation (1) to calculate the corresponding k values [15]. The effect of molten pool stirring on k was analyzed, and the results are presented in Figure 2a. The k values associated with rotating stationary scrap rods were larger than those associated with a stationary scrap rod and increased with increasing rotating speed. The k values increased by approximately 1.1 times when the rotating speed increased from 141 to 423 r/min. At T = 1573 K, k increased from 7.50×10^{-5} to 1.54×10^{-4} m/s when the rotation speed was increased from 141 to 423 r/min. The corresponding k values at T = 1673 K were 8.42×10^{-5} and 1.72×10^{-4} m/s, respectively. The effect of the rotation speed on k was the largest for the same T and amount of added scrap (Figure 2b, area enclosed by the red dotted line). Therefore, melting pool stirring played a major role in carbon mass transfer. This was ascribed to the great agitation caused by the high rotation speed promoting the convective mass transfer of carbon from the molten iron pool to the surface of the solid scrap [16,17].

The effect of T on k was also analyzed, and the results are presented in Figure 2a (area enclosed by the pink dashed line). For the stationary specimen, as T was increased from 1573 to 1773 K, k increased from 3.14×10^{-5} to 5.31×10^{-5} m/s, respectively. This was attributed to the high T value under natural convection conditions creating a large temperature gradient between the hot metal and the scrap and subsequently generating a large thermal driving force [16,17]. Therefore, heat exchange between the hot metal and solid scrap was promoted to enhance carbon diffusion.

The effect of scrap size on k was analyzed, and the results are illustrated in Figure 2b (areas enclosed by the black and green dashed lines). Under the same T and molten pool stirring conditions, k decreased from 3.14×10^{-5} to 1.97×10^{-5} m/s upon increasing the scrap size from Φ10 to Φ15 mm, respectively, and increased from 3.14×10^{-5} to 3.799×10^{-5} m/s upon increasing the carbon content of scrap from 0.1675 to 0.49 wt%, respectively.

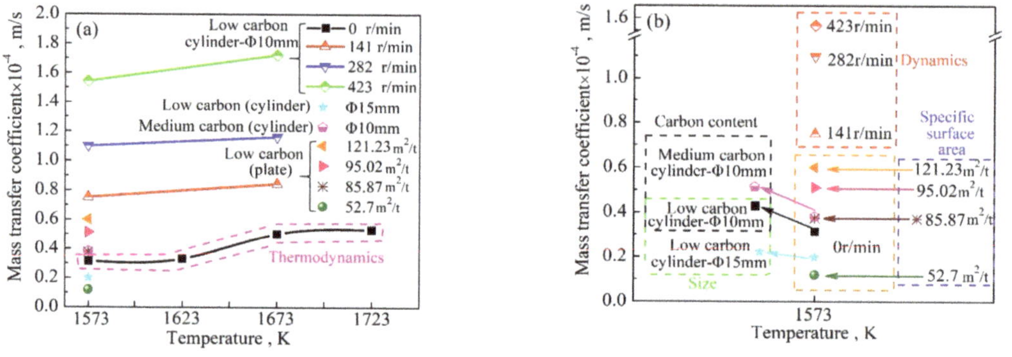

Figure 2. (**a**) Temperature dependence of the mass transfer coefficient under different experimental conditions (**b**) Magnified area of the graph in (**a**).

The effect of the scrap specific surface area (S_{sp}) on k was analyzed, and the results are presented in Figure 2b (area enclosed by the blue dashed line). Under the same T and molten pool stirring conditions, k increased from 1.19×10^{-5} to 5.99×10^{-5} m/s upon increasing S_{sp} from 52.7 to 121.23 m^2/t, respectively.

Using the aforementioned analysis results, we concluded that the rotation speed affected k the most, followed by T, S_{sp}, and C_s. Therefore, the effect of each factor on k decreased as follows: molten pool stirring rate > bath temperature > scrap type. These results were consistent with those of the subsequent theoretical evaluation model.

3. Qualitative Evaluation of Factors Affecting Mass Transfer Coefficient

3.1. Correlation Analysis of the Factors Affecting Mass Transfer Coefficient

The factors affecting k were used as explanatory (independent) variables, and k was used as the explained (dependent) variable. For the steelmaking process, the effects of top/bottom-blowing causing agitation in the molten pool during conventional steelmaking can be converted into mixing power. In this study, the relation between the imposed rotational speed and the speed induced in the metal bath has been explained as follows: firstly, the work generated by the rotation of the scrap rod can be determined using the formula for the rigid-body moment of inertia. The rotating speed was used to calculate the mixing power, as follows:

$$J = \frac{1}{2} m_s r^2 \; ; \; E = \frac{1}{2} J \omega^2 \tag{2}$$

where E is the work owing to rotation (W), J is the moment of inertia (kg·m^2), and r is the radius of the scrap metal rod (m). Furthermore, ω is the angular velocity (rad/s), which can be calculated as follows: $\omega = 2\pi n$, where n is the rotating speed in the experiment (r/s), and m_s is the total mass of hot metal (1 kg in this study). The stirring energy per unit mass of molten steel (ε; W/t) can be expressed as follows:

$$\varepsilon = E/m \tag{3}$$

where m is the total mass of the hot metal (t).

According to the above equations, the rotating speeds used in the previous experiment, that is, 141, 282, and 423 r/min, were converted into stirring powers of 0.0249, 0.0997, and 0.2243 W/t, respectively. The raw data from the previous experiments performed by the authors are summarized in Table 2.

Table 2. Raw data of the evaluation model.

Serial Number	Bath Temperature	Stirring Energy Density	Specific Surface Area	Carbon Content	Mass Transfer Coefficient
-	K	W/t	m^2/t	wt%	$\times 10^{-4}$ m/s
1	1573	0	56.03	0.168	0.314
2	1623	0	56.03	0.168	0.333
3	1673	0	56.03	0.168	0.501
4	1723	0	56.03	0.168	0.531
5	1573	0.0249	56.03	0.168	0.750
6	1573	0.0997	56.03	0.168	1.097
7	1573	0.2243	56.03	0.168	1.540
8	1673	0.0249	56.03	0.168	0.842
9	1673	0.0997	56.03	0.168	1.159
10	1673	0.2243	56.03	0.168	1.720
11	1573	0	56.03	0.491	0.379
12	1573	0	52.7	0.180	0.131
13	1573	0	95.02	0.180	0.452
14	1573	0	121.2	0.180	0.569
15	1573	0	85.87	0.180	0.375

Correlation analysis is used to test the significance of the relationship between two variables. The Pearson's correlation coefficient (r) can be expressed as follows:

$$r = \frac{S_{xy}}{S_x S_y},\ S_{xy} = \frac{\sum_{i=1}^{n}(x_i - \overline{x})(y_i - \overline{y})}{n - 1},\ S_x = \sqrt{\frac{\sum(x_i - \overline{x})^2}{n - 1}},\ \text{and}\ S_y = \sqrt{\frac{\sum(y_i - \overline{y})^2}{n - 1}} \tag{4}$$

where S_{xy} is the sample covariance, S_x and S_y are the sample standard deviations of x and y, respectively, N is the number of samples, and $\overline{x}$, $\overline{y}$ are the average values of x and y, respectively. Positive and negative correlations are indicated by $r > 0$ and $r < 0$, respectively. The magnitude of the absolute value indicates the strength of the correlation; that is, the larger the absolute value, the stronger the correlation.

Correlation analysis of the original data was performed using the Statistical Product and Service Solutions software to ensure that the independent variables were correlated with the evaluation variables, and the results are presented in Table 3.

Table 3. Correlation analysis results.

	Bath Temperature	Stirring Energy Density	Scrap Specific Surface Area	Scrap Carbon Content
Mass transfer coefficient	0.533	0.960	0.362	0.016

The analyzed factors (ε, T, S_{sp}, and C_s) were positively correlated to k (Table 3). The r-value of ε and k was the largest, followed by that of T and k. According to previous studies, the mass transfer of carbon from the hot metal pool to the scrap surface was the limiting step of the melting process. Stirring promoted mass transfer and convective heat transfer in the bath. The melting rate increased with increasing stirring rate [2,3,9,16,17]. Liu et al. [18] analyzed the effects of different bottom-blowing systems on ice melting and concluded that the stirring effect of the bottom-blowing on the molten pool accelerated the mass transfer of carbon from the hot metal to the scrap surface. Furthermore, the convective heat transfer between the molten pool and scrap was enhanced to promote scrap melting. The aforementioned results were consistent with those of our correlation analysis. Therefore, the qualitative results of the correlation analysis provided a direction for further exploration of the quantitative relationship between the melting process parameters and several factors.

3.2. Qualitative Evaluation of Mass Transfer Coefficient Using the Entropy Weight Method

It is beneficial to promote the scrap to melt to determine the degree of influence of each factor on the mass transfer coefficient [19]. The entropy weight method, an objective assignment method, can utilize the information provided by the data to determine the objective weights and to remove the subjective influence. It is more advantageous in comparison with other subjective assignment methods. The main steps of the entropy weight method are as follows [20–22]:

The evaluation matrix composed of m evaluation schemes and n indicators is defined as $X = (x_{ij})_{m\times n}$, $i = 1,2,\ldots,m$; $j = 1,2,\ldots,n$. The index standardization method is defined in Equation (5):

$$P_{ij}= x_{ij}/\sum_{i=1}^{m} x_{ij} \tag{5}$$

where P_{ij} is the standardized index data. This indicates the proportion of the i-th scheme under the j-th indicator in this index. The incommensurability among indicators is effectively eliminated through standardization. The entropy of each evaluation index is expressed as Equation (6):

$$E_j= (\sum_{i=1}^{m} P_{ij}lnP_{ij})/lnm \tag{6}$$

In particular, $P_{ij}ln\, P_{ij} = 0$ when $P_{ij} = 0$. ω_j is the weight of each indicator which can be expressed as Equation (7):

$$\begin{cases} \omega_j= (1-E_j)/\sum_{j=1}^{n}(1-E_j) \\ \sum_{j=1}^{n}\omega_j= 1 \end{cases} \tag{7}$$

r_i is defined as the comprehensive evaluation score of the i-th system. As shown in Equation (8), it is obtained using the linear weighted comprehensive evaluation formula.

$$r_i = \sum_{j=1}^{n} p_{ij}\omega_j \tag{8}$$

Finally, the influence of each factor on scrap melting is determined by comparing the comprehensive scores. As shown in Table 4, the comprehensive evaluation scores of r_1–r_{15} can be obtained using Equations (5)–(8).

Table 4. Comprehensive evaluation scores.

Influencing Factors	Factors	Serial Number	Value		Rank
Bath temperature	Bath temperature	r_1	0.0453	0.0453	2
		r_2	0.0359		
		r_3	0.0541		
		r_4	0.0573		
Stirring of the molten pool	Rotating speed	r_5	0.1279	0.1279	1
		r_6	0.1184		
		r_7	0.1663		
		r_8	0.0909		
		r_9	0.1251		
		r_{10}	0.1857		
Scrap type	Carbon content	r_{11}	0.0411	0.0411	3
		r_{12}	0.0141		
	Scrap specific surface area	r_{13}	0.0488		
		r_{14}	0.0614		
		r_{15}	0.0405		

The comprehensive evaluation scores of the affecting factors were ranked as follows: molten pool stirring > bath temperature > scrap type (Table 4). Therefore, the molten pool kinetics affected scrap melting the most, followed by the thermodynamics of the molten pool and scrap selection. Consequently, the following suggestions were evaluated in this study. The dynamics and thermodynamics of the molten pool can be improved to accelerate scrap melting and promote the steelmaking process. Similar processes are used in conventional steelmaking to promote scrap melting. Top/bottom-blowing was used to improve the kinetic conditions. Procedures such as injecting pulverized coal or top-blowing oxygen were implemented to accelerate the exothermic reaction of carbon and oxygen. T was increased to improve the thermodynamic conditions.

4. Quantitative Evaluation of Mass Transfer Coefficient Using Explicit Functions

4.1. Explicit Function under Natural Convection Conditions

The schematic diagram of cylindrical scrap melting is presented in Figure 3.

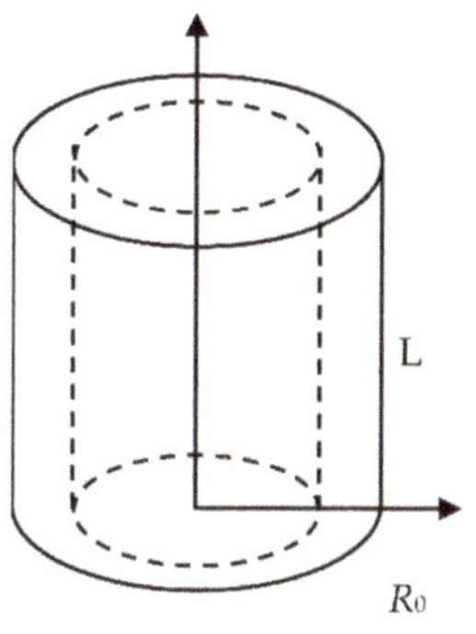

Figure 3. Schematic diagram of cylindrical scrap melting.

For a stationary specimen, the flow of liquid conformed to natural convection conditions. According to the dimensionless correlations of mass transfer under natural convection conditions ($Sh = A(GrSc)^B$) [23], k can be calculated as follows:

$$k = A(\frac{g\Delta\rho}{\mu})^B L^{3B-1} (D_C)^{1-B} \tag{9}$$

where A and B are constants obtained by fitting the experimental results. g is the gravitational acceleration in m/s^2. $\Delta\rho$ is the density difference in kg/m^3. μ is the hydrodynamic viscosity in $N{\cdot}s/m^2$. L, the specific length of the scrap, is the immersion depth of the cylindrical scrap under stationary conditions. And D_C is the diffusion coefficient in m^2/s.

According to previous studies [7,8,13,23–26], D_C is a function of the temperature and carbon concentration. For example, Shukla et al. [13] indicated that D_C can be calculated as follows: $D_C = 10^{-4} \times \exp(-(12{,}100/T + 2.568) + (1320/T - 0.554) \times C_b)$. Therefore, D_C can be expressed as $D_C = f(T, C_b)$, where T is the diffusion temperature (K), which is equal to the bath temperature. D_C increased upon increasing T and C_b.

S_{sp} for a cylindrical scrap specimen can be calculated as follows:

$$S_{sp} = \frac{2}{\rho_s}(\frac{1}{L} + \frac{1}{R_0}) \tag{10}$$

Therefore, by combining Equations (9)–(11), k can be expressed as follows:

$$k = A(\frac{g\Delta\rho}{-})^B (\frac{S_{sp}\rho_s}{2} - \frac{1}{R_0})^{1-3B} (D_C)^{1-B} \tag{11}$$

Equation (11) gives an explicit relationship between k, T, C_s, and S_{sp} under natural convection conditions, providing a basis for quantitative discussion. Under the experimental conditions in this study, A = 0.144 and B = 0.325 [15].

4.2. Explicit Function under Forced Convection Conditions

The dimensionless correlation of k under forced convection conditions is the Chilton–Colburn analogy: ($J_D = \left(\frac{k}{U}\right)(Sc)^{0.644} = ARe^B$) [8]. Therefore, k can be expressed as follows:

$$k = A(\pi)^{B+1}\left(\frac{\rho_b}{\mu}\right)^{B+0.644}\left(D_0 e^{-\frac{Q}{RT}}\right)^{0.644} L^{2B+1} n^{B+1} \quad (12)$$

Under forced convection conditions, S_{sp} can be expressed as follows:

$$S_{sp} = \frac{2(L+2h)}{Lh\rho_s} \quad (13)$$

where h is the height of the cylindrical scrap (m) and L is the average diameter of the cylindrical scrap ($L = 2R_0$)

In addition, Equation (14) was derived from the relationship between ε and the rotation speed.

$$n = \frac{2\left(\frac{m}{m_s}\varepsilon\right)^{1/2}}{\pi L} \quad (14)$$

Combining Equations (12)–(14), k can be expressed as follows:

$$k = A(2)^{B+1}\left(\frac{\rho_b}{\mu}\right)^{B+0.644}\left(D_0 e^{-\frac{Q}{RT}}\right)^{0.644}\left(\frac{4h}{S_{sp}h\rho_s - 2}\right)^{B}\left(\frac{m}{m_s}\varepsilon\right)^{\frac{B+1}{2}} \quad (15)$$

Equation (15) gives an explicit relationship between k, T, ε, C_s, and S_{sp} under forced convection conditions, providing a basis for quantitative discussion. Under the experimental conditions in this study, $A = 0.133$ and $B = -0.356$ [15].

4.3. Validation of Explicit Functions under Natural and Forced Convection Conditions

The dependence of k on T under natural convection conditions is illustrated in Figure 4. The k values determined using the evaluation model were consistent with the experimental values; k increased with increasing T, confirming the accuracy of the explicit function under natural convection conditions.

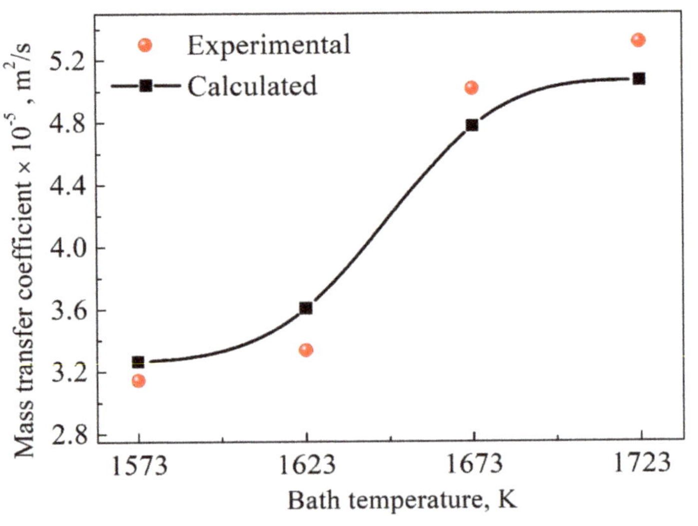

Figure 4. Validation of explicit function under natural convection conditions.

The dependence of k on ε under forced convection conditions is illustrated in Figure 5. The k values determined using the evaluation model were consistent with the experimental values; k increased with increasing ε, confirming the accuracy of the explicit function under forced convection conditions.

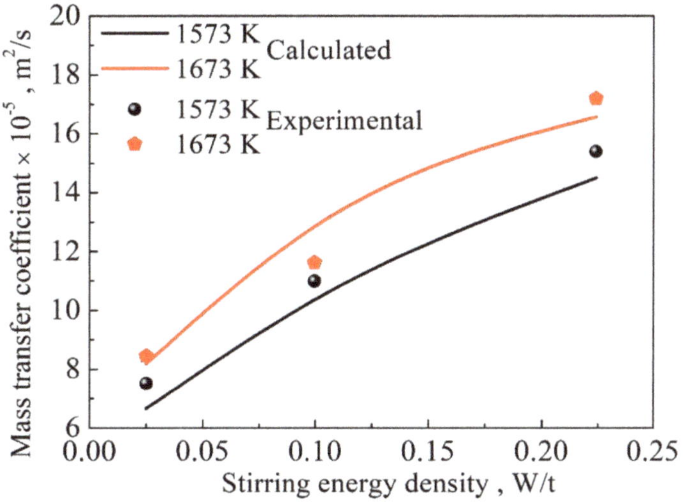

Figure 5. Validation of the explicit function under forced convection conditions.

5. Conclusions

The k values under different scrap melting conditions were obtained by laboratory-scale experiments. At T of 1573 and 1673 K, when the rotation speed was increased from 141 to 423 r/min, k increased from 7.50×10^{-5} to 1.54×10^{-4} m/s and from 8.42×10^{-5} to 1.72×10^{-4} m/s, respectively. For the stationary specimen, k increased from 3.14×10^{-5} to 5.31×10^{-5} m/s as T was increased from 1573 to 1723 K, respectively, from 3.14×10^{-5} to 3.799×10^{-5} m/s when C_s increased from 0.1675 to 0.49 wt%, respectively, and from 1.19×10^{-5} to 5.99×10^{-5} m/s when S_{sp} increased from 52.7 to 121.23 m^2/t, respectively.

The effects of various factors on k were evaluated using correlation analysis. The stirring energy density had the greatest correlation with the mass transfer coefficient, followed by the bath temperature.

The results of the entropy weight method indicated that the effects of each factor on k decreased as follows: molten pool stirring rate > bath temperature > scrap type.

The explicit functions between k and the factors affecting k under natural and forced convection conditions were described as follows: under natural convection,

$$k = A\left(\frac{g\Delta\rho}{\mu}\right)^{B}\left(\frac{S_{sp}\rho_s}{2} - \frac{1}{R_0}\right)^{1-3B}(D_C)^{1-B}$$

Under forced convection, the explicit function was as follows:

$$k = A(2)^{B+1}\left(\frac{\rho_b}{\mu}\right)^{B+0.644}\left(D_0 e^{-\frac{Q}{RT}}\right)^{0.644}\left(\frac{4h}{S_{sp}h\rho_s - 2}\right)^{B}\left(\frac{m}{m_s}\varepsilon\right)^{\frac{B+1}{2}}$$

Author Contributions: M.G., J.-T.G., and Y.-L.Z., conceived and designed the experiments; M.G. performed the experiments; M.G. analyzed the data; M.G. and Y.-L.Z. contributed to the writing and editing of the manuscript. All authors have read and agreed to the published version of the manuscript.

Funding: This research was funded by the National Key R&D Program of China, grant number 2019YFC1905701. This research was funded by Key projects of NSFC, grant number U1960201.

Institutional Review Board Statement: Not applicable.

Informed Consent Statement: Not applicable.

Data Availability Statement: Not applicable.

Conflicts of Interest: The authors declare that they have no conflict of interest.

References

1. Brabie, L.C.; Kawakami, M. Kinetics of Steel Scrap Melting in Molten Fe-C Bath. *High Temp. Mater. Process.* **2000**, *19*, 241–256. [CrossRef]
2. Yang, W.Y.; Zhang, X.G.; Yang, Y.T. The Study Overview of Utilizing Scrap in Convertor Steelmaking. In Proceedings of the 8th China Iron and Steel Annual Conference, Yantai, China, 25–28 October 2011.
3. Yang, W.Y.; Jiang, X.F.; Li, L. Thermal Simulation Experiments on Scrap Melting. *Iron Steel* **2017**, *52*, 27–35.
4. Kosaka, M.; Minowa, S. Dissolution of Steel Cylinder into Liquid Fe-C Alloy. (Stirring of Liquid Iron in H. F. Furnace-I). *Tetsu-to-Hagane* **1966**, *52*, 537–539, 1429–1432.
5. Isobe, K.; Maede, H.; Ozawa, K.; Umezawa, K.; Saito, C. Analysis of the Scrap Melting Rate in High Carbon Molten Iron. *Tetsu-to-Hagane* **1990**, *76*, 2033–2040. [CrossRef]
6. Wei, G.; Zhu, R.; Tang, T.; Dong, K. Study on the Melting Characteristics of Steel Scrap in Molten Steel. *Ironmak. Steelmak.* **2019**, *46*, 609–617. [CrossRef]
7. Li, R.H. *The Basis of Mass Transfer*; Beijing University Press: Beijing, China, 1987; pp. 195–251.
8. Don, W.P. *Chemical Engineers' Handbook*, 7th ed; Science Press: Beijing, China, 2001; p. 5.
9. Wright, J.K. Steel Dissolution in Quiescent and Gas Stirred Fe/C Melts. *Metall. Mater. Trans. B* **1989**, *20*, 363–374. [CrossRef]
10. Kim, Y.U.; Pehlke, R. Mass Transfer During Dissolution of a Solid into Liquid in the Iron—Carbon System. *Metall. Mater. Trans. A* **1974**, *5*, 2527–2532. [CrossRef]
11. Kosaka, M.; Minowa, S. Dissolution of Steel Cylinder into Liquid Fe-C Alloy. *Tetsu-to-Hagane* **1967**, *53*, 983–997. [CrossRef]
12. Kosaka, M.; Minowa, S. Mass-Transfer from Graphite Cylinder into Liquid Fe-C Alloy. *Tetsu-to-Hagane* **2010**, *53*, 1467–1477. [CrossRef]
13. Shukla, A.K.; Deo, B.; Robertson, D.G.C. Scrap Dissolution in Molten Iron Containing Carbon for the Case of Coupled Heat and Mass Transfer Control. *Metall. Mater. Trans. B* **2013**, *44*, 1407–1427. [CrossRef]
14. Szekely, J.; Chuang, Y.K.; Hlinka, J.W. The Melting and Dissolution of Low-Carbon Steels in Iron—Carbon Melts. *Metall. Mater. Trans. B* **1972**, *3*, 2825–2833. [CrossRef]
15. Gao, M. Fundamentals of Scrap Melting in the Steelmaking Process. Ph.D. Thesis, University of Science and Technology Beijing, Beijing, China, 2021.
16. Sun, H.; Liu, Y.; Lin, C.; Muh-Jung, L.U. Experimental Observation of Spherical Scrap Melting in Hot Metal. In Proceedings of the 6th International Congress on the Science & Technology of Steelmaking, Beijing, China, 12–14 May 2015; pp. 136–139.
17. Pehlke, R.D.; Goodell, P.D.; Dunla, R.W. Kinetics of Steel Dissolution in Molten Pig Iron. *Trans. Metall. Soc. AIME* **1964**, *233*, 1420–1427.
18. Liu, Y.S.; Gao, J.P. Water Model Study on the Feature of Scrap Melting in the Oxygen Top-Bottom Combined Blowing Process. *J. Iron Steel Res.* **1990**, *2*, 1–7.
19. Kruskopf, A. A Model for Scrap Melting in Steel Converter. *Metall. Mater. Trans. B* **2015**, *46*, 1195–1206. [CrossRef]
20. Yang, G.; Deng, S.; Xu, A.J.; Dai, X.Q. Analysis and Calculation of Steel Scrap Melting in Multifunctional Hot Metal Ladle. *J. Cent. South Univ. (Sci. Tech.)* **2019**, *5*, 1021–1027.
21. Wu, M.H. Simulation Study on the Stirring of the Converter Bath by the Bottom Blowing. *Hebei Metall.* **1987**, *4*, 3–17.
22. Zuo, D.W. Research and Practice on Reducing Tapping Temperature of BOF. *Hunan Metall.* **2003**, *2*, 32–35.
23. Wanibe, Y.; Shoji, T.; Fujisawa, T.; Sakao, H. Temperature-Dependency of Interdiffusion in Molten Fe-C Alloy. *Iron Steel Inst. Jpn. Trans.* **1982**, *22*, 560–565. [CrossRef]
24. Florian Penz, M.; Johannes, S. Diffusive steel scrap melting in carbon-saturated hot metal—Phenomenological investigation at the solid-liquid interface. *Materials* **2019**, *12*, 1358. [CrossRef]
25. Florian Penz, M.; Johannes, S. A Review of Steel Scrap Melting in Molten Iron-Carbon Melts. *Steel Res. Int.* **2019**, *90*, 1900124. [CrossRef]
26. Florian Penz, M.; Johannes, S. Dissolution of Scrap in Hot Metal under Linz Donawitz (LD) Steelmaking Conditions. *Metals* **2018**, *8*, 1078. [CrossRef]

metals

MDPI

Article

Data-Driven Modelling and Optimization of Energy Consumption in EAF

Simon Tomažič *, Goran Andonovski, Igor Škrjanc and Vito Logar

Faculty of Electrical Engineering, University of Ljubljana, 1000 Ljubljana, Slovenia;
goran.andonovski@fe.uni-lj.si (G.A.); igor.skrjanc@fe.uni-lj.si (I.Š.); vito.logar@fe.uni-lj.si (V.L.)
* Correspondence: simon.tomazic@fe.uni-lj.si; Tel.: +386-1-4768-760

Abstract: In the steel industry, the optimization of production processes has become increasingly important in recent years. Large amounts of historical data and various machine learning methods can be used to reduce energy consumption and increase overall time efficiency. Using data from more than two thousand electric arc furnace (EAF) batches produced in SIJ Acroni steelworks, the consumption of electrical energy during melting was analysed. Information on the consumed energy in each step of the electric arc process is essential to increase the efficiency of the EAF. In the paper, four different modelling approaches for predicting electrical energy consumption during EAF operation are presented: linear regression, k-NN modelling, evolving and conventional fuzzy modelling. In the learning phase, from a set of more than ten regressors, only those that have the greatest impact on energy consumption were selected. The obtained models that can accurately predict the energy consumption are used to determine the optimal duration of the transformer profile during melting. The models can predict the optimal energy consumption by selecting pre-processed training data, where the main steps are to find and remove outlier batches with the highest energy consumption and identify the influencing variables that contribute most to the increased energy consumption. It should be emphasised that the electrical energy consumption was too high in most batches only because the melting time was unnecessarily prolonged. Using the proposed models, EAF operators can obtain information on the estimated energy consumption before batch processing depending on the scrap weight in each basket and the added additives, as well as information on the optimal melting time for a given EAF batch. All models were validated and compared using 30% of all data, with the fuzzy model in particular providing accurate prediction results. It is expected that the use of the developed models will lead to a reduction in energy consumption as well as an increase in EAF efficiency.

Keywords: electric arc furnace; energy consumption; profile optimization; modelling; machine learning; steelmaking; regression; fuzzy modelling; evolving modelling

Citation: Tomažič, S.; Andonovski, G.; Škrjanc, I.; Logar, V. Data-Driven Modelling and Optimization of Energy Consumption in EAF. *Metals* **2022**, *12*, 816. https://doi.org/10.3390/met12050816

Academic Editors: Thomas Echterhof, Ko-Ichiro Ohno and Ville-Valtteri Visuri

Received: 30 March 2022
Accepted: 6 May 2022
Published: 9 May 2022

1. Introduction

Current market demands for steel quality, price and production times require the introduction of several technological innovations in electric arc furnace (EAF) steelmaking. Electric Arc Furnaces (EAF) are improving very rapidly. Twenty years ago, the performance of today's EAFs would have seemed impossible. Thanks to an impressive number of innovations, the melting time in the most efficient furnaces (with a capacity of 100–130 t) has been reduced to 30–40 min. Electrical energy consumption was decreased by 1.8 times, from 630 to 340 kWh/t and hourly productivity increased by six times, from 40 to 240 t/h. The share of electrical energy in total energy consumption per melt fell to 50%. Electrode consumption was reduced by about six times [1,2]. It can be assumed that such performance should be normal for most steelworks in the near future.

In modern furnaces, the fundamental processes are melting the solid scrap and heating the liquid bath. The productivity of today's furnaces therefore depends mainly on these high-energy processes. To set these processes in motion, heat must be obtained from electrical or chemical energy and then passed to the regions of the solid charge or liquid bath [3,4]. The heating technology, furnace designs and other EAF equipment are evolving very fast. Every year, new technical solutions are offered and widely advertised. Steelmakers are struggling to find their way through the flood of innovations. According to the latest trends, modern steelworks should meet four essential requirements in the following way:

- By producing different types of steel in the desired quality, the specified process requirements are met.
- By reducing the manufacturing costs, the specified economic requirements are met, which means that the profitability and competitiveness of the products can be increased.
- By limiting excessive pollution, which is regulated by government regulations, the specified environmental requirements are met.
- By limiting physically and mentally demanding work that is unacceptable for the population of a given country above a certain level of social development, the specified health and safety requirements are met.

The total costs of the EAF can be divided into the cost of scrap and ferroalloys, which account for about 70%, and the so-called operating costs, which account for the remaining 30% of the total cost. The operating costs can be further divided into the costs of electrical energy, fuel and electrodes, which account for about 40% of the operating costs [1,5,6]. The total costs can be reduced in the following ways:

- By reducing the consumption of loaded materials, refractory materials, energy sources, etc. per ton of product;
- By speeding up and increasing production and thus reducing the costs of maintenance, personnel and other specific production costs;
- By finding cheaper input materials and energy sources.

Over the last fifty years, the main objective of EAF development has been to increase productivity. During this period, almost all innovations introduced were dedicated to this problem. Apart from the cost of scrap, productivity represents a crucial factor on which the overall steelmaking economy depends to a large extent [7]. When productivity increases, labour and maintenance costs usually decrease, as do the costs of electrodes, energy sources, refractory materials, electrical energy and other operating costs [8,9]. The proposed EAF innovations, in addition to their positive contributions, also bring some drawbacks. For example, the use of oxygen-gas burners and the introduction of carbon injection for slag foaming enable a drastic reduction in electrical energy consumption, but, on the other hand, increase carbon dioxide emissions [5,6]. Due to environmental protection, the use of biomass (and biofuel produced from renewable biomass) as a renewable energy source in the electric arc furnace is also becoming increasingly important [10,11].

The electrical energy consumption can be controlled by the electrical mode, which is determined by the programme for changing the electrical parameters (current, voltage, arc power, etc.) of the EAF's circuit during the melting process. These parameters can be changed over a wide range due to the special design of the furnace transformer. The control of the transformer voltage levels during the melting process ("on-load") can be done either manually by the operator or fully automatically. The biggest challenge in EAF operation, i.e., determining the optimum melting programmes, times and batch quantities, is thus still left to the operator and his experience. Since the control of the melting process is based on indirect measurements (e.g., arc stability, energy consumption, power-on time, etc.) and not on the actual conditions in the EAF (e.g., bath temperature, melting stage, bath composition), EAF operation is suboptimal (lower raw material and energy efficiency, lower steel quality and increased CO_2 emissions), which consequently means higher operating costs [2,11,12].

With extensive use of oxygen and carbon during the melting process, chemical heat plays a major role in reducing electrical energy consumption and increasing EAF productivity as the bath absorbs a large amount of chemical heat, which is released during oxidation of carbon, iron and its alloys such as Mn, Si, etc [2,6].

Higher oxygen consumption usually occurs during bath blowing, as it depends on the use of carbon powder, which is added into the bath at the same time as oxygen. The impressive results achieved by the additional oxygen consumption cannot be achieved without the carbon injection. The latter reduces the iron oxides and thus prevents an undesirable reduction in yield. Otherwise, the amount of oxidised iron increases drastically the more oxygen is blown into the bath. In addition, the injected carbon leads to the release of CO and CO_2, which causes the slag to foam. Immersing the arc in foamy slag provides a large increase in efficiency in the use of electrical energy [2,13].

This study addresses the optimization of electric arc furnace (EAF) to increase its efficiency and thus reduce electrical energy consumption. This can be achieved by defining optimal control profiles for the EAF, i.e., transformer power, oxygen balancing, and carbon addition [13,14]. The optimization is based on a data-driven approach where different models (from linear models to evolving fuzzy models) [15–17] and statistical analyses [12] have been performed. The models can be run online in parallel with the actual EAF process and help the operator to control the EAF. Many authors have shown through simulations that optimised operating profiles allow significant reductions in production times and operating costs [2,13,18,19].

Advanced technological solutions such as post-combustion, off-gas [20] and slag [21] heat recovery, oxygen lancing [22], gas burners [23], bottom stirring, adjustable alternating current transmission systems and high-power transformers [24], have reduced energy consumption. Nevertheless, additional process improvements can be made in EAF through heat recovery systems [25], various additives [5,26] and optimal control with operating profiles [2].

The melting profiles are usually selected in advance by the operator based on the maximum energy input. The predefined profiles have the disadvantage that they do not take into account the variations in EAF conditions. Therefore, adaptive control of the EAF (via oxygen and carbon input) is required to achieve suitable conditions and also slag properties. The latter enables to protect the water-cooled panels and walls, reduce energy consumption and contribute to the correct steel composition [2]. Due to the lack of measurements, the operator has limited insight into the EAF process. Consequently, the predefined timed inputs (charging, oxygen lancing and carbon injection) may differ from the optimal times that ensure higher EAF efficiency. Many authors [2,13,18] have conducted studies to investigate EAF efficiency through optimised control. However, very few of them have considered the optimisation of energy sources over the entire tap-to-tap interval. The reason for this could be insufficiently defined optimisation objectives and rough EAF models that are not accurate enough to be used in the optimisation procedure.

The aim of this study is to find key influential factors from which energy consumption in EAF is estimated using the proposed predictive models. These can be used in a simulator to improve the EAF process in such a way that less electrical energy is consumed and the production of a certain type of steel is possible in a shorter time than with the existing process. The total energy (electrical and chemical) consumed in the EAF process is distributed between the three products (steel, slag and off-gas) and the various losses. Only the energy that is delivered to the steel bath can be considered as useful energy.

The paper is organised as follows: Section 2 describes, first, the dataset used and the preprocessing steps applied on it; second, the procedure for selecting the key input variables; and, third, four different modelling approaches (based on machine learning and fuzzy methods) for predicting the electrical energy consumption of EAF. Section 3 discusses the experimental results, comparing all the developed models. A discussion and concluding remarks are given in Sections 4 and 5.

2. Materials and Methods

This section presents the methods needed to build models for predicting electrical energy consumption. These models will be used as part of the operator advisory system to assist the operator in managing the EAF. This prevents the operator from frequently selecting suboptimal settings in the semi-automatic furnace control mode that result in lower steel yield and quality and higher energy and material consumption.

2.1. Data Description and Pre-Processing

The operation of the EAF is monitored by measuring all variables and parameters that could affect energy consumption and overall efficiency. All parameters and variables are stored and organised separately for each batch. Some of the measurements are recorded event-based at specific times, while others are recorded continuously. All important variables from the charging and melting phases are listed in Table 1. The charging recipe is determined by the scrap weight in each basket and the hotheel at the beginning. These data are aggregated for all baskets used. In the melting phase, there are several parameters that affect the total energy consumption and the overall efficiency of the process. The most important criterion and the focus of this article is the electrical energy consumption per total weight of scrap (kWh/t), which is presented in this article as a percentage of the maximum electrical energy consumption per total weight of scrap.

Table 1. List of input variables at the charging and melting phase of the EAF process.

Charging		Melting	
Description	**Unit**	**Description**	**Unit**
Total scrap weight	[kg]	Melting time	[s]
Hotheel start	[kg]	Delays	[s]
Scrap weight in basket 1	[kg]	Temperature	[°C]
Scrap weight in basket 2	[kg]	Total oxygen	[Nm^3]
Scrap weight in basket 3	[kg]	Total carbon	[kg]
Type of charged scrap		Hotheel end	[kg]
		Slag weight	[kg]

In the development of the electrical energy consumption prediction models, the first required step is the preprocessing and filtering of data (removal of a part of the data). Since the data are stored in different databases and with different sampling times during the operation of the EAF, it is necessary to resample and synchronise the data before starting the analysis phase. Since historical data from completed batches are often incomplete, these batches must be removed from the modelling process during filtering. The data cleaning procedure to eliminate all corrupted data should also be applied to efficiently identify and remove outliers (e.g., unusually long tap-to-tap time spans, i.e., more than four hours, extremely high power consumption, etc.). The steps of data pre-processing cannot be performed completely automatically, since in some special cases the knowledge and experience of the staff (especially the EAF operator) must also be taken into the account. Each batch may consist of two, three, or four baskets of raw material. Since in the available database the melting process was most frequently performed with three baskets, only these batches were used in all further analyses. After filtering the data, the first, the second, and the third baskets have an average capacities of 46 t, 36 t, and 18 t, respectively. Each individual charge takes about three minutes. The melting of the scrap after charging with the first, second, and third baskets takes about 17 min, 11 min, and 20 min on average, and the average delay per batch is 13 min.

2.2. Selection of the Key Input Variables

The operation of the EAF is a subject to several factors that affect the final product quality and energy consumption. Deviations from potentially optimal performance can be influenced by all parameters and settings during the charging and melting phases. Therefore, the most influential independent variables must be identified from historical data, as this information is necessary for the development of the models for energy consumption prediction. In the study of Glavan et al. [27], it has been shown that the input variable selection (IVS) approach can efficiently find the most important input variables from a big database for modelling and prediction purposes. The IVS approach is based on the analysis of historical data and combines a data mining approach with various selection criteria [28]. The selection of input variables has a great impact on the prediction performance, the effectiveness of the model and the better understanding of the system. Therefore, the IVS represents an important step for model identification. The authors in [27] tested and compared different methods from the literature for variable selection. They also evaluated each method to find out the most suitable methods for model-based prediction problems. Finally, the authors selected the following methods as the most effective: partial correlation measure (Pcorr) [29], partial mutual information (PMI), linear-in-the-parameters (LIP) [30], non-negative Garrote (NNGarr) [31], variable importance in projection (PLS VIP) [32], distance correlation (dCorr) [33], and least absolute shrinkage and selection operator (LASSO) [34]. All of these methods, briefly discussed in [27], were used in this study. The influential factors from all methods were averaged to determine the order of the most influential variables. In the following, all the machine learning methods that were used to obtain the predictive models for estimation of the electrical energy consumption are briefly described.

2.3. Machine Learning Methods

2.3.1. Linear Regression

The linear model has been a mainstay of statistics for the past 30 years and remains one of the most important tools [35,36]. Linear regression is a linear approach for modelling the relationship between a scalar response and one or more explanatory variables (also known as dependent and independent variables). In linear regression, the relationships are modelled using linear predictor functions, whose unknown model parameters are estimated from the data. Such models are called linear models. Using linear regression, the single output of the model $\hat{y}_j$ can be determined in the following way:

$$\hat{y}_j = \hat{\beta}_0 + \sum_{i=1}^{p} x_i \hat{\beta}_i, \tag{1}$$

where $x_j^T = [x_1, x_2, \ldots, x_p]_j$ is a regression vector (where $j = 1, \ldots, m$; m is the total number of test samples and p is the total number of independent variables) and $\hat{\beta}^T = [\hat{\beta}_1, \hat{\beta}_2, \ldots, \hat{\beta}_p]$ is the vector of linear coefficients. The term $\hat{\beta}_0$ is the intercept, which in machine learning is also called the bias [35]. It is often convenient to include the constant variable 1 in the vector x_j and include $\hat{\beta}_0$ in the vector of coefficients $\hat{\beta}$, and then write the linear model in vector form as an inner product:

$$\hat{y}_j = x_j^T \hat{\beta}. \tag{2}$$

There are many different methods of fitting the linear model to a set of training data. By far the most popular is the least squares method. In this approach, the coefficients β are chosen to minimise the residual sum of squares:

$$RSS(\beta) = \sum_{i=1}^{n} (y_i - x_i^T \beta)^2, \tag{3}$$

where n is the total number of training samples. $RSS(\beta)$ is a quadratic function of the parameters, and hence its minimum always exists but may not be unique. The solution is most easily characterised in matrix notation:

$$RSS(\beta) = (y - X\beta)^T (y - X\beta), \tag{4}$$

where X is an $n \times p$ matrix with each row an input vector x_i^T, and y is an n-vector of the outputs in the training set. Differentiating w.r.t. β the normal equations can be written as follows:

$$X^T(y - X\beta) = 0. \tag{5}$$

If $X^T X$ is non-singular, then the unique solution is given by:

$$\hat{\beta} = (X^T X)^{-1} X^T y, \tag{6}$$

and the fitted value at the i-th input x_i is $\hat{y}_i = x_i^T \hat{\beta}$.

2.3.2. K-Nearest Neighbour Method

K-nearest neighbours (k-NN) algorithms [37,38] are non-parametric supervised machine learning algorithms commonly used in the field of pattern recognition. The k-NN algorithms can be used for both classification and regression. In both cases, the input to the algorithm consists of the labelled training dataset:

$$D = \{(x_i, y_i)\} \text{ for } i = 1, \ldots, n, \tag{7}$$

where n is the number of samples in the dataset, x_i is the regression vector and y_i is the class label or a continuous output variable. To make a prediction (class label or continuous target variable), the k-NN algorithms find the k nearest neighbours of a query point $\hat{x}_j$ and compute the class label (i.e., classification) or continuous target variable (i.e., regression) based on the k nearest (most "similar") points. Since the prediction is based on a comparison of a query point with data points (regression vectors) in the training dataset, k-NN is also categorised as an instance-based (or "memory-based") method.

In k-NN regression, the output prediction is based on the labels of the k nearest neighbours. The output value $\hat{y}$ is usually the average of the values of k nearest neighbours:

$$\hat{y}_j = \frac{1}{k} \sum_{i=1}^{k} y_i. \tag{8}$$

For both classification and regression, a distance-weighted k-NN algorithm [38] can also be used, which assigns weights to the contributions of the neighbours, so that the closer neighbours contribute more to the average than the more distant ones. For example, a common weighting scheme is to assign a weight of $w_i = 1/d_i$ to each neighbour, where d_i is the distance to the i-th nearest neighbour.

$$\hat{y}_j = \frac{\sum_{i=1}^{k} w_i y_i}{\sum_{i=1}^{k} w_i}. \tag{9}$$

The best choice of k depends upon the data. In general, larger values of k reduce the effects of noise on classification but make the boundaries between classes less clear. A good k can be selected by various heuristic techniques. By changing the value of k, the complexity of a k-NN model is affected. In practise, a good trade-off must be found between high bias (the model is not complex enough to fit the data well when k is too large) and high variance (the model fits the training data too closely when k is too small).

For k-NN algorithms, many distance metrics or measures can be used to select k nearest neighbours. There is no "best" distance measure, and the choice is highly context- or problem-dependent. For continuous features, the most common distance metric is the Euclidean distance. Another popular choice is the Manhattan distance, which puts less emphasis on the differences between "distant" feature vectors or outliers than the Euclidean distance. The Mahalanobis distance would be another good choice for a distance metric, as it takes into account the variance of the different feature vectors as well as the covariance among them.

One of the main advantages of k-NN is that it is relatively easy to implement and interpret. Moreover, with its approach to approximate complex global functions locally, it can be a powerful predictive "model". Another advantage is that k-NN has some strong consistency results. As the amount of data approaches infinity, the two-class k-NN algorithm is guaranteed to yield an error rate no worse than twice the Bayes error rate (the minimum achievable error rate given the distribution of the data). The drawback is that k-NN is very sensitive to the curse of dimensionality [38] and is expensive to compute with an $O(n)$ prediction step. Therefore, various data structures have been developed to improve the computational performance of k-NN in prediction. In particular, the idea is to identify the k nearest neighbours more intelligently. Instead of matching each training sample in the training set to a given query point vector, various approaches have been developed to partition the search space as efficiently as possible and reduce the number of distance evaluations actually performed. Data structures such as KD-trees and Ball-trees are often used for this purpose, as they can make k-NN substantially more efficient.

2.4. Takagi–Sugeno Fuzzy Modeling

Fuzzy logic was developed in 1965 as an extension of the classical (Boolean) logic. The classical logic assigns to a variable or a statement the value of 1 for "true" or the value of 0 for "false", fuzzy logic allows the value assignment at an interval between [0, 1]. The reason for this can be found in the observation of the way of human thinking when deciding on the very approximate estimates of various facts that they present to themselves in the form of rules. To address such a concept, a mechanism for recording knowledge based on rules in the form of approximate reasoning based on fuzzy logic has been introduced. First, some basic concepts of fuzzy logic and approximative reasoning, which are necessary for the understanding of fuzzy models, are introduced. Fuzzy logic records relationships, knowledge and decisions in the form of rules. For conjunction of the linguistic statements, the conjunction operator (t-norm) "min" is used. The combination of the affiliation of all linguistic expressions determines degree of rule fulfillment or rule firing strength because it expresses how well the premise matches the given values of input variables. For the entire fuzzy system, only fulfillment degrees greater than zero are important. It must be guaranteed that the rules complete the entire possible input space to avoid situations where no rule gets activated for certain input values. In the case of non-explicit local affiliation functions, this problem does not exist because all rules are always fulfilled, although with very small values. After the degree of fulfillment of an individual rule is calculated, the contributions of individual consequent parts have to be determined and assembled to obtain the output of the fuzzy system. This is called accumulation. Usually, the output of the fuzzy system is a fuzzy set that needs to be transformed into a sharp form for further work. This is called the process of defuzzification. Of course, this is not necessary if a sharp value is chosen for the consequent part, or if the result is used for qualitative estimations. In general, three basic types of fuzzy systems exist, i.e., a linguistic or Mamdani, a special fuzzy system or a singleton, and a Takagi–Sugeno Fuzzy System. In our case, the focus is on the Takagi–Sugeno (TS) Fuzzy approach, which provides excellent interpretability and the best fuzzy modelling results [39]. The first step of fuzzy modelling is the fuzzification, where the degree of membership for all linguistic statements $\mu_{ij}(x_j)$ ($i = 1, \ldots, M$ and $j = 1, \ldots, p$) is calculated, where M is the number of fuzzy system rules and p is the number of inputs x_j. The TS fuzzy rule $\mathcal{R}_i$ can be written as follows:

$$\mathcal{R}_i : \text{ IF } x_1 = \mathsf{A}_{i1} \text{ AND } \ldots \text{ AND } x_p = \mathsf{A}_{ip} \text{ THEN } y = f_i(x_1, x_2, \ldots, x_p) \tag{10}$$

where A_{ij} represents a fuzzy set for the variable x_j and y is the output. By aggregation, the individual linguistic statements into the level of activation of the rule (with respect to the operators between them) are composed. The output of the TS fuzzy model is defined as:

$$\hat{y} = \frac{\sum\limits_{i=1}^{M} f_i(x)\,\mu_i(x)}{\sum\limits_{i=1}^{M} \mu_i(x)} \tag{11}$$

where x is the input vector and $f_i(x) = w_{i0} + w_{i1}x_1 + w_{i2}x_2 + \ldots + w_{ip}x_p$ is linear regression function. If the fuzzy model is written in a conjunctive form and min function is used for the t-norm, then the degree of fulfillment of the rule is:

$$\mu_i(x) = \min(\mu_{ij}(x_j)). \tag{12}$$

In this study, Gaussian membership functions were used:

$$\mu_{ij}(x_j) = \exp\left(-\frac{1}{2}\frac{(x_j - c_{ij})^2}{\sigma_{ij}^2}\right), \tag{13}$$

where σ_{ij}^2 is variance and c_{ij} is expected value of the Gaussian function (belonging to A_{ij} fuzzy set).

In fuzzy models, the nonlinear parameters in the premise (i.e., the parameters in the causal part of the rule that define the membership functions, their positions and widths) and the linear parameters (w_{ip}) in the consequent part of the rules can be optimized. The latter can be easily estimated using the least squares method. The parameters in the causal part of the rule correspond to the parameters on a hidden layer of neural networks and are non-linear. The optimization of the rule structure is a combination problem that can be solved by a selection of linear subsets or by a nonlinear global optimization, for example, by an optimization with a genetic algorithm (GA) or a particle swarm optimization (PSO) [40].

2.5. Evolving the Cloud-Based Prediction Model

Due to the refinement of the technological process of melting in the EAF, the data collected from the new batches are increasingly different and the consequently developed models are predicting electrical energy consumption worse and worse. The evolving modelling approach is appropriate for the purpose of constantly updating models also during the process of melting. In this paper, an online evolving fuzzy identification method (based on data clouds) [41], which represents an upgrade according to the Takagi–Sugeno fuzzy modelling, is used. By upgrade means the ability to evolve the structure of the model online and to adapt the parameters of each local model during the process.

In evolving modelling, the structure of the fuzzy model is identified online using the evolving mechanisms, i.e., principles for adding and removing fuzzy rules. The rule-based form of i-th rule is defined as:

$$\mathcal{R}^i : \text{IF} \quad (x_f(k) \sim X^i) \quad \text{THEN} \quad y_i(k) = f_i(x_f(k)) \tag{14}$$

where $x_f(k) = [u_1(k), u_2(k) \ldots, u_p(k)]$ represents the input (regression) vector, X^i stands for the i-th data-cloud, $y_i(k)$ represents the output of that fuzzy rule, and $f_i(x_f(k))$ represents an arbitrary function. In our case, the NARX model is used and therefore the output function is defined as:

$$f_i(k) = \theta_i^T \psi(k) \tag{15}$$

where $\psi(k) = \left[x_f(k),\ 1\right]^T$ stands for the extended regressor and θ_i^T is vector of local parameters of i-th fuzzy rule, which are calculated using the recursive Weighted Least Squares method (rWLS) as presented in [41]. The final value of the output is calculated as follows:

$$y(k) = \sum_{i=1}^{c} \beta_i \theta_i^T \psi_k \tag{16}$$

where c is the number of data-clouds (fuzzy rules), and β_i stands for normalized relative density, which is defined as relation between the current data sample $x_f(k)$ and the i-th fuzzy rule X^i. Normalized relative density is calculated as follows:

$$\beta_i(k) = \frac{\gamma_i(k)}{\sum\limits_{j=1}^{c} \gamma_j(k)}, \qquad i = 1, \ldots, c \tag{17}$$

where $\gamma_i(k)$ stands for the local density of the data $x_f(k)$ and is calculated as:

$$\gamma_i(k) = \frac{1}{1 + \|x_f(k) - \mu_i(k)\|^2 + \sigma_i(k) - \|\mu_i(k)\|^2} \tag{18}$$

In Equation (18), $\mu_i(k)$ and $\sigma_i(k)$ denote mean value vector and mean-square length of the data vector from i-th cloud, respectively. Please refer to [41] for more details about the whole evolving algorithm including the evolving mechanisms of adding and removing (data-clouds) fuzzy rules.

3. Results

The data used within the methods for key input variables selection and for the validation of the developed models were collected from the actual EAF in the SIJ Acroni company. From the collected database, 577 different batches were selected with the filtering. In the stage of predictive models development, the whole dataset was divided into training (404 batches) and testing (173 batches) subsets (70% of the data for training and 30% for testing). For each batch, 13 input variables were recorded, which are listed in Table 1. For each batch, the loading recipe (marked from 1 to 12) and melting program (marked from 1 to 15) are also selected according to the required properties of the steel produced.

3.1. Results of the Selection of Key Input Variables

Using the methods presented in Section 2.2 (Pcorr, PMI, LIP, NNGarr, PLS VIP, and LASSO), the most influential variables for predictive models were found. Since the results vary widely from one method to another, average influential factors were calculated to be more generally usable regardless of modelling method. Figure 1 shows the sorted results of finding the most influential variables considering all data in the database. The boxes in the figure show the average values, the median values and the intervals within the 25th and 75th percentiles.

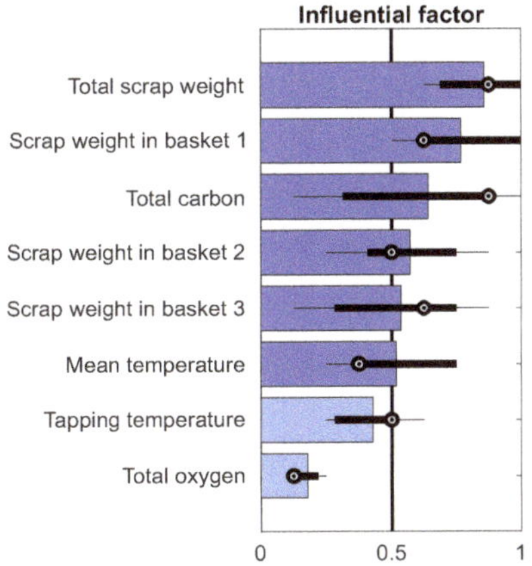

Figure 1. The average influential factors for all independent input variables.

Table 2 shows the average values for each influential factor. According to the obtained results, it is reasonable to include the following variables in further consideration: total scrap weight, scrap weight in individual baskets, total carbon, average temperature during melting, tapping temperature and total oxygen. The significance of the individual input variables can also be partially inferred from Figure 2, where linear models describe the relationships between the various input variables and the total electrical energy consumption (as a percentage of the maximum value (kWh/t)). In determining the most influential variables, the dispersion or data distribution plays a major role. Figure 2 shows one of the most influential variables and one of the least influential variables in each case. The simultaneous use of multiple independent variables to predict electrical energy consumption can change the influential factor of a single variable (due to the interconnectedness of the variables). Therefore, it is difficult to conclude from Figure 2 why total scrap weight is more important than the total carbon variable.

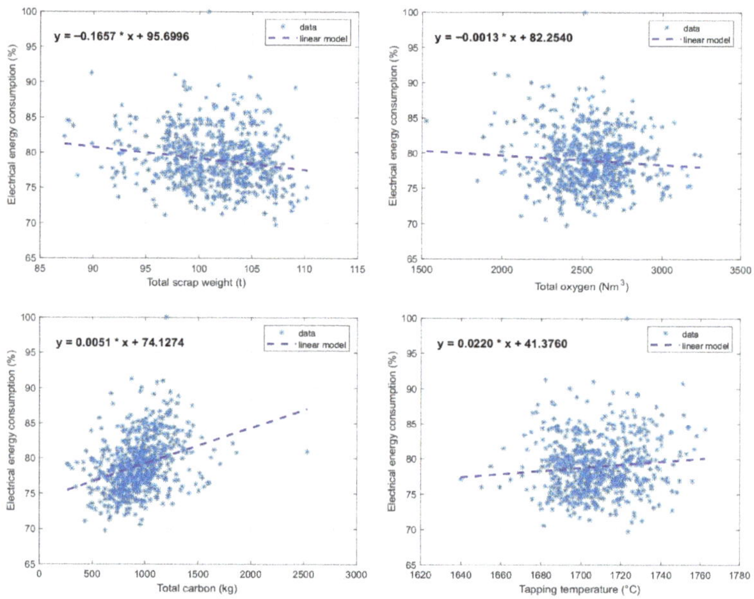

Figure 2. Linear models of electrical energy consumption (as a percentage of the maximum value) as a function of total scrap weight (**top left**), total oxygen (**top right**), total carbon (**bottom left**), and tapping temperature (**bottom right**), respectively.

Table 2. The average influential factors for all independent input variables.

Variable	Influential Factor
Total scrap weight	0.8571
Scrap weight in basket 1	0.7679
Total carbon	0.6429
Scrap weight in basket 2	0.5714
Scrap weight in basket 3	0.5357
Mean temperature	0.5179
Tapping temperature	0.4286
Total oxygen	0.1786

When modelling electrical energy consumption, reducing the dimensionality of the input space is also very important; otherwise, the (fuzzy) model structure may become too complex and the large number of model parameters may be difficult to determine.

If the modelling method also includes an optimization phase of the model parameters, the modelling process can become very slow and inefficient. On the other hand, considering only a limited number of the most influential variables can lead to worse prediction results as some of the information is lost. Therefore, different variations of combined input variables were also considered. Using the methods presented in Section 2.2, the following combined input variables (Figure 3) were selected as the most influential: the quotient of tapping temperature and total scrap weight, the quotient of mean temperature and scrap weight in the first two baskets, chemical energy (calculated from total carbon and total oxygen as proposed in [42]), the quotient of total oxygen and total carbon, scrap weight in the third basket. The average influential factors for all combined input variables are listed in Table 3. Figure 4 shows that the use of only one combined input variable does not drastically improve the prediction of electrical energy consumption, but as mentioned earlier, the main advantage of selecting the most influential variables is shown only when all input variables are used together in the exact combination.

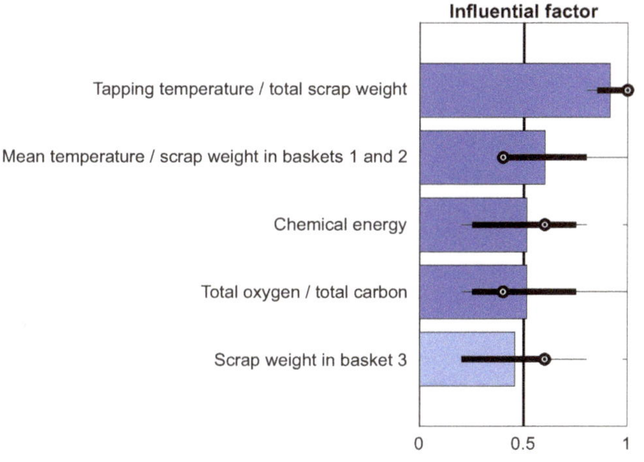

Figure 3. The average influential factors for the five most influential combined input variables.

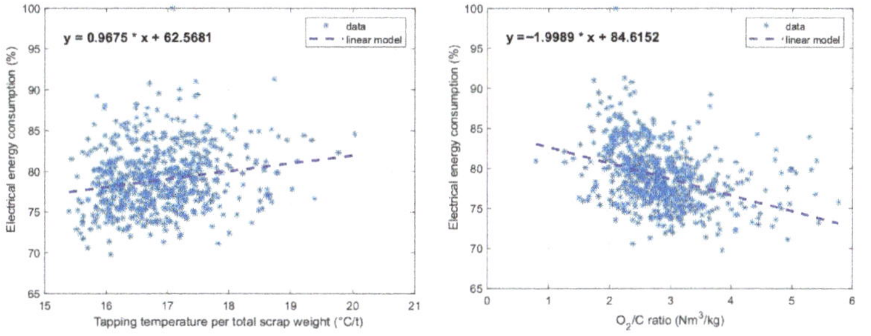

Figure 4. Linear models of electrical energy consumption as a function of the quotient of tapping temperature and total scrap weight (**left**) and the quotient of total oxygen and total carbon (**right**).

Table 3. The average influential factors for the five most influential combined input variables.

Variable	Influential Factor
Tapping temperature/total scrap weight	0.9143
Mean temperature/scrap weight in baskets 1 and 2	0.6000
Chemical energy	0.5143
Total oxygen/total carbon	0.5143
Scrap weight in baskets 3	0.4571

3.2. Analysis of Models for Energy Consumption Prediction

This subsection presents the comparative results of predicting electrical energy consumption with the static models explained in Section 2. Each model is used to predict the total electrical energy consumption of the current batch as a function of the key input variables listed in Table 3. All models are compared using the root–mean–square error (RMSE), which is a measure of the differences between the values (electrical energy consumption in percentages) predicted by a model $\hat{y}_i$ and the observed values y_i:

$$\mathrm{RMSE} = \sqrt{\frac{1}{m}\sum_{i=1}^{m}(\hat{y}_i - y_i)^2}, \tag{19}$$

where m is the number of all test batches.

Figure 5 shows the results of predicting electrical energy consumption with the k-NN model (left) and the linear regression model (right) compared to the electrical energy consumption measurements. In the figure, the line shows the ideal (completely accurate) prediction of electrical energy consumption according to the test samples. The k-NN model was constructed to consider Mahalanobis distance and the six nearest neighbours. The output of the k-NN model is calculated according to Equation (9), which means that the nearer neighbour has more influence on the output than the farther neighbour. Compared to the prediction results of the k-NN model, the linear regression model achieves slightly better results (see Table 4) in terms of R^2 (coefficient of determination) and RMSE, although this model is simpler.

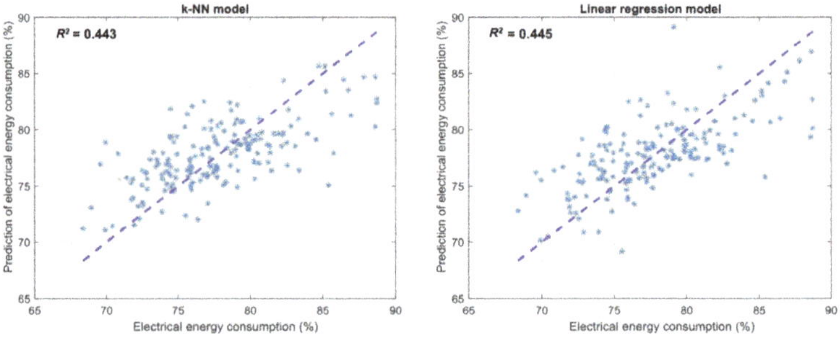

Figure 5. Prediction of electrical energy consumption with the k-NN model (**left**) and the linear regression model (**right**) compared to measurements of electrical energy consumption.

Artificial intelligence algorithms, i.e., evolving and fuzzy modelling approaches proposed in this work, achieve better prediction results than machine learning methods (k-NN and linear regression), as expected. Figure 6 shows the results of predicting electrical energy consumption with the evolving model (left) and the fuzzy model (right) compared to the electrical energy consumption measurements. When looking at Figures 5 and 6, it is difficult to decide which model is the best because the differences are quite small. Therefore, all RMSE and R^2 results for each method are presented in Table 4. From the table, it can be concluded that the best results were obtained with the conventional fuzzy method and the evolving method proposed in this paper. In the conventional fuzzy modelling, the PSO optimization method was used to determine the optimal structure (number, distribution and width of Gaussian membership functions) of the fuzzy logic system that gives the best prediction results.

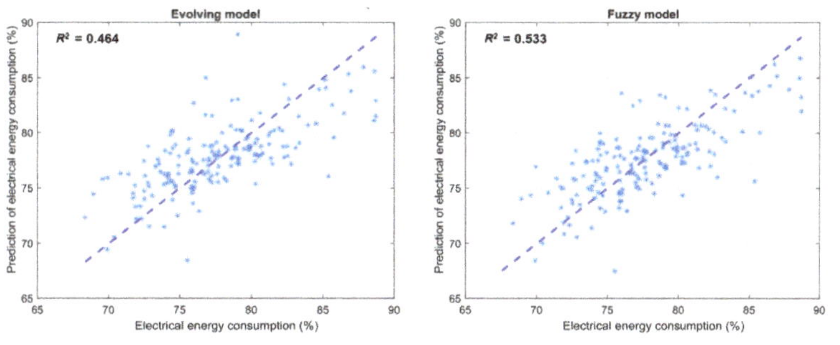

Figure 6. Prediction of electrical energy consumption with the evolving model (**left**) and fuzzy model (**right**) compared to measurements of electrical energy consumption.

Table 4. Comparison of RMSE results for the prediction models.

Method	RMSE (%)	R^2
k-NN method	3.177	0.443
Linear regression	3.171	0.445
Evolving model	3.118	0.464
Fuzzy model	2.910	0.533

All the developed models can also be compared with the calculation of the cumulative distribution functions, which are shown in Figure 7. From this graph, for example, it is easy to see that 90% of all errors are less than 5% (of the maximum electrical energy consumption) when the k-NN model is used. Thus, a steeper curve represents a better model.

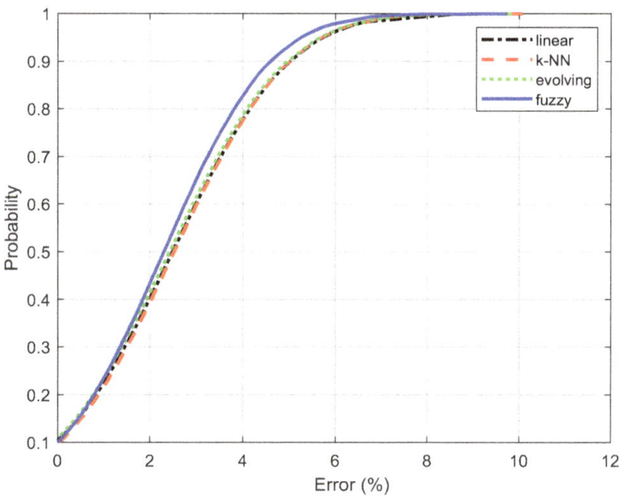

Figure 7. Cumulative distribution functions for the electrical energy prediction errors.

The comparison between the results considering all input variables and only the most influential variables (selected variables) shows that reducing the independent variables can improve the fitting results by at least 20% according to the RMSE of linear regression. The effect of reducing the input space is even more evident when evolving or a fuzzy modelling approach is used, since in these cases model complexity translates into more challenging optimization conditions due to the large number of input variables. The larger number of optimization parameters slows down the training process and may lead to suboptimal results. Fuzzy membership functions may not be optimally defined and distributed, and consequently the model may be over-fitted to the training dataset. However, over-fitted models fail quickly when applied to new batches that differ slightly from those in the past.

The prediction results of all developed models can be drastically improved (by at least 20% according to the RMSE of linear regression) if melting time is also used as an independent input variable. Although up to 80% of authors of all published papers dealing with the EAF energy consumption prediction have used melting time as an input variable to achieve better results, this approach is completely incorrect as the melting time is not known in advance. If melting time was known in advance, advanced models would actually be unnecessary because the melting time is almost entirely proportional to electrical energy consumption (see Figure 8). In Figure 8, two different linear models are shown according to the maximum transformer tap level (in the profile), which is either seven or eight for all melting programmes. The linear models show that the melting programmes with the maximum transformer tap level eight have a slightly higher energy consumption than the melting programmes with the maximum transformer tap level seven, but the slope is almost the same in both cases. The obtained models, shown in Figure 8, are used to predict the melting time from the electrical energy consumption prediction (obtained with the fuzzy model). This information is essential for the EAF operator as he can try different scenarios in the simulator and determine the optimal time to complete the batch. This is one of the possible ways to partially reduce the electrical energy consumption without intervening in the EAF itself because until now, in most cases, electrical energy consumption was only too high due to an unnecessary prolongation of the melting time. This is because it is difficult for the operator to determine exactly when the material is completely melted.

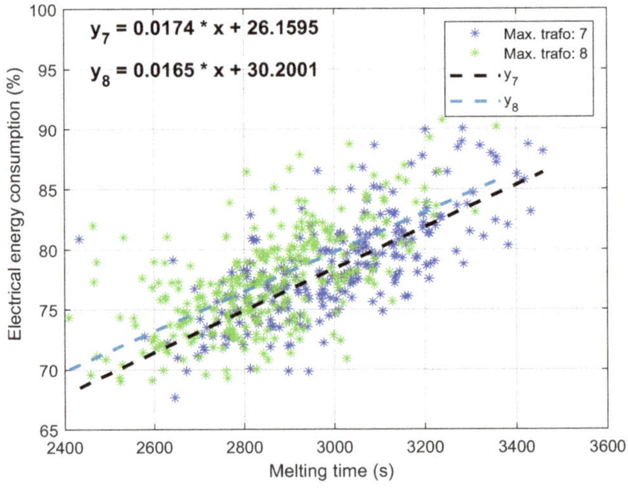

Figure 8. Electrical energy consumption as a function of melting time for two different maximum transformer tap levels in the profile.

4. Discussion

Technological processes in the steel industry have improved greatly in recent decades. Further optimization of the processes is possible by introducing digital tools that advise operators on setting parameters and help control production (also in terms of equipment maintenance). In this study, the focus is on the optimization of electrical energy consumption through the analysis of existing historical data and the construction of prediction models. The latter allows the operator to perform preliminary simulations through an advisory tool that determines the electrical energy consumption according to the selected conditions.

The operator can thus test the optimal values for the materials added, the amount of carbon and oxygen added, the melting temperature and, above all, the final melting time. From all of the influential variables, total melting time is the one that total energy consumption depends the most on it but should not be considered as an input variable, which is a common mistake. Although the transformer profiles (in historical data) that define the EAF electrical parameters (current, voltage, arc power) have two different final values for the power levels in the existing melting programs (7 and 8), these values have an almost negligible impact on the final consumption compared to the final melting time (which is also defined with the transformer profiles). The developed models can predict the electrical energy consumption quite accurately since the error is less than 5% (of the maximum energy consumption) for 90% of all errors. Converting the electrical energy consumption to the final melting time is also very straightforward since consumption and melting time are proportional to each other. The choice of input variables is critical to developing applicable models, especially when a large number of variables are available. Without algorithms to analyze influential factors, the types of charged materials would certainly be chosen as input variables, as well as the amount of slag or delays during the process. As the results show, the total mass has the greatest influence on the prediction of energy consumption, although the consumption is normalized with respect to the total mass (in kWh/t). When determining the key variables, their simultaneous consideration is crucial because the influential factors are distributed differently than when only one input variable is considered at a time. For successful model construction, it is also critical to eliminate bad measurements (outlier filtering) that occur in batches with many interruptions and extended melting time due to faults at the EAF. Poor measurements are also possible due to incorrectly recorded charged materials (quantities and types), but not all such anomalies in the measurements can be detected. All the developed models are comparable with each other in terms of the prediction error (RMSE) and the coefficient of determination R^2, which means that, if the combined variables are chosen appropriately, the linear methods also work effectively. Each of the methods has its advantages and disadvantages. For example, the evolving method, although it does not give the best results, may be best suited for online updating of models during the process itself, which may improve the prediction for the current batch. The conventional fuzzy method is computationally the most demanding because it involves a PSO optimization, but it provides the best prediction results. On the other hand, linear regression is the simplest since it does not require parameter adjustments, while the k-NN method is the fastest since it does not require a training phase.

5. Conclusions

This paper presents the results of a study in which preprocessed historical data from the real EAF process were used to identify the influential variables that have the greatest impact on electrical energy consumption during melting. The results show that the root mean square error in predicting electrical energy consumption can be reduced by at least 20% with proper selection of the influential variables. Four different prediction models were constructed from the filtered data, using linear regression, k-NN, evolving, and fuzzy modelling methods. When comparing the errors in the prediction of electrical energy consumption, the fuzzy model was found to be the most accurate, as the root mean square error has the lowest value and the coefficient of determination has the highest value. The developed models will be used within the advisory tool, which will help the EAF operator to adjust the parameters correctly during the melting process and, in this way, improve the efficiency of the EAF.

Author Contributions: Conceptualization, S.T., I.Š. and V.L.; Data curation, V.L.; Formal analysis, S.T. and G.A.; Methodology, S.T., G.A., I.Š. and V.L.; Project administration, I.Š.; Software, S.T., G.A. and V.L.; Visualization, S.T. and V.L.; Writing—original draft, S.T., G.A., I.Š. and V.L.; Writing—review & editing, V.L. All authors have read and agreed to the published version of the manuscript.

Funding: The work presented in this paper is funded by the European Union's Horizon 2020 research and innovation programme, the SPIRE initiative, under Grant No. 869815, the INEVITABLE project ("Optimization and performance improving in metal industry by digital technologies") and by the Slovenian Research Agency Programme: Modelling, simulation and control of processes (P2-0219).

Institutional Review Board Statement: Not applicable.

Informed Consent Statement: Not applicable.

Data Availability Statement: Not applicable.

Conflicts of Interest: The authors declare no conflict of interest.

References

1. Toulouevski, Y.N.; Zinurov, I.Y. Modern Steelmaking in Electric Arc Furnaces: History and Development. In *Innovation in Electric Arc Furnaces: Scientific Basis for Selection*; Springer: Berlin/Heidelberg, Germany, 2013; pp. 1–24. [CrossRef]
2. Saboohi, Y.; Fathi, A.; Škrjanc, I.; Logar, V. Optimization of the Electric Arc Furnace Process. *IEEE Trans. Ind. Electron.* **2019**, *66*, 8030–8039. [CrossRef]
3. Carlsson, L.S.; Samuelsson, P.B.; Jönsson, P.G. Predicting the Electrical Energy Consumption of Electric Arc Furnaces Using Statistical Modeling. *Metals* **2019**, *9*, 959. [CrossRef]
4. Kovačič, M.; Stopar, K.; Vertnik, R.; Šarler, B. Comprehensive Electric Arc Furnace Electric Energy Consumption Modeling: A Pilot Study. *Energies* **2019**, *12*, 2142. [CrossRef]
5. Sung, Y.; Lee, S.; Han, K.; Koo, J.; Lee, S.; Jang, D.; Oh, C.; Jang, B. Improvement of Energy Efficiency and Productivity in an Electric Arc Furnace through the Modification of Side-Wall Injector Systems. *Processes* **2020**, *8*, 1202. [CrossRef]
6. Echterhof, T. Review on the Use of Alternative Carbon Sources in EAF Steelmaking. *Metals* **2021**, *11*, 222. [CrossRef]
7. Ahmed, W.; Moazzam, M.; Sarkar, B.; Ur Rehman, S. Synergic Effect of Reworking for Imperfect Quality Items with the Integration of Multi-Period Delay-in-Payment and Partial Backordering in Global Supply Chains. *Engineering* **2021**, *7*, 260–271. [CrossRef]
8. Mahapatra, A.S.; N Soni, H.; Mahapatra, M.S.; Sarkar, B.; Majumder, S. A Continuous Review Production-Inventory System with a Variable Preparation Time in a Fuzzy Random Environment. *Mathematics* **2021**, *9*, 747. [CrossRef]
9. Bhuniya, S.; Pareek, S.; Sarkar, B. A supply chain model with service level constraints and strategies under uncertainty. *Alex. Eng. J.* **2021**, *60*, 6035–6052. [CrossRef]
10. Sarkar, B.; Mridha, B.; Pareek, S. A sustainable smart multi-type biofuel manufacturing with the optimum energy utilization under flexible production. *J. Clean. Prod.* **2022**, *332*, 129869. [CrossRef]
11. Yadav, D.; Kumari, R.; Kumar, N.; Sarkar, B. Reduction of waste and carbon emission through the selection of items with cross-price elasticity of demand to form a sustainable supply chain with preservation technology. *J. Clean. Prod.* **2021**, *297*, 126298. [CrossRef]
12. Carlsson, L.S.; Samuelsson, P.B.; Jönsson, P.G. Using Statistical Modeling to Predict the Electrical Energy Consumption of an Electric Arc Furnace Producing Stainless Steel. *Metals* **2020**, *10*, 36. [CrossRef]
13. Logar, V.; Fathi, A.; Škrjanc, I. A Computational Model for Heat Transfer Coefficient Estimation in Electric Arc Furnace. *Steel Res. Int.* **2016**, *87*, 330–338. [CrossRef]
14. Meier, T.; Logar, V.; Echterhof, T.; Škrjanc, I.; Pfeifer, H. Modelling and Simulation of the Melting Process in Electric Arc Furnaces—Influence of Numerical Solution Methods. *Steel Res. Int.* **2016**, *87*, 581–588. [CrossRef]
15. Núñez, A.; De Schutter, B.; Sáez, D.; Škrjanc, I. Hybrid-fuzzy modeling and identification. *Appl. Soft Comput.* **2014**, *17*, 67–78. [CrossRef]
16. Dovžan, D.; Logar, V.; Škrjanc, I. Implementation of an Evolving Fuzzy Model (eFuMo) in a Monitoring System for a Waste-Water Treatment Process. *IEEE Trans. Fuzzy Syst.* **2015**, *23*, 1761–1776. [CrossRef]
17. Škrjanc, I.; Iglesias, J.A.; Sanchis, A.; Leite, D.; Lughofer, E.; Gomide, F. Evolving fuzzy and neuro-fuzzy approaches in clustering, regression, identification, and classification: A Survey. *Inf. Sci.* **2019**, *490*, 344–368. [CrossRef]
18. Fathi, A.; Saboohi, Y.; Škrjanc, I.; Logar, V. Comprehensive Electric Arc Furnace Model for Simulation Purposes and Model-Based Control. *Steel Res. Int.* **2017**, *88*, 1600083. [CrossRef]
19. Hay, T.; Visuri, V.V.; Aula, M.; Echterhof, T. A Review of Mathematical Process Models for the Electric Arc Furnace Process. *Steel Res. Int.* **2021**, *92*, 2000395. [CrossRef]
20. Lee, B.; Sohn, I. Review of Innovative Energy Savings Technology for the Electric Arc Furnace. *JOM* **2014**, *66*, 1581–1594. [CrossRef]
21. Barati, M.; Esfahani, S.; Utigard, T. Energy recovery from high temperature slags. *Energy* **2011**, *36*, 5440–5449. [CrossRef]
22. Lee, B.; Ryu, J.W.; Sohn, I. Effect of Hot Metal Utilization on the Steelmaking Process Parameters in the Electric Arc Furnace. *Steel Res. Int.* **2015**, *86*, 302–309. [CrossRef]
23. Kirschen, M.; Risonarta, V.; Pfeifer, H. Energy efficiency and the influence of gas burners to the energy related carbon dioxide emissions of electric arc furnaces in steel industry. *Energy* **2009**, *34*, 1065–1072. [CrossRef]
24. Bisio, G.; Rubatto, G.; Martini, R. Heat transfer, energy saving and pollution control in UHP electric-arc furnaces. *Energy* **2000**, *25*, 1047–1066. [CrossRef]

25. Meier, T.; Hay, T.; Echterhof, T.; Pfeifer, H.; Rekersdrees, T.; Schlinge, L.; Elsabagh, S.; Schliephake, H. Process Modeling and Simulation of Biochar Usage in an Electric Arc Furnace as a Substitute for Fossil Coal. *Steel Res. Int.* **2017**, *88*, 1600458. [CrossRef]
26. Gandt, K.; Meier, T.; Echterhof, T.; Pfeifer, H. Heat recovery from EAF off-gas for steam generation: Analytical exergy study of a sample EAF batch. *Ironmak. Steelmak.* **2016**, *43*, 581–587. [CrossRef]
27. Glavan, M.; Gradišar, D.; Atanasijević-Kunc, M.; Strmčnik, S.; Mušič, G. Input variable selection for model-based production control and optimisation. *Int. J. Adv. Manuf. Technol.* **2013**, *68*, 2743–2759. [CrossRef]
28. Van De Wal, M.; De Jager, B. Review of methods for input/output selection. *Automatica* **2001**, *37*, 487–510. [CrossRef]
29. May, R.; Dandy, G.; Maier, H. Review of Input Variable Selection Methods for Artificial Neural Networks. In *Artificial Neural Networks-Methodological Advances and Biomedical Applications*; BoD—Books on Demand: Norderstedt, Germany, 2011.
30. Li, K.; Peng, J.X. Neural input selection-A fast model-based approach. *Neurocomputing* **2007**, *70*, 762–769. [CrossRef]
31. Breiman, L. Better subset regression using the nonnegative garrote. *Technometrics* **1995**, *37*, 373–384. [CrossRef]
32. Chong, I.G.; Jun, C.H. Performance of some variable selection methods when multicollinearity is present. *Chemom. Intell. Lab. Syst.* **2005**, *78*, 103–112. [CrossRef]
33. Székely, G.J.; Rizzo, M.L.; Bakirov, N.K. Measuring and testing dependence by correlation of distances. *Ann. Stat.* **2007**, *35*, 2769–2794. [CrossRef]
34. Tibshirani, R. Regression Shrinkage and Selection via the Lasso. *J. R. Stat. Soc.* **1996**, *58*, 267–288. [CrossRef]
35. Freedman, D. *Statistical Models: Theory and Practice*; Cambridge University Press: Cambridge, UK, 2009.
36. Montgomery, D.C.; Peck, E.A.; Vining, G.G. *Introduction to Linear Regression Analysis*; John Wiley & Sons: Hoboken, NJ, USA, 2012.
37. Friedman, J.H.; Bentley, J.L.; Finkel, R.A. An Algorithm for Finding Best Matches in Logarithmic Expected Time. *ACM Trans. Math. Softw.* **1977**, *3*, 209–226. [CrossRef]
38. Chen, G.H.; Shah, D. Explaining the Success of Nearest Neighbor Methods in Prediction. *Found. Trends Mach. Learn.* **2018**, *10*, 1–250. [CrossRef]
39. Takagi, T.; Sugeno, M. Fuzzy identification of systems and its applications to modeling and control. *IEEE Trans. Syst. Man Cybern.* **1985**, *SMC-15*, 116–132. [CrossRef]
40. Kennedy, J.; Eberhart, R. Particle swarm optimization. In Proceedings of the ICNN'95-International Conference on Neural Networks, Perth, WA, Australia, 27 November–1 December 1995; Volume 4, pp. 1942–1948. [CrossRef]
41. Andonovski, G.; Mušič, G.; Blažič, S.; Škrjanc, I. On-line Evolving Cloud-based Model Identification for Production Control. *IFAC-PapersOnLine* **2016**, *49*, 79–84. [CrossRef]
42. Blažič, A.; Škrjanc, I.; Logar, V. Soft sensor of bath temperature in an electric arc furnace based on a data-driven Takagi–Sugeno fuzzy model. *Appl. Soft Comput.* **2021**, *113*, 107949. [CrossRef]

metals

MDPI

Article

Thermophysical Model for Online Optimization and Control of the Electric Arc Furnace

Sudi Jawahery [1,*], Ville-Valtteri Visuri [2,3], Stein O. Wasbø [1], Andreas Hammervold [1], Niko Hyttinen [2] and Martin Schlautmann [4]

1 Cybernetica AS, Leirfossvegen 27, 7038 Trondheim, Norway; stein.wasbo@cybernetica.no (S.O.W.); andreas.hammervold@cybernetica.no (A.H.)
2 Research and Development, Outokumpu Stainless Oy, Terästie, 95490 Tornio, Finland; ville-valtteri.visuri@outokumpu.com (V.-V.V.); niko.hyttinen@outokumpu.com (N.H.)
3 Process Metallurgy Research Unit, University of Oulu, 90014 Oulu, Finland
4 VDEh-Betriebsforschungsinstitut GmbH, Sohnstraße 69, 40237 Düsseldorf, Germany; Martin.Schlautmann@BFI.de
* Correspondence: sudi.jawahery@cybernetica.no

Abstract: A dynamic, first-principles process model for a steelmaking electric arc furnace has been developed. The model is an integrated part of an application designed for optimization during operation of the furnace. Special care has been taken to ensure that the non-linear model is robust and accurate enough for real-time optimization. The model is formulated in terms of state variables and ordinary differential equations and is adapted to process data using recursive parameter estimation. Compared to other models available in the literature, a focus of this model is to integrate auxiliary process data in order to best predict energy efficiency and heat transfer limitations in the furnace. Model predictions are in reasonable agreement with steel temperature and weight measurements. Simulations indicate that industrial deployment of Model Predictive Control applications derived from this process model can result in electrical energy consumption savings of 1–2%.

Keywords: electric arc furnace; mathematical modeling; model predictive control

Citation: Jawahery, S.; Visuri, V.-V.; Wasbø, S.O.; Hammervold, A.; Hyttinen, N.; Schlautmann, M. Thermophysical Model for Online Optimization and Control of the Electric Arc Furnace. *Metals* **2021**, *11*, 1587. https://doi.org/10.3390/met11101587

Academic Editor: Geoffrey Brooks

Received: 20 August 2021
Accepted: 27 September 2021
Published: 5 October 2021

1. Introduction

Electrical arc furnaces (EAF) perform a primary steelmaking process that converts recycled steel scrap into liquid steel, which can be refined further in downstream processes. The EAF is a refractory-lined vessel that is filled with steel scrap at the start of each new heat. Through holes in the vessel roof, graphite electrodes (a single electrode in DC furnaces and and three electrodes in AC furnaces) are lowered and used to conduct a high-voltage electric arc that supplies electrical energy to melt the scrap metal. Gas burners are mounted along the outer vessel sidewalls. During the course of a heat, the burners can operate in two different modes: (1) by providing pure oxygen for refining, or (2) by providing a mixture of oxygen and either liquefied natural gas (LNG) or propane to burn for extra heating. The burners are typically operated in fuel combustion mode during the early process stages, while refining takes place towards the end of the heat. The use of gas burners has been shown to decrease batch time and reduce electrical power consumption. To protect the vessel and furnace equipment from sustaining damage due to radiation from the electric arc and heated metal, cooling water heat exchange panels are mounted along the upper parts of the vessel's sidewalls and the roof [1].

A heat is typically run as either a one-, two- or three-basket heat. This means the vessel is charged with scrap metal one, two or three times during a heat, with the first basket always being charged before the electric arc and gas burners are turned on. The baskets can vary significantly, both in size and in the type of scrap being charged. Carbon and additional slag-forming materials are also added to the furnace in order to achieve the desired slag-phase composition and foaming. At the end of each heat, the slag phase and liquid steel are tapped separately from the furnace, and a new heat is ready to begin [1].

Numerous mathematical models have been proposed for predicting the course of the EAF process. Recently, Hay et al. [2] presented a comprehensive review of mathematical models proposed to date. They concluded that while there are still several development areas, modern models can predict the main dynamic changes in distribution of species and energy with reasonable accuracy. Furthermore, it was suggested that fundamental models are now sufficiently fast to be used for model predictive control (MPC).

Some relevant studies [3–7] on the application of MPC for the EAF process are summarized in the following sentences. The model by Bekker et al. [3] is intended for controlling the offgas system and manipulates two variables (fan force and slip-gap) to adjust three outputs: the relative furnace pressure, offgas temperature and offgas CO mass fraction. Of these three variables, the relative furnace pressure was regulated, while the offgas temperature and offgas CO mass fraction were only limited. Extending the model by Bekker et al. [3], Oosthuizen et al. [4] presented a slag foaming model and introduced the rate of direct reduced iron (DRI) addition as an additional manipulated input variable. Later, Oosthuizen et al. [5] proposed a related MPC algorithm based on economic objectives. The MPC proposed by MacRosty and Swartz is formulated in terms of an economic performance objective. More specifically, the model adjusts the arc power, oxygen flow from the burner, natural gas flow from the burner, oxygen injection, carbon injection, and mass of the second charge to minimize the total costs of the EAF process. Shyamal [7] proposed a shrinking horizon MPC algorithm, which was coupled with multi-rate moving horizon estimation (MHE) for real-time model calibration. The model was directed at real-time energy management and employed time-varying electricity prices for decision making. Shyamal [7] also proposed a real-time dynamic advisory system, which was based on multi-tiered optimization of the estimated states from MHE. It is worth noting that the MPC algorithm employed by Oosthuizen et al. [4,5] is linear, while those employed by Bekker et al. [3], MacRosty and Swartz [6] and Shyamal [7] are non-linear.

A model comprising monitoring and prediction of thermal and metallurgical heat state evolution in the EAF has been developed by BFI [8–10]. This dynamic EAF process model uses event driven and cyclically measured process data to calculate the temperature, weights and analyses of the steel and slag phases in the furnace. The model considers these phases without spatial resolution and uses ordinary differential equations in time and algebraic equations to describe the process state. The same model kernel can be used to monitor the current heat state from actual process data and to predict its further evolution based on related practice data for the remaining treatment steps. In order to monitor the thermal process state, the BFI model calculates the current energy content of the melt based on a cyclically evaluated overall energy balance. The energy into the balance is the sum of the electrical energy supplied and the chemical energy released by reactions. The energy leaving the balance takes into account the losses to cooling water, offgas, radiation and convection. The bath temperature is obtained from the difference of the current energy content and the energy requirement for meltdown, which is in turn calculated from the reference enthalpies (i.e., specific enthalpies at reference temperature) of the charged materials (scrap and slag formers), where the hot heel is also taken into account.

The monitoring of the metallurgical process state in the BFI model comprises the cyclic calculation of the weight and the composition for the metal bath and the slag phase. For this purpose, the input by the charged materials as well the effects of the different oxidation and reduction reactions (decarburization, dephosphorization and slagging/reduction of metallic elements) are considered. The latter are based on appropriate first-order differential equations where the reaction rate of an element or oxide is given by its content in steel or slag multiplied by the oxygen or reduction agent input rate and an adapted oxidation or reduction efficiency, respectively.

The aim of this work was to formulate a new dynamic model of an EAF that could be used to optimize the electric power profile and electric arc operation. For that purpose, it goes a step beyond the mentioned state-of-the art approaches and uses a more detailed modelling of different control volumes with liquid and solid phases. The model in this work is non-linear, allowing for the representation of complex interrelated phenomena, including estimation of the visibility of the electric arc and arc efficiency for melting and heating. The MPC application of the proposed model uses a finite receding horizon, where the duration of the heat is not specified, but the predictions horizon becomes shorter as the process nears the completion criteria. To validate the viability of the new approach, the model and related MPC were tested for an industrial scale EAF in stainless steel-making. While Visuri et al. [11] presented some preliminary results, extensive results along with accompanying interpretation and explanation are discussed in this article.

2. Materials and Methods

The process model developed for the EAF is based on physical modeling principles such as mass and energy balances. When creating a process model, it is essential that the model can provide the necessary information for solving the problem at hand without becoming computationally cumbersome. In this case, the model is designed to be used as a basis for real-time optimization. Hence, the focus of the modeling efforts is to ensure that the model is fast enough to be recalculated multiple times in each time sample. The model is developed as a continuous-time model that is integrated over selected time steps using the forward Euler method for numerical integration. The numerical smoothness of the model has been emphasized for two reasons: (1) so that computationally fast explicit integration schemes can be applied without losing accuracy, and (2) so that optimization problems formulated with model output can be more easily designed to be convex [12]. Further, a Kalman filter (KF) has been designed to ensure that the process model follows the efficiency of the real process. Measurement outputs are calculated by the model and compared with process measurements. The residuals between the model predictions and real measurements are fed into the KF, which updates state variables and selected parameters for estimation [13].

2.1. Control Volumes and State-Space Variables

Figure 1 shows the process state variables that are included in the model. Energy supplied by three electrodes is used to directly heat the contents of the inner solid and liquid masses. As a result, the inner control volume has the highest temperatures in the model and is therefore colored in red, with bordering masses (the outer control volume, gas) colored in orange gradients. The temperatures, total masses and masses of individual components in the solid and liquid phases of the inner and outer steel control volumes are modeled. The temperature, total masses and individual component masses of the solid and liquid slag phases are also modeled. The component masses are enumerated in Table 1. The environment is modeled by including dynamic states for the temperatures of the roof, side panels and gas that fills the space not occupied by steel in the furnace. The roof and side panels are both in contact with cooling water streams for which the temperature measurement is recorded. Calculation of the cooling water temperature variation as predicted by the model allows for real-time comparison to process measurements. The temperature of the process offgas is recorded downstream in a duct that extracts fumes from the furnace. The offgas temperature in the duct is modeled accordingly and also compared to real-time data.

In order to maintain the model's focus on the energy balance, the modeled slag masses exchange heat and mass only with the steel and not with the furnace environment. This assumption reduces the model complexity and allows the parameter estimation discussed in Section 2.5 to more directly impact the states of interest, namely the solid and liquid steel. Heat transfer between steel and the slag masses is then tuned to indirectly account for the interactions of slag with the environment. Slag properties are taken from Jiao et al. [14], and properties for the furnace materials are taken from Fruehan [1].

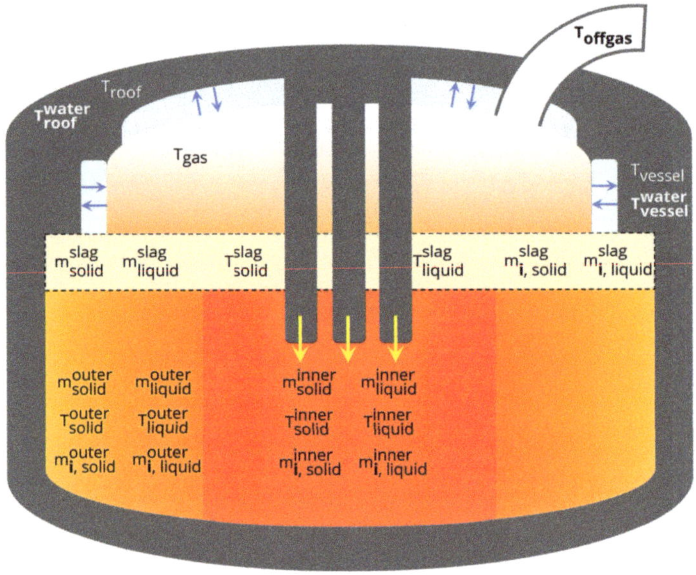

Figure 1. State variables in the EAF process model. T^{water}_{roof}, T^{water}_{vessel} and T_{offgas} can be compared to real-time process data.

Table 1. Modeled components in the steel and slag phases.

Dissolved Component	Phase(s)	Reactive with O_2 in Model?	Equilibrium Reaction in Model?
Fe	Liquid, Solid	Yes	Yes
C	Liquid, Solid	Yes	Yes
Cr	Liquid, Solid	Yes	Yes
Si	Liquid, Solid	Yes	Yes
Al	Liquid, Solid	Yes	Yes
Mn	Liquid, Solid	No	Yes
FeO	Slag	No	Yes
SiO_2	Slag	No	Yes
Cr_2O_3	Slag	No	Yes
Al_2O_3	Slag	No	Yes
MnO	Slag	No	Yes

Because the focus of this model is to achieve a dynamic energy balance rather than a detailed mass balance, not all components recorded in process data are included in the model. The components listed in Table 1 represent the components whose non-oxide element represent more than 2% of charged mass and whose reactions have the potential to significantly affect the energy balance.

2.2. Chemical Reactions

Gas burners supply flows of LNG and oxygen that react to release energy:

$$\{CH_4\} + 2\{O_2\} \rightarrow 2\{H_2O\} + \{CO_2\}. \quad (1)$$

Oxygen that is not fully consumed by the reaction with LNG, for example during the refining phase, can react with CO gas and components in the liquid steel to form liquid slag components and gases:

$$\begin{aligned}
\{CO\} + \frac{1}{2}\{O_2\} &\rightarrow \{CO_2\} \\
Fe + \frac{1}{2}\{O_2\} &\rightarrow (FeO) \\
[C] + \frac{1}{2}\{O_2\} &\rightarrow \{CO\} \\
[C] + \{O_2\} &\rightarrow \{CO_2\} \\
[Si] + \{O_2\} &\rightarrow (SiO_2) \\
2[Cr] + \frac{3}{2}\{O_2\} &\rightarrow (Cr_2O_3) \\
2[Al] + \frac{3}{2}\{O_2\} &\rightarrow (Al_2O_3).
\end{aligned} \tag{2}$$

The oxygen for the reactions in Equation (Equation (2)) is divided between the CO gas and the liquid steel components. Oxygen is allocated to the inner and outer liquid steel proportionally based on their masses. Within each control volume, the oxygen consumption in Equation (Equation (2)) is proportional to the mass fractions of Fe, C, Si, Cr and Al.

The model's explicit treatment of oxygen consumption by reactions in (Equation (2)) does not account for the activity coefficients of the different components. However, thermodynamic equilibrium is enforced by the inclusion of reversible reactions between the liquid steel and slag phases. Including these equilibrium reactions in the model achieves component mass fraction ratios that are consistent with the equilibrium constants given in Appendix A.2, which have been adapted from Turkdogan [15]. The following steel–slag equilibrium reactions take place, where both the forward and backward reactions are modeled:

$$\begin{aligned}
(FeO) + [C] &\leftrightarrow Fe + \{CO\} \\
(FeO) + [Mn] &\leftrightarrow Fe + (MnO) \\
(MnO) + [C] &\leftrightarrow [Mn] + \{CO\} \\
2(FeO) + [Si] &\leftrightarrow 2Fe + (SiO_2) \\
2(MnO) + [Si] &\leftrightarrow 2[Mn] + (SiO_2) \\
3(FeO) + 2[Cr] &\leftrightarrow 3Fe + (Cr_2O_3) \\
3(SiO_2) + 4[Al] &\leftrightarrow 3[Si] + 2(Al_2O_3).
\end{aligned} \tag{3}$$

Equations (1)–(3) form a pared down version of the reactions modeled by Logar et al. [16], with the exception of the reactions involving Al and Al_2O_3.

The energy released and consumed by the reactions in Equations (1)–(3) is distributed between different masses in the furnace model. Energy released by Equation (1) is split between the steel and the gas, where the efficiency of LNG burning to heat steel changes during the process as described by Logar et al. [17]. All of the energy released by Equations (2) and (3) heats the steel. The steel-heating reaction energy is first divided between the inner and outer steel proportionally based on their volumes, as defined by the dimensions given in Appendix A.4. The model distributes the total reaction energy to the entire furnace contents because distributing energy based on the control volume of the reactants produces physically unreasonable results. For example, using all of the reaction energy to heat the inner steel when only the inner steel contains liquid phase reactants leads to excessively high temperatures and unreasonable temperature gradients. Within each control volume, the reaction energy is divided between the solid and liquid phases proportionally based on mass.

2.3. Heat Transfer

Heat transfer in the model is simplified such that all energy supplied to and lost by the furnace contents is exchanged exclusively with the steel masses. The heating of other furnace contents, namely slag, is then accounted for indirectly by tuning the heat transfer between the steel and slag masses. Heat transfer between some masses are modeled linearly, while others are accounted for only by radiation.

Convection and conduction between select masses in the furnace are modeled as linear heat transfer:

$$Q_{ij} = k_{ij} A_{ij} \left(T_i - T_j\right). \tag{4}$$

In Equation (4), the amount of heat flowing from mass i to mass j is proportional to the temperature difference $\left(T_i - T_j\right)$, heat transfer coefficient k_{ij} and area for heat transfer A_{ij}. A comprehensive list of all masses involved in linear heat transfer along with the corresponding coefficients and areas is available in Table 2.

Table 2. Prefactors for linear heat transfer between different masses in the furnace. The subscript letters (s, l, c, b, r, v, g) refer to (solid steel, liquid steel, solid slag, liquid slag, roof, vessel, gas), respectively. An empty table entry indicates that linear heat transfer between the two masses is omitted from the model.

	Inner Solid	Outer Solid	Inner Liquid	Outer Liquid	Solid Slag	Liquid Slag	Roof	Vessel	Gas
Inner Solid	-	$k_{ss}\ A_{ss}^{cross}$	$k_{sl}\ A_{sl}^{inner}$	$k_{sl}\ A_{sl}^{cross}$	$k_{cs}\ A_{cs}^{inner}$	$k_{bs}\ A_{bs}^{inner}$	-	-	$k_{sg}\ A_{bs}^{inner}$
Outer Solid	$k_{ss}\ A_{ss}^{cross}$	-	$k_{sl}\ A_{ls}^{cross}$	$k_{sl}\ A_{sl}^{outer}$	$k_{cs}\ A_{cs}^{outer}$	$k_{bs}\ A_{bs}^{outer}$	-	-	$k_{sg}\ A_{bs}^{outer}$
Inner Liquid	$k_{sl}\ A_{sl}^{inner}$	$k_{sl}\ A_{ls}^{cross}$	-	$k_{ll}\ A_{ll}^{cross}$	$k_{cl}\ A_{cl}^{inner}$	$k_{bl}\ A_{bl}^{inner}$	-	-	$k_{lg}\ A_{bl}^{inner}$
Outer Liquid	$k_{sl}\ A_{sl}^{cross}$	$k_{sl}\ A_{sl}^{outer}$	$k_{ll}\ A_{ll}^{cross}$	-	$k_{cl}\ A_{cl}^{outer}$	$k_{bl}\ A_{bl}^{outer}$	-	-	$k_{lg}\ A_{bl}^{outer}$
Solid Slag	$k_{cs}\ A_{cs}^{inner}$	$k_{cs}\ A_{cs}^{outer}$	$k_{cl}\ A_{cl}^{inner}$	$k_{cl}\ A_{cl}^{outer}$	-	-	-	-	-
Liquid Slag	$k_{bs}\ A_{bs}^{inner}$	$k_{bs}\ A_{bs}^{outer}$	$k_{bl}\ A_{bl}^{inner}$	$k_{bl}\ A_{bl}^{outer}$	-	-	-	-	-
Roof	-	-	-	-	-	-	-	-	$k_{gr}\ A_r$
Vessel	-	-	-	-	-	-	-	-	$k_{gv}\ A_v$
Gas	$k_{sg}\ A_{bs}^{inner}$	$k_{sg}\ A_{bs}^{outer}$	$k_{lg}\ A_{bl}^{inner}$	$k_{lg}\ A_{bl}^{outer}$	-	-	$k_{gr}\ A_r$	$k_{gv}\ A_v$	-

To model the area for heat transfer between solid and liquid phases within each steel control volume, the scrap metal is assumed to linearly transition from being a single slab at the beginning of the process (solid mass fraction $x_{solid} \approx 1$) to small solid particles at the end of the process (solid mass fraction $x_{solid} \approx 0$) in a manner that resembles the melting phenomenon described by González et al. [18]. This transition is modeled as:

$$\begin{aligned} r_{particle}^{max} &= \left(\frac{3m_{solid}}{4\rho_{solid}\pi}\right)^{1/3} \\ r_{particle} &= \left(r_{particle}^{max} - r_{particle}^{min}\right) x_{solid} + r_{particle}^{min}. \end{aligned} \tag{5}$$

Equation (5) means that the maximum particle size in either control volume is a function of the instantaneous mass of solid m_{solid} and the solid fraction, where m_{solid} refers either to the model states m_{solid}^{outer} or m_{solid}^{inner} as depicted in Figure 1. The maximum particle radius is first calculated by assuming the whole mass of the solid to be a single spherical particle. This assumption is then corrected for by relating the actual particle radius size to the solid mass fraction. When $x_{solid} < 1$, the mass of solid is assumed to be broken up, leading to smaller particle radii. The radii of solid particles decrease until they reach the model constant $r_{particle}^{min} = 10\,\text{cm}$, at which point the particle's mass is assumed to be purely virtual. The area for solid–liquid heat transfer is then modeled as a function of both the liquid mass fraction x_{liquid}, the solid mass and $r_{particle}$:

$$A_{sl} = \frac{3x_{liquid} m_{solid}}{\rho_{solid} r_{particle}}. \tag{6}$$

The liquid fraction factor in Equation (6) accounts for the liquid coverage of the solid particles: at small x_{liquid}, the entire surface area of the solid particles may not be in contact with liquid metal. Equation (6) is applied directly to calculate $A_{\text{sl}}^{\text{inner}}$ and $A_{\text{sl}}^{\text{outer}}$. The area for heat transfer between unlike phases in different control volumes is also calculated from Equation (6), but the result is scaled by a factor of 3 to account for reduced mixing between the control volumes and the substitution of terms depends on the specific combination of phases being modeled ($A_{\text{sl}}^{\text{cross}} = \frac{x_{\text{liquid}}^{\text{outer}} m_{\text{solid}}^{\text{inner}}}{\rho_{\text{solid}} r_{\text{particle}}^{\text{inner}}}$; $A_{\text{ls}}^{\text{cross}} = \frac{x_{\text{liquid}}^{\text{inner}} m_{\text{solid}}^{\text{outer}}}{\rho_{\text{solid}} r_{\text{particle}}^{\text{outer}}}$).

To model the area for heat transfer between like phases in different control volumes, the total area separating the control volumes is first calculated and then scaled with the appropriate phase fractions:

$$\begin{aligned} h_{\text{inner}} &= \frac{\frac{m_{\text{solid}}^{\text{inner}}}{\rho_{\text{solid}}} + \frac{m_{\text{liquid}}^{\text{inner}}}{\rho_{\text{liquid}}}}{A_{\text{inner}}} \\ h_{\text{outer}} &= \frac{\frac{m_{\text{solid}}^{\text{outer}}}{\rho_{\text{solid}}} + \frac{m_{\text{liquid}}^{\text{outer}}}{\rho_{\text{liquid}}}}{A_{\text{outer}}} \\ A_{\text{cross}} &= \pi d_{\text{inner}} \frac{h_{\text{inner}} + h_{\text{outer}}}{2} \\ A_{(\text{ss}/\text{ll})}^{\text{cross}} &= x_{(\text{solid}/\text{liquid})}^{\text{inner}} x_{(\text{solid}/\text{liquid})}^{\text{outer}} A_{\text{cross}}. \end{aligned} \tag{7}$$

The inner and outer control volume areas and diameters are defined by the model dimensions given in Appendix A.4.

The areas for heat transfer between steel and solid slag are calculated based on the solid slag mass, the specific area of slag A_{s} given by Bekker et al. [19] and the appropriate phase fractions:

$$\begin{aligned} A_{\text{c}}^{(\text{inner}/\text{outer})} &= m_{\text{c}} A_{\text{s}} \frac{A_{(\text{inner}/\text{outer})}}{A_{\text{inner}} + A_{\text{outer}}} \\ A_{\text{c}(\text{s}/\text{l})}^{(\text{inner}/\text{outer})} &= x_{(\text{solid}/\text{liquid})}^{(\text{inner}/\text{outer})} A_{\text{c}}^{(\text{inner}/\text{outer})}. \end{aligned} \tag{8}$$

The areas for heat transfer between steel and liquid slag are calculated based on the metal bath surface area because the liquid slag forms as a layer on top of the steel. These areas are also applicable for the heat transfer between steel and the surrounding gas:

$$A_{\text{b}(\text{s}/\text{l})}^{(\text{inner}/\text{outer})} = x_{(\text{solid}/\text{liquid})}^{(\text{inner}/\text{outer})} A_{(\text{inner}/\text{outer})}. \tag{9}$$

Although the roof and side panel cooling water does not exchange heat directly with the steel, the linear heat flux from the furnace to the cooling water must be calculated in order to predict the outlet water temperature and compare to process data:

$$Q_{(\text{r}/\text{v})\text{w}} = k_{(\text{r}/\text{v})\text{w}} A_{(\text{r}/\text{v})} \left(T_{(\text{r}/\text{v})} - T_{(\text{r}/\text{v})}^{\text{W}} \right). \tag{10}$$

Radiation between steel and the furnace surfaces is included in the model. Table 3 lists the equations for heat flux from the steel to the furnace roof and vessel $Q_{(\text{s}/\text{l})(\text{r}/\text{v})}^{(\text{inner}/\text{outer})}$.

Table 3. Radiative heat transfer between different masses in the furnace. The subscript letters (s, l, b, r, v) refer to (solid steel, liquid steel, liquid slag, roof, vessel), respectively.

	Roof	Vessel
Inner Solid	$\sigma_{SB}\, A_{bs}^{inner}\, VF_{r}^{inner} \left(\varepsilon_s T_s^{inner^4} - \varepsilon_r T_r^{\,4}\right)$	$\sigma_{SB}\, A_{bs}^{inner}\, VF_{v}^{inner} \left(\varepsilon_s T_s^{inner^4} - \varepsilon_r T_v^{\,4}\right)$
Outer Solid	$\sigma_{SB}\, A_{bs}^{outer}\, VF_{r}^{outer} \left(\varepsilon_s T_s^{outer^4} - \varepsilon_r T_r^{\,4}\right)$	$\sigma_{SB}\, A_{bs}^{outer}\, VF_{v}^{outer} \left(\varepsilon_s T_s^{outer^4} - \varepsilon_r T_v^{\,4}\right)$
Inner Liquid	$\sigma_{SB}\, A_{bl}^{inner}\, VF_{r}^{inner} \left(\varepsilon_l T_l^{inner^4} - \varepsilon_r T_r^{\,4}\right)$	$\sigma_{SB}\, A_{bl}^{inner}\, VF_{v}^{inner} \left(\varepsilon_l T_l^{inner^4} - \varepsilon_r T_v^{\,4}\right)$
Outer Liquid	$\sigma_{SB}\, A_{bl}^{outer}\, VF_{r}^{outer} \left(\varepsilon_l T_l^{outer^4} - \varepsilon_r T_r^{\,4}\right)$	$\sigma_{SB}\, A_{bl}^{outer}\, VF_{v}^{outer} \left(\varepsilon_l T_l^{outer^4} - \varepsilon_r T_v^{\,4}\right)$

σ_{SB} is the Stefan–Boltzmann constant. The view factors VF_v^{inner} and VF_v^{outer} are calculated based on equations for disks (inner steel) and annular rings (outer steel) embedded in the base of a cylinder to the cylinder column (vessel) [20]. Because no radiative heat transfer from steel to gas or between steel phases is included in the model, the roof is the only other surface that absorbs steel radiation, and the view factors between the steel control volumes and the roof are solved for ($VF_r^{inner} = 1 - VF_v^{inner}$; $VF_r^{outer} = 1 - VF_v^{outer}$).

Radiative heat transfer from the roof to the furnace vessel is given by:

$$Q_{rv} = \sigma_{SB} A_r VF_v^r \left(\varepsilon_r T_r^{\,4} - \varepsilon_v T_v^{\,4}\right). \tag{11}$$

The view factor VF_v^r in Equation (11) is calculated based on an equation for the base of a cylinder (roof) to the cylinder column (vessel) [20].

2.4. Solid–Liquid Phase Change

Melting and freezing are a mass transfer mechanism between the solid and liquid phases in the inner steel, outer steel and slag. Melting is assumed to take place gradually as the solid temperature increases in a range centered around a defined melting/liquidus temperature T_m. Similarly, freezing takes place as the liquid temperature decreases in the same range. When both solid and liquid temperatures are within this range, melting and freezing take place simultaneously with temperature-dependent rates.

The melting and freezing mechanisms are illustrated in Figure 2. During the melting process, the solid temperature increases above the lower boundary for the phase change region. Liquid mass begins to accumulate and the liquid temperature changes quickly from its original virtual value. Eventually, the solid mass disappears, and additional energy inputs heats the liquid phase. Analogously, the liquid temperature decreases below the upper boundary for the phase change region to start the freezing process. Solid mass accumulates and the solid temperature changes quickly from its original virtual value. Eventually, the liquid mass disappears, and the solid mass continues to cool.

The rates of melting and freezing are given by:

$$\begin{aligned} r_{melt} &= k_{phase} m_s \frac{\max(0,\, (T_{solid} + dT_m) - T_m)}{2dT_m} \\ r_{freeze} &= k_{phase} m_l \frac{\max\left(0,\, T_m - \left(T_{liquid} - dT_m\right)\right)}{2dT_m}. \end{aligned} \tag{12}$$

The heat of fusion determines the amount of energy exchanged between the two phases:

$$\begin{aligned} Q_{melt} &= \Delta H_{fusion} r_{melt} \\ Q_{freeze} &= -\Delta H_{fusion} r_{freeze}. \end{aligned} \tag{13}$$

The terms for the heat of melting and freezing of each phase contribute to the energy balance equations found in Equation (A21).

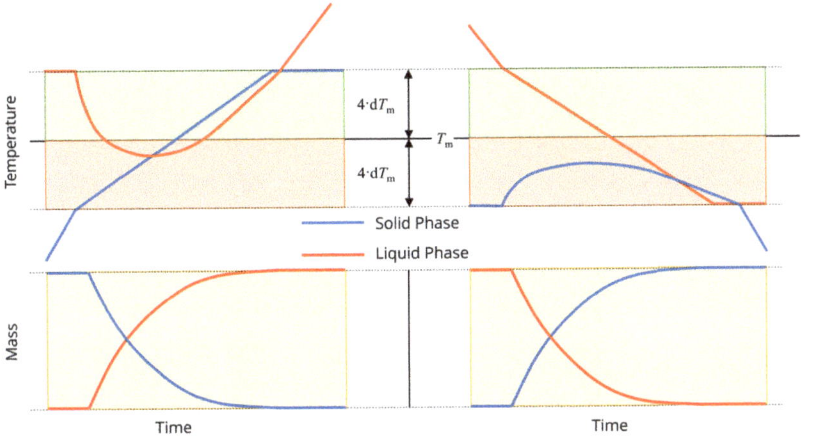

Figure 2. Illustration of (**left**) melting and (**right**) freezing. T_m indicates the steel melting temperature and dT_m is a constant used for calculating the rates of melting and freezing.

2.5. Arc Efficiency

A novel model for electrical energy efficiency based on arc visibility has been implemented. Observations from process data and literature indicate that arc efficiency, defined as the percentage of supplied electrical energy that heats steel, changes during the course of a heat [21]. While visibility of the arc is just one of many factors that affects arc efficiency [21], a description of arc coverage is a natural extension of the dynamic model states and can provide valuable insight into the current and future states of the process.

Figure 3 denotes the dimensions used by the arc efficiency model. Depending on the electrode height and the heights of the scrap (solid), liquid and slag phases, the electric arc length l_{arc} can be completely covered or partially exposed. The length of the electric arc l_{arc} is a constant while the position of the arc in the furnace is determined by the height of the electrode $h_{electrode}$, which changes during the boredown and melting stages of the process. The heights of the phases and the total bath are given by:

$$h_{(solid/liquid/slag)} = \frac{m_{(solid/liquid/slag)}}{A_{furnace}\rho_{(solid/liquid/slag)}} \quad (14)$$
$$h_{bath} = h_{scrap} + h_{liquid} + h_{slag}.$$

The visibility of the arc can then be written:

$$vis_{arc} = \max\left(\min\left(\frac{h_{electrode} - h_{bath}}{l_{arc}}, 1\right), 0\right). \quad (15)$$

The arc visibility is zero when the full length of the arc is below the cumulative height of the bath components (scrap h_{scrap}, liquid h_{liquid} and slag h_{slag}). The visibility is combined with additional model parameters in order to write the model for arc energy losses that includes estimation parameter k_{loss}, which can be used to tune the model to better follow individual heats:

$$x_{arc}^{loss} = k_{loss}\left(vis_{arc} + k_{basket}^{loss}\right). \quad (16)$$

Using process data to calculate the electrode height $h_{electrode}$ is challenging because the electrodes are consumed during the process, causing the length of the electrodes to vary from heat to heat. Instead of using process data to determine to electrode position, the electrode height is modeled with an equation that captures typical boredown behavior as a function of the total instantaneous solid and liquid mass and the total electrical energy supplied to the furnace kWh:

$$h_{\text{electrode}} = \exp(-\text{kWh}/\text{kWh}_{\text{basket}}) + \frac{m_{\text{solid}} + m_{\text{liquid}}}{A_{\text{furnace}}\rho_{\text{liquid}}}. \quad (17)$$

The density of slag ρ_{slag} is modeled as a function of the overall liquid fraction, consistent with the observation that foaminess increases as meltdown progresses [22]:

$$\rho_{\text{slag}} = 120 + 1380\exp(-x_l/x_{\text{basket}}). \quad (18)$$

Equations (16)–(18) each include a term with the subscript *basket*. These terms are model-fitting constants that are fit for the cases of one-, two- and three-basket heats.

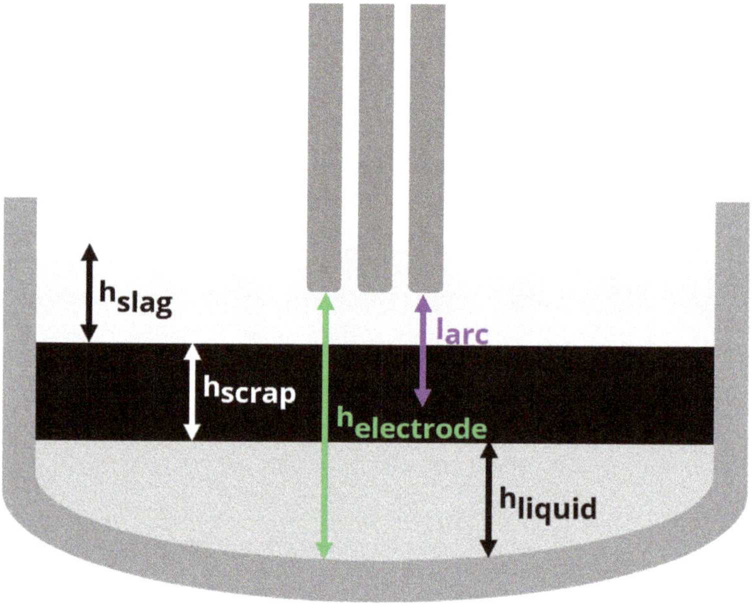

Figure 3. Variable and static dimensions used to calculate arc visibility for the arc efficiency model.

3. Results and Discussion

3.1. Model Behavior

Figure 4 shows typical model meltdown behavior. Heats typically result in the production of approximately 140 tonnes of liquid steel. The inner solid mass is heated directly by the electric arc and always begins to melt first, while the outer solid mass begins to melt before the inner solid is completely liquefied. The melting process is interrupted by pauses in process operation and the addition of the second basket, at which point all mass solidifies before continuing to be heated and re-melted. Each time a new heat begins, the furnace is emptied (solid steel, liquid steel, slag and component masses are reset to initial states, along with all temperatures except for the furnace roof and vessel).

The process begins when the power and gas burners are turned on. The electrodes are located in the center of the furnace and supply power to the most closely situated charged material. The arc power is therefore used to heat the inner steel control volume as described in Appendix A.1. As a result, the inner solid temperature rises much faster than the outer solid temperature. Heat transfer limitations described in Section 2.3 govern the rate at which arc energy is dissipated from the inner to the outer control volume. The furnace surroundings are heated by energy losses from the burners and arc as well as by heat transfer from the steel. Eventually, the inner solid becomes hot enough to melt and an inner liquid mass begins to appear. At this point in the process, additional planned baskets will usually be added, cooling the furnace contents. Time delays in adding baskets also cause the furnace contents to cool undesirably. Power continues to be supplied to the furnace, and the outer solid finally begins to melt before the inner solid has fully disappeared.

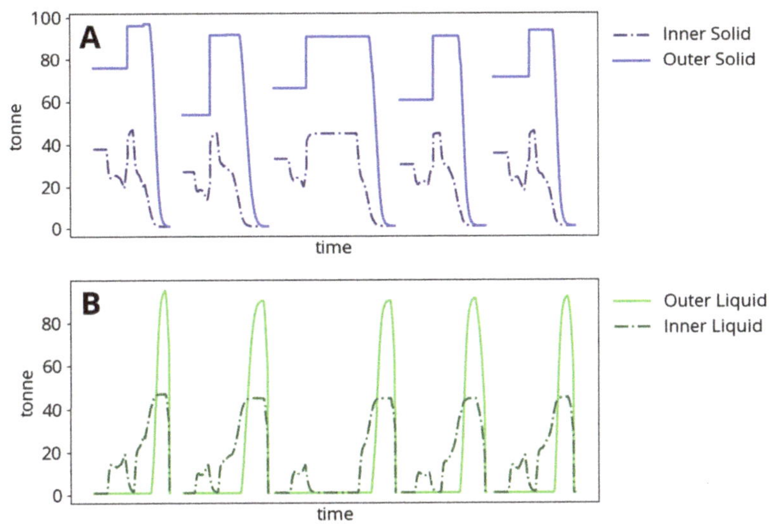

Figure 4. Steel meltdown dynamics and change in (**A**) solid mass and (**B**) liquid mass during five consecutive two-basket heats.

Figure 5 shows model agreement with process data from 250 heats for liquid steel temperature and weight after tapping. The model does not account for a hot heel, which can vary from heat to heat and may explain some of the observed scatter in weight agreement. The hot heel discrepancy may also affect the scatter in temperature agreement. Both the weight and temperature agreement are within reasonable expectations for model behavior and measurement accuracy.

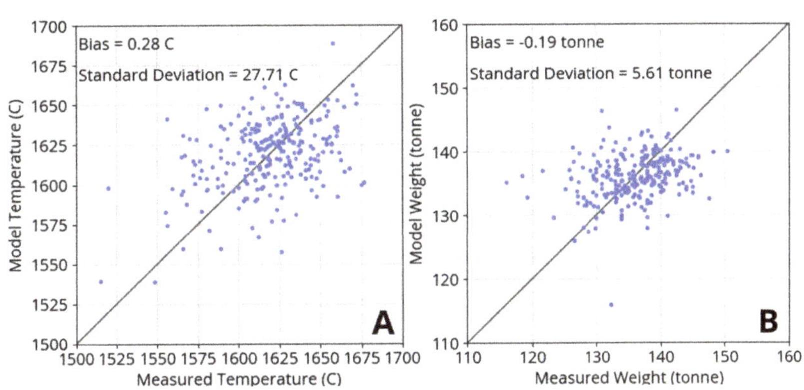

Figure 5. Agreement of model predictions with process measurements from 250 heats for (**A**) liquid steel temperature and (**B**) liquid steel weight after tapping. Model biases for temperature and weight are negligible while residual standard deviations reflect scatter.

3.2. Recursive Estimation of Arc Efficiency

In order for MPC to be effective, the underlying model describing the process needs to be able to accurately predict the behavior of the optimization targets as the heat progresses. For the purpose of optimization for energy savings, the model not only has to follow the present efficiency, but must also be able to predict the future efficiency with sufficient accuracy. The goal of developing the arc energy loss model described in Section 2.5 is therefore to accurately predict efficiency at upcoming stages in the meltdown process such that the power profile can be optimized accordingly. Arc power can be adjusted down and up during low and high efficiency periods, respectively, in order to minimize energy losses to the environment.

The arc efficiency model includes basket-dependent constants as described in Section 2.5. In order to fit the constants k^{loss}_{basket}, kWh_{basket} and x_{basket}, a precursor study using process data was performed in which the percentage of arc power loss to the furnace roof and vessel (x^{loss}_{arc} in Equation (16)) was recursively estimated with a KF [13]. The average result of this study for one-, two- and three- basket heats is shown in Figure 6, where a higher arc energy loss proportion corresponds to a lower arc efficiency. The discernible features denoted in Figure 6 allow us to propose a physical explanation that we can later use to model arc efficiency. These features are most apparent for one-basket heats, as multiple-basket heat results are impacted by the variable proportions of electric energy added per basket. High arc losses at the beginning of the heats are explained by arc exposure during boredown. As boredown continues, the arc is covered by scrap and the loss proportion decreases until the scrap begins to melt and re-exposes the arc. The arc is, however, only temporarily exposed, as the appearance of a liquid steel phase is followed quickly by a liquid slag phase, and the reactions between liquid steel and slag components lead to a larger, foamier slag phase. This foamy slag phase provides the arc with coverage as the heat nears completion.

Efficiency model constants k^{loss}_{basket}, kWh_{basket} and x_{basket} are fit to best reproduce the average data for one-, two- and three-basket heats presented in Figure 6. The arc efficiency model effectively reproduces the one- and two-basket heat data, but performs less well when compared to the three-basket heat data. The discrepancy between the model and the three-basket heat data could be due to poor statistics, as far fewer three-basket heats were recorded in the data series.

While the basket-dependent parameters do enable the model to capture average process behavior, there still exists a variable degree of energy losses between heats with the same number of baskets. These differences between heats could be due to many factors not currently accounted for by the efficiency model, including variable density of baskets, electrode tip wear and hot heel. The strategy employed for addressing this variation is to include scaling parameter k_{loss} in the model for arc efficiency given by Equation (16). Unlike the efficiency model constants, which are tuned for all heats with the same number of planned charged baskets, the k_{loss} term in Equation (16) is a designated parameter for recursive estimation. Variations in efficiency from heat to heat are to be expected and can be followed using a KF [13]. Figure 7 shows an example of the impact of recursive estimation on the same five consecutive two-basket heats plotted in Figure 4. The ballistic simulation uses a k_{loss} value that best matches the average for all heats with the same number of baskets. The ballistic simulation follows the trend in cooling water process data well, but is prone to sometimes overpredicting heat losses and the resulting cooling water outlet temperature. Recursive estimation of k_{loss} fares better: after overcoming initial errors that come from reinitializing the heat, the estimation results match both the trend and level of cooling water temperature data.

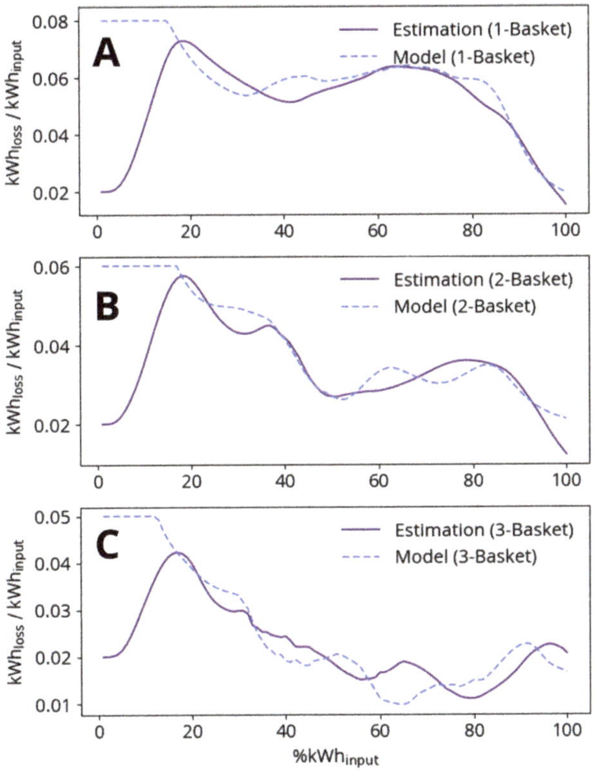

Figure 6. Comparison of arc efficiency estimation and model for (**A**) one-basket heats, (**B**) two-basket heats and (**C**) three-basket heats. The x-axis is the normalized progress of each heat as measured by the percentage of total electric energy added to the furnace. Low $\%kWh_{input}$-results are not meaningful as the efficiency estimation requires several samples to change from the initial guess factor of 0.02.

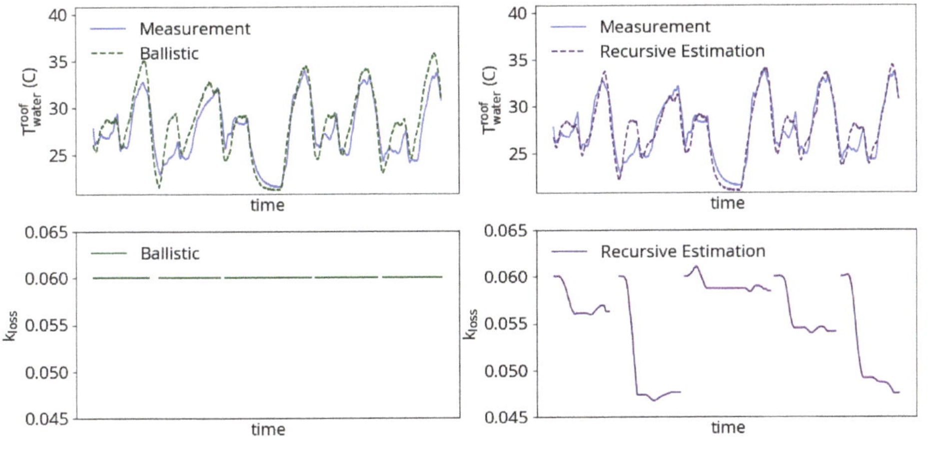

Figure 7. Example of arc loss coefficient k_{loss} estimation during five consecutive two-basket heats. (**left**) A constant k_{loss} produces model results that follow the cooling water dynamics but sometimes overpredict the outlet temperature. (**right**) Estimation of k_{loss} produces more accurate model results.

3.3. Industrial Use and Application

Model predictive control (MPC) is an advanced method of process control that determines a sequence of process inputs that optimizes a predicted process output at specified time points in the future [23]. Online MPC routines re-evaluate the current process state at each successive sampling time, allowing the optimization to adapt to process disturbances. The process control scheme referred to in this work can be more specifically described as Non-linear model predictive control (NMPC) because the predicted response to proposed inputs are calculated based on a non-linear process model.

MPC simulations were performed based on logged plant data from 250 heats with the Cybernetica^{TM} Cenit software. The goal of these MPC simulations is to optimize the electrical power input in order to increase the overall efficiency of the arc power. In order to adapt the process data for the MPC study, the basket contents and schedule of charges are preserved according to logged data without the exact schedule being preemptively revealed to the MPC. Logged power input is overwritten by the closed-loop simulation. Because the operation of the gas burners should be in sync with the accumulated electrical energy added to the furnace, logged LNG and oxygen flows are replaced with the gas burner recipe used in plant operation. Logged time delays and pauses in electrical power supply are preserved in the MPC simulations.

In the Cybernetica^{TM} Cenit implementation of MPC, the optimization takes the form of minimizing an objective function. Because the EAF is operated as a batch process, the process outputs that contribute to the objective function are evaluated at the end of the batch (the time at which the model predicts the furnace contents are fully melted). The MPC algorithm seeks to simultaneously minimize the total batch time and maximize the efficiency of the electric arc based on the objective function J:

$$J = \frac{1}{2}\Delta U^{\mathrm{T}}\,\mathbf{S}\,\Delta U + R^{\mathrm{T}}(Z - Z_{\mathrm{max}}). \tag{19}$$

The optimization criteria J is a scalar calculated from the sum of the right-hand side terms in Equation (19). U represents manipulated variable (MV) process inputs and ΔU is the vector of changes to manipulated inputs proposed by the optimization. Changes to MVs are weighted by the penalties contained by the diagonal of matrix $\mathbf{S}$. Z is vector of control variables (CV) calculated by the model, and Z_{max} is a vector of soft maximum constraints for each CV. The violation of each constraint is weighted linearly by vector R, and the term is only evaluated for the largest constraint violation in the prediction horizon.

For the EAF optimization simulations, the elements of the U-vector are:

- U_{1-15} or MV_{1-15}: Shift from the nominal power profile during optimization interval ($i = 1$ to 15) (MW).

The MPC optimizes deviation U from a nominal power profile in order to propose a more efficient power profile solution. The MV in this optimization problem is therefore change to power input rather than the power input itself. The prediction horizon is divided into 15 intervals, each of which is shifted by a separate U_i. The nominal power profile is given by plant recipes for one-, two- and three-basket heats. The elements of the Z-vector are:

1. Z_1 or CV_1: Batch time (seconds)
2. Z_2 or CV_2: Energy losses from the electric arc (kWh).

The elements of Z_{max} are in this case set to zero in order to direct the optimization to minimize both of the CVs.

Because the arc efficiency changes as the solid scrap melts down and a slag phase forms, as discussed in Section 3.2, adjusting the power levels over the course of the heat can lead to a more optimal power profile. Figure 8 shows an illustration of power profile optimization. The MPC scheme uses a finite receding horizon, meaning that the as the heat proceeds the output power profile will have shorter remaining duration. The criteria for ending the output power is the model prediction of full meltdown of the solid mass in the furnace. The example batches following the nominal and optimized input power trajectories are predicted to end at different times, and the optimized trajectory is projected to incur fewer energy losses directly from the arc to the environment.

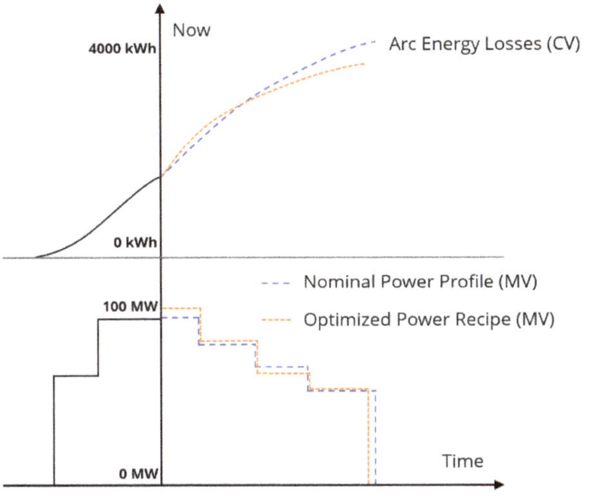

Figure 8. An illustration of MPC optimization for batch time and electric arc efficiency. Historical MV and model CV predictions are shown as solid black lines. Within the prediction horizon, optimized MV (MPC output) and CV are shown in orange, while nominal MV and the resulting CV are shown in blue.

Figure 9 shows the energy savings predicted by the MPC simulations for one-, two- and three-basket heats. The average predicted savings are:

- One-basket heats: 15.75 kWh/tonne per heat (2315 kWh per heat, based on an average charged weight of 147 tonne);
- Two-basket heats: 13.32 kWh/tonne per heat (1945 kWh per heat, based on an average charged weight of 146 tonne);
- Three-basket heats: 6.78 kWh/tonne per heat (983 kWh per heat, based on an average charged weight of 145 tonne).

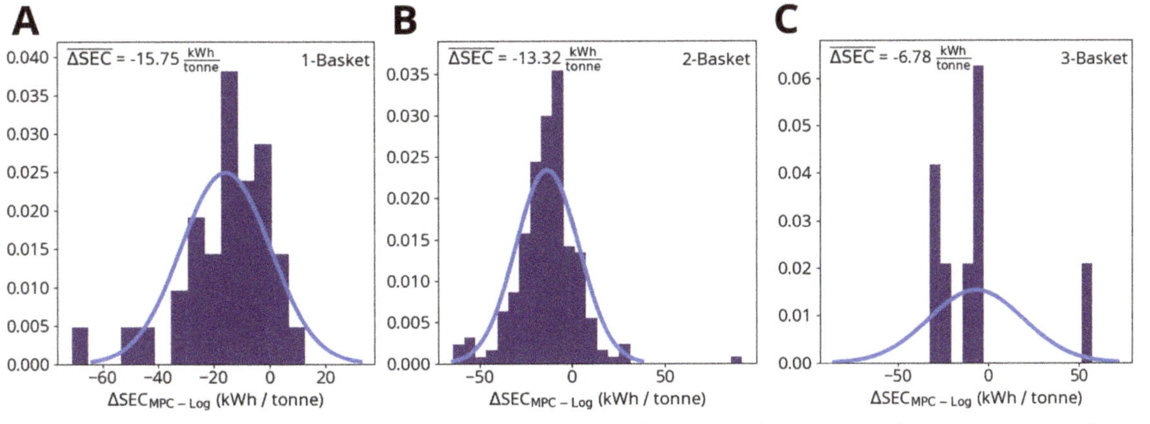

Figure 9. Change in specific energy consumption (SEC) due to MPC for (**A**) one-basket heats, (**B**) two-basket heats and (**C**) three-basket heats. ΔSEC is calculated by subtracting the logged SEC up to the point where the model predicts steel meltdown from the MPC simulation result.

These results only account for the process energy savings up to the point when the model predicts that all the scrap metal has melted. This means that if the model predicts that the logged power profiles continue to heat the furnace after all the scrap metal is liquefied, which happens frequently, the additional energy savings beyond the point of full meltdown are neglected. While the model predicts that accounting for turning off the furnace earlier can cut down energy usage per heat by up to 15 kWh/tonne, these additional savings are not included in the average predicted savings listed above or in the data shown in Figures 9 and 10. The purpose of this savings criteria is to evaluate the MPC scheme's potential for efficiency improvements independently from the model's accuracy of end-point prediction.

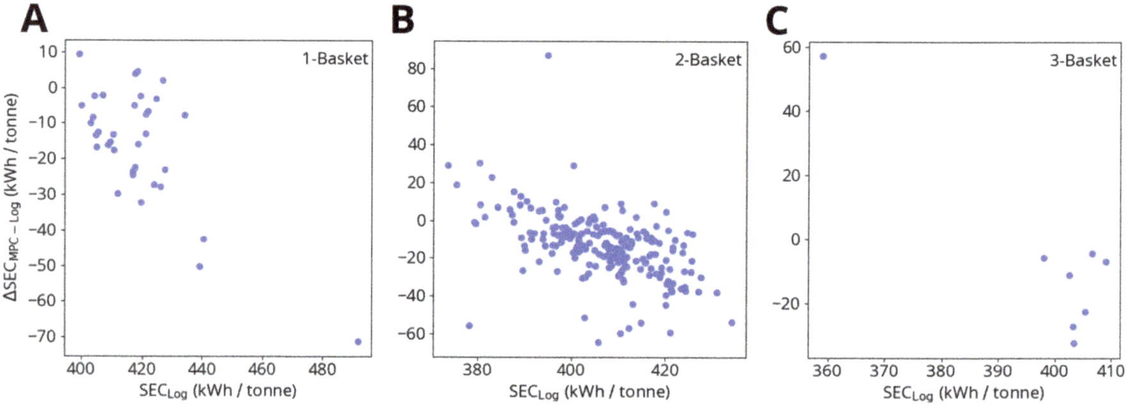

Figure 10. Logged specific energy consumption (SEC) up to the point of melting vs. change in SEC due to MPC for (**A**) one-basket heats, (**B**) two-basket heats and (**C**) three-basket heats. ΔSEC is calculated as described in Figure 9.

While the MPC simulations predict the largest energy savings for one-basket heats, the overall findings for one-, two- and three-basket heats are similar. Within these subgroups, the electrical energy savings are not uniform for all heats: Figure 10 shows that the MPC savings (ΔSEC) are correlated with the magnitude of the logged specific electrical energy consumption (SEC). As SEC approaches approximate threshold values of 400 kWh/tonne (one-basket heats) and 390 kWh/tonne (two- and three-basket heats) ΔSEC steadily decreases, potentially indicating that the MPC is not able to save significant amounts of energy beyond a given lower bound.

The decreased energy demand for melting the furnace contents while following MPC recommendations raises the question of whether the the optimized power profile significantly affects the endpoint state of the liquid steel. The process and model dynamics are such that fully melting the scrap metal within the time frame of a standard batch requires heating the liquid steel well above the melting temperature. In practice, temperature measurements are made only after tapping, making a temperature target difficult to verify, and the assumption that typical meltdown temperatures are high enough for downstream processing has been shown to be sufficient. Because the optimization scheme uses a meltdown criteria rather than an explicit temperature target, the energy-saving recommendations could conceivably result in lower temperatures that are not optimal for further processing and refining of the steel. To investigate this question, model predictions of liquid steel temperatures from optimized and logged power profile simulations are examined and compiled in Table 4. Two process data-derived temperatures are presented: T_{Log}^{Melt} is the liquid steel temperature at the moment the model predicts the steel is melted, while T_{Log}^{Full} is the liquid steel temperature when the logged power is shut off. Because the MPC simulations shut off power once all the steel has melted, T_{Log}^{Melt} is the most appropriate quantity for comparison to T_{MPC}.

Table 4. Temperature of fully-liquefied steel prior to tapping as predicted by the model when following logged process data and MPC simulations. Both means and standard deviations are presented.

	T_{Log}^{Full}	T_{Log}^{Melt}	T_{MPC}
1-Basket	1688.5 °C ± 27.8 °C	1684.5 °C ± 27.9 °C	1693.0 °C ± 13.0 °C
2-Basket	1693.3 °C ± 21.6 °C	1687.6 °C ± 23.4 °C	1703.7 °C ± 17.1 °C
3-Basket	1717.9 °C ± 2.1 °C	1717.2 °C ± 2.2 °C	1702.5 °C ± 0.9 °C

The results presented in Table 4 show that, at the point of meltdown, the liquid steel actually reaches higher temperatures in the MPC simulations than in the simulations with logged data. These results point to an interesting finding: according to the model, the optimized power profile leads not only to lower process energy demand due to reduced heating of the environment directly by the electric arc, but also to more efficient heating of the steel by the energy that is able to be absorbed. MPC meltdown temperatures are also higher than predicted temperatures at the point of logged power shut off, indicating that following the optimized profile should not cause the liquid steel to be too cold at tapping such as to cause problems for downstream plant processes.

4. Conclusions

The EAF is a challenging process to model and optimize from a precision point of view: there are significant uncertainties associated with materials and electrodes that can be difficult to resolve using an automated approach. This work aimed at formulating a first-principles mathematical model in terms of ordinary differential equations for the state variables of the EAF that can be adapted to process data using recursive parameter estimation. The resulting model described in this article is of manageable size yet sufficiently detailed and adaptable to be useful for process optimization.

An MPC-based optimization application based on this model has been running online using data for a 140 tonne EAF furnace since August 2020, demonstrating that the model is fast enough for industrial deployment. The predicted metal temperatures and weights were found to be in reasonably good agreement with the measured values. Results indicate that MPC-based process operation leads to both a reduction in total energy usage as well as more efficient dissipation and heating of the steel by the consumed energy.

While several MPC studies for EAF processes have been reported in the literature [3,6], this study is, to the best of the authors' knowledge, the first where the efficiency of the power input and electric arc has been the target for MPC optimization. The framework for optimizing power input with MPC opens up the possibility of incorporating a more complex model of the electrical power in order to achieve more effective and sensitive optimization of arc efficiency [21]. Additional optimization scenarios can be considered in further work: an economic optimization of burner vs. electrical power can be implemented using the same framework, and new data and dynamic states can be added to the model in order to optimize for operational costs such as equipment life cycle and electrode consumption [24].

Author Contributions: Conceptualization, S.O.W., A.H., N.H., V.-V.V. and M.S.; methodology, S.J., S.O.W., V.-V.V. and M.S.; software, S.J., S.O.W. and A.H.; validation, S.J., V.-V.V. and N.H.; formal analysis, S.J.; investigation, S.J. and N.H.; resources, S.O.W. and N.H.; data curation, S.J. and S.O.W.; writing—original draft preparation, S.J.; writing—review and editing, S.O.W., V.-V.V. and M.S.; visualization, S.J. and A.H.; supervision, S.O.W. and V.-V.V.; project administration, S.O.W. and V.-V.V.; funding acquisition, S.O.W. and M.S. All authors have read and agreed to the published version of the manuscript.

Funding: This research was funded by the European Commission, Directorate-General for Research and Innovation, through the project "Model-based optimisation for efficient use of resources and energy" (MORSE), grant number 768652. https://www.spire2030.eu/morse (accessed on 26 September 2021).

Institutional Review Board Statement: Not applicable.

Informed Consent Statement: Not applicable.

Data Availability Statement: Restrictions apply to the availability of these data. The data presented in this study are available on request from the corresponding author with the permission of authors/partners at Outokumpu Stainless Oy.

Acknowledgments: The authors acknowledge helpful discussions with Eero Auer and Esa Puukko from Outokumpu regarding the collection, processing and results of EAF data. The tragic death of Ralf Pierre, a much appreciated expert in EAF modeling and very helpful partner in many fruitful discussions, has left the authors deeply saddened.

Conflicts of Interest: The authors declare no conflict of interest. The funders had no role in the design of the study; in the collection, analyses, or interpretation of data; in the writing of the manuscript, or in the decision to publish the results.

Abbreviations

The following abbreviations are used in this manuscript:

EAF	Electric Arc Furnace
MPC	Model Predictive Control
NMPC	Non-linear Model Predictive Control
SEC	Specific Energy Consumption
KF	Kalman Filter
LNG	Liquefied Natural Gas
VF	View Factor
CV	Controlled Variable
MV	Manipulated Variable
DRI	Direct Reduced Iron

Nomenclature

The nomenclature for constants and variables used in the main text and appendices of this manuscript are listed:

A_{ij}	Area for heat transfer between mass i and mass j	m^2
d_i	Diameter of i	m
$\frac{\mathrm{d}}{\mathrm{d}t}$	Derivative operator	
ϵ_j	Radiation emissivity of surface j	
F_i	Molar rate of change of component i	$\frac{\mathrm{mol}}{\mathrm{s}}$
$C_{\mathrm{p},\,i}$	Heat capacity of i	$\frac{\mathrm{J}}{\mathrm{kg\,K}}$
h_k	Height of mass k	m
H_i	Enthalpy of component i	$\frac{\mathrm{J}}{\mathrm{kg}}$
k_{ij}	Heat transfer coefficient between type i and type j	$\frac{\mathrm{W}}{\mathrm{m^2K}}$
m_i^j	Mass of component i in phase j	kg
M_i	Molar mass of component i	$\frac{\mathrm{g}}{\mathrm{mol}}$
Q_{ij}	Heat flowing from mass i to mass j	W
P_k	Power from source k	MW
p_i	Pressure of phase i	Pa
r_i	Reaction rate of reaction i	$\frac{\mathrm{kg}}{\mathrm{s}}$
r_{particle}	Radius of particle	m
ρ_i	Density of i	$\frac{\mathrm{kg}}{\mathrm{m}^3}$
σ_{SB}	Stefan-Boltzmann constant for radiation	$5.670374 \times 10^{-8} \frac{\mathrm{W}}{\mathrm{m^2K^4}}$
t	Time	s
T_j	Temperature of phase j	K
v_i	Stoichiometric coefficient of component i in a chemical reaction	
V_k	Volume of mass k	m^3

W_i	Mass rate of change of component i	$\frac{kg}{s}$
x_i^j	Mass fraction of component i in phase j	
x_j	Area fraction of control volume j	
subscript: b	Liquid slag	
subscript: c	Solid slag	
subscript: g	Gas	
subscript: l	Liquid steel	
subscript: r	Roof	
subscript: s	Solid steel	
subscript: v	Vessel	

Appendix A. Extended Model Details

Appendix A.1. Electrical and Chemical Power

The power input supplied by the electric arc P_{arc} is a logged process input. The arc power is partitioned between several model masses:

$$\begin{aligned} P_{\text{arc}}^{\text{gas}} &= x_{\text{arc}}^{\text{gas}} P_{\text{arc}} \\ P_{\text{arc}}^{\text{loss}} &= x_{\text{arc}}^{\text{loss}} P_{\text{arc}} \\ P_{\text{arc}}^{\text{steel}} &= \left(1 - x_{\text{arc}}^{\text{gas}} - x_{\text{arc}}^{\text{loss}}\right) P_{\text{arc}} \end{aligned} \tag{A1}$$

The calculation and estimation of $x_{\text{arc}}^{\text{loss}}$ is described in Section 2.5, while $x_{\text{arc}}^{\text{gas}}$ is a model constant for the fraction of arc power used to heat the gas phase. $P_{\text{arc}}^{\text{loss}}$ is used to heat two pieces of furnace equipment: the furnace vessel and roof. The further partitioning of $P_{\text{arc}}^{\text{loss}}$ is determined by model constant $x_{\text{v}}^{\text{loss}}$:

$$\begin{aligned} P_{\text{arc}}^{\text{vessel}} &= x_{\text{vessel}}^{\text{loss}} P_{\text{arc}}^{\text{loss}} \\ P_{\text{arc}}^{\text{roof}} &= \left(1 - x_{\text{vessel}}^{\text{loss}}\right) P_{\text{arc}}^{\text{loss}} \end{aligned} \tag{A2}$$

The molar LNG flow rate into the furnace F_{LNG} is given by conversion from logged gas flow rates according to standard temperature and pressure (STP; $T = 0\ ^\circ\text{C}$, $p = 100$ kPa). The molar flow rate of oxygen F_{O_2} is similarly calculated by conversion from logged standard flow rates. Given that enough oxygen is present, the LNG is assumed to combust completely, allowing us to write an expression for the LNG power P_{LNG}:

$$P_{\text{LNG}} = \left(1.0 \times 10^{-3}\right) \min\left(F_{\text{LNG}}, \frac{F_{O_2}}{2}\right) M_{\text{LNG}} H_{\text{LNG}} \tag{A3}$$

M_{LNG} and H_{LNG} refer to the the molar mass and enthalpy of combustion for the specified composition of LNG. P_{LNG} heats both the steel and gas phases, with the gas phase being heated to a greater extend as meltdown progresses:

$$\begin{aligned} x_{\text{LNG}}^{\text{gas}} &= 0.25\left(\tanh\left(5x_{\text{liquid}}^{\text{overall}} - 2.5\right) + 2\right) \\ P_{\text{LNG}}^{\text{gas}} &= x_{\text{LNG}}^{\text{gas}} P_{\text{LNG}} \\ P_{\text{LNG}}^{\text{steel}} &= \left(1 - x_{\text{LNG}}^{\text{gas}}\right) P_{\text{LNG}} \end{aligned} \tag{A4}$$

Oxygen consumption by the combustion of LNG can be calculated:

$$F_{O_2}^{\text{LNG}} = 2 \times \min\left(F_{\text{LNG}}, \frac{F_{O_2}}{2}\right) \tag{A5}$$

Any oxygen not consumed by LNG combustion remains available for further chemical reactions. Oxygen not consumed by LNG can be used for combustion with CO:

$$F_{O_2}^{CO} = \min\left(F_{O_2} - F_{O_2}^{LNG}, \frac{k_{CO} n_{CO}}{2} \right) \tag{A6}$$

n_{CO} denotes the accumulated moles of CO from refining reactions present in the furnace and k_{CO} refers to a model limiting rate constant for the CO combustion reaction. The power released by CO combustion is given:

$$p_{CO} = F_{O_2}^{CO} H_{CO} \tag{A7}$$

H_{CO} refers to the enthalpy of combustion for CO. The amount of oxygen available for refining reactions can be calculated based on the mass of liquid steel species:

$$\begin{aligned} m_{\text{refine}}^{\text{inner}} &= \sum_{i=\text{Fe, C, Si, Cr}}^{\text{inner}} m_{i,\text{ liquid}}^{\text{inner}} \\ m_{\text{refine}}^{\text{outer}} &= \sum_{i=\text{Fe, C, Si, Cr}}^{\text{outer}} m_{i,\text{ liquid}}^{\text{outer}} \\ F_{O_2}^{\text{refine}} &= \max\left(F_{O_2} - F_{O_2}^{LNG} - F_{O_2}^{CO}, 0 \right) \end{aligned} \tag{A8}$$

$m_{i,\text{ liquid}}$ denotes the mass of species i within a given control volume and k_{refine} refers to a model limiting rate constant of refining for the liquid steel components. $F_{O_2}^{\text{refine}}$ is partitioned to the inner and outer control volumes and used for the reactions described in Appendix A.2:

$$\begin{aligned} F_{O_2}^{\text{refine, inner}} &= \frac{m_{\text{refine}}^{\text{inner}}}{m_{\text{refine}}^{\text{inner}} + m_{\text{refine}}^{\text{outer}}} F_{O_2}^{\text{refine}} \\ F_{O_2}^{\text{refine, outer}} &= \frac{m_{\text{refine}}^{\text{outer}}}{m_{\text{refine}}^{\text{inner}} + m_{\text{refine}}^{\text{outer}}} F_{O_2}^{\text{refine}} \end{aligned} \tag{A9}$$

The power from all refining reactions is calculated from the sum of reactions described in Appendix A.2:

$$P_{\text{refine}} = \sum_{r_{\text{oxygen}}} \left(F_{O_2,\text{ inner}}^{r_{\text{oxygen}}} + F_{O_2,\text{ outer}}^{r_{\text{oxygen}}} \right) H_{O_2}^{r_{\text{oxygen}}} + \sum_{r_{\text{equilibrium}}} \left(F_{\mathbf{XO},\text{ inner}}^{r_{\text{equilibrium}}} + F_{\mathbf{XO},\text{ outer}}^{r_{\text{equilibrium}}} \right) H_{\mathbf{XO}}^{r_{\text{equilibrium}}} \tag{A10}$$

H_{O_2} refers to the enthalpy of each refining reaction involving oxygen and H_{XO} refers to the enthalpy of each steel-slag equilibrium reaction per *mol* of **XO**, where **XO** is the oxide species listed for each reaction in Tables A3 and A4.

The total chemical power to the steel is given:

$$P_{\text{chemical}} = P_{LNG}^{\text{steel}} + P_{\text{refine}} + P_{CO} \tag{A11}$$

The total steel-heating power is partitioned between the inner and outer control volumes, with all of P_{arc}^{steel} being used to heat the inner steel:

$$\begin{aligned} P_{\text{inner}} &= x_{\text{inner}} \left(P_{LNG}^{\text{steel}} + P_{\text{refine}} + P_{CO} \right) + P_{\text{arc}}^{\text{steel}} \\ P_{\text{outer}} &= x_{\text{outer}} \left(P_{LNG}^{\text{steel}} + P_{\text{refine}} + P_{CO} \right) \end{aligned} \tag{A12}$$

x_{inner} and x_{outer} are the area fractions of the inner and outer control volumes, respectively, as calculated from the model dimensions given in Appendix A.4. The total power to the gas phase is given:

$$P_{\text{gas}} = P_{LNG}^{\text{gas}} + P_{\text{arc}}^{\text{gas}} \tag{A13}$$

Appendix A.2. Reaction Kinetics

The rates of oxygen consumption by the different reactions listed in Equation (2) depend on the species mass fractions. The rates of consumption and production of the other species involved can be calculated from the oxygen consumption rate. The consumption and production rates of different species are given in Table A1 and hold for both the inner and outer control volumes, where the steel mass fractions and oxygen available for refining can vary.

Table A1. Oxygen consumption and species rate of change for each steel reaction with oxygen. F denotes molar rates of change, v denotes stoichiometric coefficients and x denotes mass fractions.

Reaction	O_2 Consumption Rate	Species i Rate of Change
$Fe + \frac{1}{2}\{O_2\} \rightarrow (FeO)$	$F_{O_2}^{Fe\rightarrow FeO} = x_{Fe}^{steel} F_{O_2}^{refine}$	$F_i^{Fe\rightarrow FeO} = \frac{v_i^{Fe\rightarrow FeO}}{v_{O_2}^{Fe\rightarrow FeO}} F_{O_2}^{Fe\rightarrow FeO}$
$[C] + \frac{1}{2}\{O_2\} \rightarrow \{CO\}$	$F_{O_2}^{C\rightarrow CO} = \frac{1}{2} x_{C}^{steel} F_{O_2}^{refine}$	$F_i^{C\rightarrow CO} = \frac{v_i^{C\rightarrow CO}}{v_{O_2}^{C\rightarrow CO}} F_{O_2}^{C\rightarrow CO}$
$[C] + \{O_2\} \rightarrow \{CO_2\}$	$F_{O_2}^{C\rightarrow CO_2} = \frac{1}{2} x_{C}^{steel} F_{O_2}^{refine}$	$F_i^{C\rightarrow CO_2} = \frac{v_i^{C\rightarrow CO_2}}{v_{O_2}^{C\rightarrow CO_2}} F_{O_2}^{C\rightarrow CO_2}$
$[Si] + \{O_2\} \rightarrow (SiO_2)$	$F_{O_2}^{Si\rightarrow SiO_2} = x_{Si}^{steel} F_{O_2}^{refine}$	$F_i^{Si\rightarrow SiO_2} = \frac{v_i^{Si\rightarrow SiO_2}}{v_{O_2}^{Si\rightarrow SiO_2}} F_{O_2}^{Si\rightarrow SiO_2}$
$2[Cr] + \frac{3}{2}\{O_2\} \rightarrow (Cr_2O_3)$	$F_{O_2}^{Cr\rightarrow Cr_2O_3} = x_{Cr}^{steel} F_{O_2}^{refine}$	$F_i^{Cr\rightarrow Cr_2O_3} = \frac{v_i^{Cr\rightarrow Cr_2O_3}}{v_{O_2}^{Cr\rightarrow Cr_2O_3}} F_{O_2}^{Cr\rightarrow Cr_2O_3}$
$2[Al] + \frac{3}{2}\{O_2\} \rightarrow (Al_2O_3)$	$F_{O_2}^{Al\rightarrow Al_2O_3} = x_{Al}^{steel} F_{O_2}^{refine}$	$F_i^{Al\rightarrow Al_2O_3} = \frac{v_i^{Al\rightarrow Al_2O_3}}{v_{O_2}^{Al\rightarrow Al_2O_3}} F_{O_2}^{Al\rightarrow Al_2O_3}$

The equilibrium reaction constants k_{eq} and dependence on species i mass fractions x_i^{steel} and x_i^{slag} are adapted from Turkdogan [15] and are given in Tables A2–A4. The rates of change of different species are given in Table A5 and hold for both the inner and outer control volumes, where the steel mass fractions and temperatures can vary.

The reference states for all species and the equilibrium constants in Table A2 are the standard state of 25 °C and 1 atm.

Table A2. Equilibrium constants for reactions. T_l denotes liquid phase temperatures.

Reaction	Equilibrium Constant
$(FeO) + [C] \leftrightarrow Fe + \{CO\}$	$\log_{10}\left(K_{eq}^{FeO\leftrightarrow CO}\right) = \frac{-5730}{\frac{1}{2}\left(T_l^{steel}+T_l^{slag}\right)+273.15} + 5.096$
$(FeO) + [Mn] \leftrightarrow Fe + (MnO)$	$\log_{10}\left(K_{eq}^{FeO\leftrightarrow MnO}\right) = 2$
$(MnO) + [C] \leftrightarrow [Mn] + \{CO\}$	$\log_{10}\left(K_{eq}^{MnO\leftrightarrow CO}\right) = \frac{-13182}{\frac{1}{2}\left(T_l^{steel}+T_l^{slag}\right)+273.15} + 8.574$
$2(FeO) + [Si] \leftrightarrow 2Fe + (SiO_2)$	$\log_{10}\left(K_{eq}^{FeO\leftrightarrow SiO_2}\right) = \frac{1510}{\frac{1}{2}\left(T_l^{steel}+T_l^{slag}\right)+273.15} + 1.72$
$2(MnO) + [Si] \leftrightarrow 2[Mn] + (SiO_2)$	$\log_{10}\left(K_{eq}^{MnO\leftrightarrow SiO_2}\right) = \frac{1510}{\frac{1}{2}\left(T_l^{steel}+T_l^{slag}\right)+273.15} + 1.27$
$3(FeO) + 2[Cr] \leftrightarrow 3Fe + (Cr_2O_3)$	$\log_{10}\left(K_{eq}^{FeO\leftrightarrow Cr_2O_3}\right) = 0.3$
$3(SiO_2) + 4[Al] \leftrightarrow 3[Si] + 2(Al_2O_3)$	$\log_{10}\left(K_{eq}^{SiO_2\leftrightarrow Al_2O_3}\right) = \frac{17065}{\frac{1}{2}\left(T_l^{steel}+T_l^{slag}\right)+273.15} - 14.465$

Table A3. Forward reaction rates for equilibrium reactions. x denotes mass fractions and k_f refers to kinetic model constants for each reaction.

Reaction	Forward Reaction Rate	Units
$(FeO)+[C]\leftrightarrow Fe+(CO)$	$F_f^{FeO\leftrightarrow CO}=(0.11\cdot 10^4)K_f^{FeO\leftrightarrow CO}x_{FeO}^{slag}x_C^{steel}$	$\frac{mol\ FeO}{s}$
$(FeO)+[Mn]\leftrightarrow Fe+(MnO)$	$F_f^{FeO\leftrightarrow MnO}=10^4\cdot K_f^{FeO\leftrightarrow MnO}x_{FeO}^{slag}x_{Mn}^{steel}$	$\frac{mol\ FeO}{s}$
$(MnO)+[C]\leftrightarrow [Mn]+\{CO\}$	$F_f^{MnO\leftrightarrow CO}=(0.017\cdot 10^4)K_f^{MnO\leftrightarrow CO}x_{MnO}^{slag}x_C^{steel}$	$\frac{mol\ MnO}{s}$
$2(FeO)+[Si]\leftrightarrow 2Fe+(SiO_2)$	$F_f^{FeO\leftrightarrow SiO_2}=(2\cdot 10^4)K_f^{FeO\leftrightarrow SiO_2}\left(x_{FeO}^{slag}\right)^2x_{Si}^{steel}$	$\frac{mol\ FeO}{s}$
$2(MnO)+[Si]\leftrightarrow 2[Mn]+(SiO_2)$	$F_f^{MnO\leftrightarrow SiO_2}=(2\cdot 10^4)K_f^{MnO\leftrightarrow SiO_2}\left(x_{MnO}^{slag}\right)^2x_{Si}^{steel}$	$\frac{mol\ MnO}{s}$
$3(FeO)+2[Cr]\leftrightarrow 3Fe+(Cr_2O_3)$	$F_f^{FeO\leftrightarrow Cr_2O_3}=(3\cdot 10^4)K_f^{FeO\leftrightarrow Cr_2O_3}x_{FeO}^{slag}x_{Cr}^{steel}$	$\frac{mol\ FeO}{s}$
$3(SiO_2)+4[Al]\leftrightarrow 3[Si]+2(Al_2O_3)$	$F_f^{SiO_2\leftrightarrow Al_2O_3}=(1.5\cdot 10^4)K_f^{SiO_2\leftrightarrow Al_2O_3}x_{SiO_2}^{slag}x_{Al}^{steel}$	$\frac{mol\ SiO_2}{s}$

Table A4. Backward reaction rates for equilibrium reactions. x denotes mass fractions, k_f refers to kinetic model constants for each reaction, p_{CO} is the partial pressure of CO and M denotes molar masses.

Reaction	Backward Reaction Rate	Units
$(FeO)+[C]\leftrightarrow Fe+\{CO\}$	$F_b^{FeO\leftrightarrow CO}=\frac{K_f^{FeO\leftrightarrow CO}}{K_{eq}^{FeO\leftrightarrow CO}}p_{CO}$	$\frac{mol\ Fe\ (FeO)}{s}$
$(FeO)+[Mn]\leftrightarrow Fe+(MnO)$	$F_b^{FeO\leftrightarrow MnO}=10^2\cdot\frac{K_f^{FeO\leftrightarrow MnO}}{K_{eq}^{FeO\leftrightarrow MnO}}x_{MnO}^{slag}$	$\frac{mol\ Fe\ (FeO)}{s}$
$(MnO)+[C]\leftrightarrow [Mn]+\{CO\}$	$F_b^{MnO\leftrightarrow CO}=10^2\cdot\frac{K_f^{MnO\leftrightarrow CO}}{K_{eq}^{MnO\leftrightarrow CO}}p_{CO}x_{Mn}^{steel}$	$\frac{mol\ Mn\ (MnO)}{s}$
$2(FeO)+[Si]\leftrightarrow 2Fe+(SiO_2)$	$F_b^{FeO\leftrightarrow SiO_2}=(2\cdot 10^2)\frac{K_f^{FeO\leftrightarrow SiO_2}}{K_{eq}^{FeO\leftrightarrow SiO_2}}x_{SiO_2}^{slag}$	$\frac{mol\ Fe\ (FeO)}{s}$
$2(MnO)+[Si]\leftrightarrow 2[Mn]+(SiO_2)$	$F_b^{MnO\leftrightarrow SiO_2}=(2\cdot 10^4)\frac{K_f^{MnO\leftrightarrow SiO_2}}{K_{eq}^{MnO\leftrightarrow SiO_2}}\left(x_{Mn}^{steel}\right)^2x_{SiO_2}^{slag}$	$\frac{mol\ Mn\ (MnO)}{s}$
$3(FeO)+2[Cr]\leftrightarrow 3Fe+(Cr_2O_3)$	$F_b^{FeO\leftrightarrow SiO_2}=\left(\frac{6\cdot 10^2\cdot M_{Cr}}{M_{Cr_2O_3}}\right)\frac{K_f^{FeO\leftrightarrow Cr_2O_3}}{K_{eq}^{FeO\leftrightarrow Cr_2O_3}}x_{Cr_2O_3}^{slag}$	$\frac{mol\ Fe\ (FeO)}{s}$
$3(SiO_2)+4[Al]\leftrightarrow 3[Si]+2(Al_2O_3)$	$F_b^{SiO_2\leftrightarrow Al_2O_3}=(1.5\cdot 10^4)\frac{K_f^{SiO_2\leftrightarrow Al_2O_3}}{K_{eq}^{SiO_2\leftrightarrow Al_2O_3}}x_{Si}^{steel}x_{Al_2O_3}^{slag}$	$\frac{mol\ Si\ (SiO_2)}{s}$

Table A5. Species rates of change for each steel–slag equilibrium reaction.

Reaction	Species i Rate of Change
$(FeO)+[C]\leftrightarrow Fe+\{CO\}$	$F_i^{FeO\leftrightarrow CO}=\frac{v_i^{FeO\leftrightarrow CO}}{v_{FeO}^{FeO\leftrightarrow CO}}\left(F_f^{FeO\leftrightarrow CO}-F_b^{FeO\leftrightarrow CO}\right)$
$(FeO)+[Mn]\leftrightarrow Fe+(MnO)$	$F_i^{FeO\leftrightarrow MnO}=\frac{v_i^{FeO\leftrightarrow MnO}}{v_{FeO}^{FeO\leftrightarrow MnO}}\left(F_f^{FeO\leftrightarrow MnO}-F_b^{FeO\leftrightarrow MnO}\right)$
$(MnO)+[C]\leftrightarrow [Mn]+\{CO\}$	$F_i^{MnO\leftrightarrow CO}=\frac{v_i^{MnO\leftrightarrow CO}}{v_{MnO}^{MnO\leftrightarrow CO}}\left(F_f^{MnO\leftrightarrow CO}-F_b^{MnO\leftrightarrow CO}\right)$
$2(FeO)+[Si]\leftrightarrow 2Fe+(SiO_2)$	$F_i^{FeO\leftrightarrow SiO_2}=\frac{v_i^{FeO\leftrightarrow SiO_2}}{v_{FeO}^{FeO\leftrightarrow SiO_2}}\left(F_f^{FeO\leftrightarrow SiO_2}-F_b^{FeO\leftrightarrow SiO_2}\right)$
$2(MnO)+[Si]\leftrightarrow 2[Mn]+(SiO_2)$	$F_i^{MnO\leftrightarrow SiO_2}=\frac{v_i^{MnO\leftrightarrow SiO_2}}{v_{MnO}^{MnO\leftrightarrow SiO_2}}\left(F_f^{MnO\leftrightarrow SiO_2}-F_b^{MnO\leftrightarrow SiO_2}\right)$
$3(FeO)+2[Cr]\leftrightarrow 3Fe+(Cr_2O_3)$	$F_i^{FeO\leftrightarrow Cr_2O_3}=\frac{v_i^{FeO\leftrightarrow Cr_2}}{v_{FeO}^{FeO\leftrightarrow Cr_2}O_3}\left(F_f^{FeO\leftrightarrow Cr_2O_3}-F_b^{FeO\leftrightarrow Cr_2O_3}\right)$

The molar rates of change due to reactions $F_i\left(\frac{mol}{s}\right)$ can be combined with the molar mass of each component $M_i\left(\frac{g}{mol}\right)$ to calculate the total mass rates of change of all dynamic state components $W_i\left(\frac{kg}{s}\right)$:

$$W_i = \left(1.0 \times 10^{-3}\right) M_i \left(\sum_{\text{oxygen}} F_i^{\text{oxygen}} + \sum_{\text{equilibrium}} F_i^{\text{equilibrium}} \right) \tag{A14}$$

Appendix A.3. Overall Heat and Mass Balances

Many of the heat and mass balances refer to the area fractions x_{inner} and x_{outer}:

$$\begin{aligned} x_{\text{inner}} &= \left(\frac{r_{\text{inner}}}{r_{\text{furnace}}} \right)^2 \\ x_{\text{outer}} &= 1 - x_{\text{inner}} \end{aligned} \tag{A15}$$

Depending on the sub- and super-scripts, $x_{(\text{solid/liquid})}^{(\text{inner/outer})}$ refers to the solid (s) or liquid (l) mass fractions in the inner or outer control volume. Similarly, $x_{i,\,(\text{solid/liquid})}^{(\text{inner/outer})}$.

In the following overall steel mass balances, u_{scrap} refers to scrap metal input to the furnace.

$$\begin{aligned} \frac{dm_{\text{solid}}^{\text{inner}}}{dt} &= x_{\text{inner}} u_{\text{scrap}} - r_{\text{melt}}^{\text{inner}} + r_{\text{freeze}}^{\text{inner}} \\ \frac{dm_{\text{solid}}^{\text{outer}}}{dt} &= x_{\text{outer}} u_{\text{scrap}} - r_{\text{melt}}^{\text{outer}} + r_{\text{freeze}}^{\text{inner}} \\ \frac{dm_{\text{liquid}}^{\text{inner}}}{dt} &= r_{\text{melt}}^{\text{inner}} - r_{\text{freeze}}^{\text{inner}} + \sum_{i,\text{ steel}} W_{i,\text{ steel}}^{\text{inner}} \\ \frac{dm_{\text{liquid}}^{\text{outer}}}{dt} &= r_{\text{melt}}^{\text{outer}} - r_{\text{freeze}}^{\text{outer}} + \sum_{i,\text{ steel}} W_{i,\text{ steel}}^{\text{outer}} \end{aligned} \tag{A16}$$

In the following overall slag mass balances, u_{slag} refers to slag input to the furnace.

$$\begin{aligned} \frac{dm_{\text{solid}}^{\text{slag}}}{dt} &= u_{\text{slag}} - r_{\text{melt}}^{\text{slag}} + r_{\text{freeze}}^{\text{slag}} \\ \frac{dm_{\text{liquid}}^{\text{slag}}}{dt} &= r_{\text{melt}}^{\text{slag}} - r_{\text{freeze}}^{\text{slag}} + \sum_{i,\text{ slag}} \left(W_{i,\text{ slag}}^{\text{inner}} + W_{i,\text{ slag}}^{\text{outer}} \right) \end{aligned} \tag{A17}$$

The species i fraction of the charged scrap x_i^{scrap} enters the scrap component balances:

$$\begin{aligned} \frac{dm_{i,\text{ solid}}^{\text{inner}}}{dt} &= x_{\text{inner}} u_{\text{scrap}} x_i^{\text{scrap}} - r_{\text{melt}}^{\text{inner}} x_{i,\text{ solid}}^{\text{inner}} + r_{\text{freeze}}^{\text{inner}} x_{i,\text{ liquid}}^{\text{inner}} \\ \frac{dm_{i,\text{ solid}}^{\text{outer}}}{dt} &= x_{\text{outer}} u_{\text{scrap}} x_i^{\text{scrap}} - r_{\text{melt}}^{\text{outer}} x_{i,\text{ solid}}^{\text{outer}} + r_{\text{freeze}}^{\text{outer}} x_{i,\text{ liquid}}^{\text{outer}} \\ \frac{dm_{i,\text{ liquid}}^{\text{inner}}}{dt} &= r_{\text{melt}}^{\text{inner}} x_{i,\text{ solid}}^{\text{inner}} - r_{\text{freeze}}^{\text{inner}} x_{i,\text{ liquid}}^{\text{inner}} + \sum_{i,\text{ steel}} W_{i,\text{ steel}}^{\text{inner}} \\ \frac{dm_{i,\text{ liquid}}^{\text{outer}}}{dt} &= r_{\text{melt}}^{\text{outer}} x_{i,\text{ solid}}^{\text{outer}} - r_{\text{freeze}}^{\text{outer}} x_{i,\text{ liquid}}^{\text{outer}} + \sum_{i,\text{ steel}} W_{i,\text{ steel}}^{\text{outer}} \end{aligned} \tag{A18}$$

The species i fraction of the charged slag x_i^{slag} enters the slag component balances:

$$\begin{aligned} \frac{dm_{i,\text{ solid}}^{\text{slag}}}{dt} &= u_{\text{slag}} x_i^{\text{slag}} - r_{\text{melt}}^{\text{slag}} x_{i,\text{ solid}}^{\text{slag}} + r_{\text{freeze}}^{\text{slag}} x_{i,\text{ liquid}}^{\text{slag}} \\ \frac{dm_{i,\text{ liquid}}^{\text{slag}}}{dt} &= r_{\text{melt}}^{\text{slag}} x_{i,\text{ solid}}^{\text{slag}} - r_{\text{freeze}}^{\text{slag}} x_{i,\text{ liquid}}^{\text{slag}} + \sum_{i_{\text{slag}}} W_{i_{\text{slag}}} \end{aligned} \tag{A19}$$

The mol balance of CO gas is given:

$$\frac{dn_{CO}}{dt} = \sum_{\text{equilibrium}} F_{CO}^{\text{equilibrium}} - 2F_{O_2}^{CO} \tag{A20}$$

The temperature balances of steel phases:

$$\begin{aligned}
\frac{dT_{\text{solid}}^{\text{inner}}}{dt} &= \frac{\begin{array}{r} x_{\text{solid}}^{\text{inner}} P_{\text{inner}} - Q_{\text{solid-liquid}}^{\text{inner-inner}} - Q_{\text{solid-solid}}^{\text{inner-outer}} - Q_{\text{solid-liquid}}^{\text{inner-outer}} - Q_{\text{solid-roof}}^{\text{inner}} - Q_{\text{solid-vessel}}^{\text{inner}} - Q_{\text{solid-gas}}^{\text{inner}} \\ - Q_{\text{solid-solid}}^{\text{inner-slag}} - Q_{\text{solid-liquid}}^{\text{inner-slag}} + Q_{\text{freeze}}^{\text{inner}} - Q_{\text{melt}}^{\text{inner}} - C_{\text{p,solid}} x_{\text{inner}} u_{\text{scrap}} \left(T_{\text{solid}}^{\text{inner}} - T_{\text{ambient}}\right) \end{array}}{C_{\text{p, solid}} m_{\text{solid}}^{\text{inner}}} \\
\frac{dT_{\text{solid}}^{\text{outer}}}{dt} &= \frac{\begin{array}{r} x_{\text{solid}}^{\text{outer}} P_{\text{outer}} - Q_{\text{solid-liquid}}^{\text{outer-outer}} + Q_{\text{solid-solid}}^{\text{inner-outer}} - Q_{\text{liquid-solid}}^{\text{inner-outer}} - Q_{\text{solid-roof}}^{\text{outer}} - Q_{\text{solid-vessel}}^{\text{outer}} - Q_{\text{solid-gas}}^{\text{outer}} \\ - Q_{\text{solid-solid}}^{\text{outer-slag}} - Q_{\text{solid-liquid}}^{\text{outer-slag}} + Q_{\text{freeze}}^{\text{outer}} - Q_{\text{melt}}^{\text{outer}} - C_{\text{p,solid}} x_{\text{outer}} u_{\text{scrap}} \left(T_{\text{solid}}^{\text{outer}} - T_{\text{ambient}}\right) \end{array}}{C_{\text{p, solid}} m_{\text{solid}}^{\text{outer}}} \\
\frac{dT_{\text{liquid}}^{\text{inner}}}{dt} &= \frac{\begin{array}{r} x_{\text{liquid}}^{\text{inner}} P_{\text{inner}} + Q_{\text{solid-liquid}}^{\text{inner-inner}} - Q_{\text{liquid-solid}}^{\text{inner-outer}} - Q_{\text{liquid-liquid}}^{\text{inner-outer}} - Q_{\text{liquid-roof}}^{\text{inner}} - Q_{\text{liquid-vessel}}^{\text{inner}} - Q_{\text{liquid-gas}}^{\text{inner}} \\ - Q_{\text{liquid-solid}}^{\text{inner-slag}} - Q_{\text{liquid-liquid}}^{\text{inner-slag}} + Q_{\text{freeze}}^{\text{inner}} - Q_{\text{melt}}^{\text{inner}} \end{array}}{C_{\text{p, liquid}} m_{\text{liquid}}^{\text{inner}}} \\
\frac{dT_{\text{liquid}}^{\text{outer}}}{dt} &= \frac{\begin{array}{r} x_{\text{liquid}}^{\text{outer}} P_{\text{outer}} + Q_{\text{solid-liquid}}^{\text{outer-outer}} + Q_{\text{solid-liquid}}^{\text{inner-outer}} + Q_{\text{liquid-liquid}}^{\text{inner-outer}} - Q_{\text{liquid-roof}}^{\text{outer}} - Q_{\text{liquid-vessel}}^{\text{outer}} - Q_{\text{liquid-gas}}^{\text{outer}} \\ - Q_{\text{liquid-solid}}^{\text{outer-slag}} - Q_{\text{liquid-liquid}}^{\text{outer-slag}} + Q_{\text{freeze}}^{\text{outer}} - Q_{\text{melt}}^{\text{outer}} \end{array}}{C_{\text{p, liquid}} m_{\text{liquid}}^{\text{outer}}}
\end{aligned} \tag{A21}$$

The temperature balances of slag phases:

$$\begin{aligned}
\frac{dT_{\text{solid}}^{\text{slag}}}{dt} &= \frac{\begin{array}{r} Q_{\text{solid-solid}}^{\text{inner-slag}} + Q_{\text{solid-solid}}^{\text{outer-slag}} + Q_{\text{liquid-solid}}^{\text{inner-slag}} + Q_{\text{liquid-solid}}^{\text{outer-slag}} \\ + Q_{\text{freeze}}^{\text{slag}} - Q_{\text{melt}}^{\text{slag}} - C_{\text{p, slag}} u_{\text{slag}} \left(T_{\text{solid}}^{\text{slag}} - T_{\text{ambient}}\right) \end{array}}{C_{\text{p, slag}} m_{\text{solid}}^{\text{slag}}} \\
\frac{dT_{\text{liquid}}^{\text{slag}}}{dt} &= \frac{\begin{array}{r} Q_{\text{solid-liquid}}^{\text{inner-slag}} + Q_{\text{solid-liquid}}^{\text{outer-slag}} + Q_{\text{liquid-liquid}}^{\text{inner-slag}} + Q_{\text{liquid-liquid}}^{\text{outer-slag}} \\ + Q_{\text{freeze}}^{\text{slag}} - Q_{\text{melt}}^{\text{slag}} \end{array}}{C_{\text{p, slag}} m_{\text{liquid}}^{\text{slag}}}
\end{aligned} \tag{A22}$$

The temperature balances of gas phases:

$$\begin{aligned}
\frac{dT_{\text{gas}}}{dt} &= \frac{\begin{array}{r} P_{\text{gas}} + Q_{\text{solid-gas}}^{\text{inner}} + Q_{\text{solid-gas}}^{\text{outer}} + Q_{\text{liquid-gas}}^{\text{inner}} + Q_{\text{liquid-gas}}^{\text{outer}} - Q_{\text{gas-roof}} - Q_{\text{gas-vessel}} \\ - F_{\text{LNG}} C_{\text{p, LNG}} \left(T_{\text{gas}} - T_{\text{ambient}}\right) - F_{O_2} C_{\text{p}, O_2} \left(T_{\text{gas}} - T_{\text{ambient}}\right) \end{array}}{C_{\text{p, gas}} n_{\text{gas}}} \\
\frac{dT_{\text{offgas}}}{dt} &= \frac{F_{\text{offgas}} C_{\text{p, gas}} \left(T_{\text{gas}} - T_{\text{offgas}}\right) + 0.05 F_{\text{air}} C_{\text{p, air}} \left(T_{\text{ambient}} - T_{\text{offgas}}\right)}{2 C_{\text{p, gas}} n_{\text{gas}}}
\end{aligned} \tag{A23}$$

n_{gas} is calculated from the ideal gas law using standard temperature and pressure (STP) conditions ($T = 0$ °C, $p = 100$ kPa) and the furnace volume given by the dimensions in Appendix A.4. $C_{\text{p, gas}}$ is calculated from the LNG and oxygen flow rates and heat capacities:

$$C_{\text{p, gas}} = \frac{F_{\text{LNG}} C_{\text{p, LNG}} + F_{O_2} C_{\text{p}, O_2}}{F_{\text{LNG}} + F_{O_2}} \tag{A24}$$

The temperature balances of furnace equipment:

$$\frac{\mathrm{d}T_{\text{roof}}}{\mathrm{d}t} = \frac{P_{\text{arc}}^{\text{roof}} + Q_{\text{solid-roof}}^{\text{inner}} + Q_{\text{solid-roof}}^{\text{outer}} + Q_{\text{liquid-roof}}^{\text{inner}} + Q_{\text{liquid-roof}}^{\text{outer}} + Q_{\text{gas-roof}} - Q_{\text{roof-water}} - Q_{\text{roof-vessel}}}{C_{\text{p, roof}} m_{\text{roof}}}$$

$$\frac{\mathrm{d}T_{\text{vessel}}}{\mathrm{d}t} = \frac{P_{\text{arc}}^{\text{vessel}} + Q_{\text{solid-vessel}}^{\text{inner}} + Q_{\text{solid-vessel}}^{\text{outer}} + Q_{\text{liquid-vessel}}^{\text{inner}} + Q_{\text{liquid-vessel}}^{\text{outer}} + Q_{\text{gas-vessel}} - Q_{\text{vessel-water}} + Q_{\text{roof-vessel}}}{C_{\text{p, vessel}} m_{\text{vessel}}} \tag{A25}$$

Appendix A.4. Model Constants and Dimensions

Table A6. Model constants and dimensions. An empty entry in the **Units** column indicates a unitless quantity.

Constant	Description	Value	Units
d_{furnace}	Furnace diameter	8.1	m
d_{inner}	Inner control volume diameter	4.65	m
h_{furnace}	Furnace height	5.2	m
h_{panel}	Cooling water panel height	2.89	m
$C_{\text{p, solid}}$	Heat capacity of solid steel	39	$\frac{\text{J}}{\text{mol K}}$
$C_{\text{p, liquid}}$	Heat capacity of liquid steel	46	$\frac{\text{J}}{\text{mol K}}$
$C_{\text{p, slag}}$	Heat capacity of slag	50	$\frac{\text{J}}{\text{mol K}}$
ρ_{solid}	Density of solid steel	2000	$\frac{\text{kg}}{\text{m}^3}$
ρ_{liquid}	Density of liquid steel	7000	$\frac{\text{kg}}{\text{m}^3}$
k_{phase}	Phase change constant	0.005	$\frac{1}{\text{s}}$
k_{CO}	Limiting constant for CO combustion	0.25	$\frac{1}{\text{s}}$
$x_{\text{arc}}^{\text{gas}}$	Fraction of arc power used to heat gas	0.05	
$x_{\text{vessel}}^{\text{loss}}$	Fraction of arc losses used to heat vessel	0.3	
k_{ss}	Heat transfer coefficient: solid steel–solid steel	400	$\frac{\text{W}}{\text{m}^2\text{K}}$
k_{sl}	Heat transfer coefficient: solid steel–liquid steel	12,000	$\frac{\text{W}}{\text{m}^2\text{K}}$
k_{ll}	Heat transfer coefficient: liquid steel–liquid steel	60,000	$\frac{\text{W}}{\text{m}^2\text{K}}$
k_{cs}	Heat transfer coefficient: solid slag–solid steel	2000	$\frac{\text{W}}{\text{m}^2\text{K}}$
k_{cl}	Heat transfer coefficient: solid slag–liquid steel	2000	$\frac{\text{W}}{\text{m}^2\text{K}}$
k_{bs}	Heat transfer coefficient: liquid slag–solid steel	5	$\frac{\text{W}}{\text{m}^2\text{K}}$
k_{bl}	Heat transfer coefficient: liquid slag–liquid steel	5	$\frac{\text{W}}{\text{m}^2\text{K}}$
k_{sg}	Heat transfer coefficient: solid steel–gas	20	$\frac{\text{W}}{\text{m}^2\text{K}}$
k_{lg}	Heat transfer coefficient: liquid steel–gas	10	$\frac{\text{W}}{\text{m}^2\text{K}}$
k_{gr}	Heat transfer coefficient: gas–roof	25	$\frac{\text{W}}{\text{m}^2\text{K}}$
k_{gv}	Heat transfer coefficient: gas–vessel	25	$\frac{\text{W}}{\text{m}^2\text{K}}$
k_{rw}	Heat transfer coefficient: roof–water	300	$\frac{\text{W}}{\text{m}^2\text{K}}$
k_{vw}	Heat transfer coefficient: vessel–water	300	$\frac{\text{W}}{\text{m}^2\text{K}}$
ε_{s}	Emissivity of solid steel	0.4	
ε_{l}	Emissivity of liquid steel	0.6	
ε_{r}	Emissivity of furnace roof	0.7	
ε_{v}	Emissivity of side panels	0.5	

References

1. Fruehan, R.J. *The Making, Shaping and Treating of Steel*; The AISE Steel Foundation: Pittsburgh, PA, USA, 1998; ISBN 9780930767020.
2. Hay, T.; Visuri, V.-V.; Aula, M.; Echterhof, T. A Review of Mathematical Process Models for the Electric Arc Furnace Process. *Steel Res. Int.* **2021**, *92*, 2000395. [CrossRef]
3. Bekker, J.G.; Craig, I.K.; Pistorius, P.C. Model Predictive Control of an Electric Arc Furnace Off-gas Process. *Control Eng. Pract.* **2000**, *8*, 445–455. [CrossRef]
4. Oosthuizen, D.J.; Craig, I.K.; Pistorius, P.C. Model Predictive Control of an Electric Arc Furnace Off-Gas Process Combined with Temperature Control. In Proceedings of the IEEE 5th Africon Conference, Cape Town, South Africa, 28 September–1 October 1999; pp. 415–420.
5. Oosthuizen, D.J.; Craig, I.K.; Pistorius, P.C. Economic Evaluation and Design of an Electric Arc Furnace Controller Based on Economic Objectives. *Control Eng. Pract.* **2004**, *12*, 235–265. [CrossRef]
6. MacRosty, R.D.M.; Swartz, C.L.E. Nonlinear Predictive Control of an Electric Arc Furnace. *IFAC Proc. Vol.* **2007**, *40*, 285–290. [CrossRef]
7. Shyamal, S. Dynamic Optimization, State Estimation and Control of Electric Arc Furnace Operation. Ph.D. Thesis, McMaster University, Hamilton, ON, Canada, 2014.
8. Kleimt, B.; Pierre, R.; Dettmer, B.; Deng, J.; Schlinge, L.; Schliephake, H. Continuous Dynamic EAF Process Control for Increased Energy and Resource Efficiency. In Proceedings of the 10th European Electric Steelmaking Conference, Graz, Austria, 25–28 September 2012.
9. Pierre, R.; Kleimt, B.; Schlinge, L.; Unamuno, I.; Arteaga, A. Modelling of EAF Slag Properties for Improved Process Control. In Proceedings of the 11th European Electric Steelmaking Conference, Venice, Italy, 25–27 May 2016.
10. Rekersdrees, T.; Snatkin, H.; Schlinge, L.; Pierre, R.; Kordel, T.; Kleimt, B.; Gogolin, S.; Haverkamp, V. Adaptive EAF Online Control Based on Innovative Sensors and Comprehensive Models. In Proceedings of the 3rd European Steel Technology & Application Days, Vienna, Austria, 26–29 June 2017.
11. Visuri, V.-V.; Jawahery, S.; Hyttinen, N.; Wasbø, S.O. Preliminary Experiences from the Application of Model Predictive Control for the EAF Process in Stainless Steelmaking. In Proceedings of the 4th European Academic Symposium on EAF Steelmaking, Online, 16–18 June 2021.
12. Borrelli, F.; Bemporad, A.; Morari, M. *Predictive Control for Linear and Hybrid Systems*; Cambridge University Press: Cambridge, UK, 2018; ISBN 9781107016880.
13. Simon, D. *Optimal State Estimation: Kalman, H Infinity, and Nonlinear Approaches*; Wiley: Hoboken, NJ, USA, 2006; ISBN 9780471708582.
14. Jiao, K.X.; Chang, Z.Y.; Chen, C.; Zhang, J.L. Thermodynamic Properties and Viscosities of CaO-SiO_2-MgO-Al_2O_3 Slags. *Metall. Mater. Trans. B* **2019**, *50*, 1012–1022. [CrossRef]
15. Turkdogan, E.T. *Fundamentals of Steelmaking*; The Institute of Materials: London, UK, 1996; ISBN 9781861250049.
16. Logar, V.; Dovžan, D.; Škrjanc, I. Modeling and Validation of an Electric Arc Furnace: Part 2, Thermo-chemistry. *ISIJ Int.* **2012**, *52*, 413–423. [CrossRef]
17. Logar, V.; Dovžan, D.; Škrjanc, I. Modeling and Validation of an Electric Arc Furnace: Part 1, Heat and Mass Transfer. *ISIJ Int.* **2012**, *52*, 402–412. [CrossRef]
18. González, O.J.P.; Ramírez-Argáez, M.A.; Conejo, A.N. Mathematical Modeling of the Melting Rate of Metallic Particles in the Electric Arc Furnace. *ISIJ Int.* **2010**, *50*, 9–16. [CrossRef]
19. Bekker, J.G.; Craig, I.K.; Pistorius, P.C. Modeling and Simulation of an Electric Arc Furnace Process. *ISIJ Int.* **1999**, *39*, 23–32. [CrossRef]
20. Logar, V.; Škrjanc, I. Modeling and Validation of the Radiative Heat Transfer in an Electric Arc Furnace. *ISIJ Int.* **2012**, *52*, 1225–1232. [CrossRef]
21. Martell-Chávez, F.; Ramírez-Argáez, M.; Llamas-Terres, A.; Micheloud-Vernackt, O. Theoretical Estimation of Peak Arc Power to Increase Efficiency in Electric Arc Furnaces. *ISIJ Int.* **2013**, *53*, 743–750. [CrossRef]
22. MacRosty, R.D.M.; Swartz, C.L.E. Dynamic Modeling of an Industrial Electric Arc Furnace. *Ind. Eng. Chem. Res.* **2005**, *44*, 8067–8083. [CrossRef]
23. Maciejowski, J.M. *Predictive Control with Constraints*; Prentice Hall: Englewood Cliffs, NJ, USA, 2002; ISBN 9780201398236.
24. Veijola, T.; Aula, M.; Jokinen, M.; Hyttinen, N.; Visuri, V.-V.; Fabritius, T. Modifying the EAF Voltage Tap Profile for Lower Electrode Tip Consumption and Increased Energy Efficiency. In Proceedings of the 4th European Steel Technology and Application Days, Düsseldorf, Germany, 24–28 June 2019.

metals MDPI

Article

Toward a Simplified Arc Impingement Model in a Direct-Current Electric Arc Furnace

Mohamad Al-Nasser [1], Abdellah Kharicha [1,*], Hadi Barati [1,2], Christoph Pichler [3], Gernot Hackl [3], Markus Gruber [3], Anton Ishmurzin [3], Christian Redl [4], Menghuai Wu [2] and Andreas Ludwig [2]

1 Christian Doppler Laboratory for Metallurgical Applications of Magnetohydrodynamics, Franz-Josef Street 18, 8700 Leoben, Austria; Mohamad.al-nasser@unileoben.ac.at (M.A.-N.); Hadi.Barati@unileoben.ac.at (H.B.)

2 Chair for Modeling and Simulation of Metallurgical Processes, Department of Metallurgy, Montanuniversität, Franz-Josef Street 18, 8700 Leoben, Austria; menghuai.wu@unileoben.ac.at (M.W.); ludwig@unileoben.ac.at (A.L.)

3 Department of Modelling and Simulation, RHI Magnesita, Magnesitstrasse 2, 8700 Leoben, Austria; Christoph.Pichler@RHIMagnesita.com (C.P.); Gernot.Hackl@RHIMagnesita.com (G.H.); Markus.Gruber@RHIMagnesita.com (M.G.); Anton.Ishmurzin@RHIMagnesita.com (A.I.)

4 INTECO Melting and Casting Technologies GmbH, Wienerstrasse 25, 8600 Bruck a.d. Mur, Austria; Christian.Redl@inteco.at

* Correspondence: Abdellah.kharicha@unileoben.ac.at

Abstract: A 2D axisymmetric two-phase model was developed to study the effect of an arc impingement on the liquid metal inside an electric arc furnace. In addition to the arc flow dynamics, the model covered the heat transfer and magneto hydrodynamics of the arc and the liquid metal. Through a parametric study, three different parameters were considered to predict the most important factors affecting the arc and overall behaviour of the process: the arc gap, the density of the gas, and the total electric current. Understanding the effect of these parameters can show the key factors altering the arc dynamics. The study showed that the total applied current was the most important parameter that influenced the impingement depth and mixing of the liquid metal. The depth of the impingement and strength of the mixing of the liquid bath were directly proportional to the current applied in the furnace. The initial arc gap distance was found to be crucial for sustaining a continuous and stable arc. The value of the gas density was very important for the velocity profile; however, it had no significant effect on the impingement depth. This showed that a constant density could be used instead of a varying gas density with temperature to increase the computational efficiency. The study assessed the effects of the aforementioned factors on the arc impingement depth, velocity magnitude, and arc stability. The conclusions acquired and challenges are also presented.

Keywords: direct current; electric arc furnace; arc impingement; arc gap; gas density; electric arc; magneto hydrodynamics; computational fluid dynamics

Citation: Al-Nasser, M.; Kharicha, A.; Barati, H.; Pichler, C.; Hackl, G.; Gruber, M.; Ishmurzin, A.; Redl, C.; Wu, M.; Ludwig, A. Toward a Simplified Arc Impingement Model in a Direct-Current Electric Arc Furnace. *Metals* **2021**, *11*, 1482. https://doi.org/10.3390/met11091482

Academic Editors: Thomas Echterhof, Ko-Ichiro Ohno and Ville-Valtteri Visuri

Received: 15 July 2021
Accepted: 10 September 2021
Published: 17 September 2021

1. Introduction

The first use of an electric arc furnace (EAF) was in the 19th century. Although a direct current (DC) in an EAF was first used in the late 1800s, the major development and industrial use was focused on alternating current (AC) furnaces. This was driven by the better efficiency and power transmission of AC power. In the past three decades, the DC-EAF has again been highly utilised in the metallurgical industry through a variety of applications such as steel scrap melting and smelting processes [1]. The shift toward DC furnaces is because they are now believed to have a better power efficiency, less electrode consumption, and a lower level of noise. DC arcs are also more stable by nature compared with AC arcs. As a result of the increasing demand for DC arc furnaces in industry, further improvements in their design and a better understanding of the method of operations are required. A typical configuration of the DC-EAF entails a large liquid metal bath cylinder with refractory walls covered by a conical roof cooled by water. The cathode electrode is mounted on the top of the metal scrap bath separated by the arc gap. A flat anode of graphite is situated directly under the metal bath in direct contact, as shown in Figure 1.

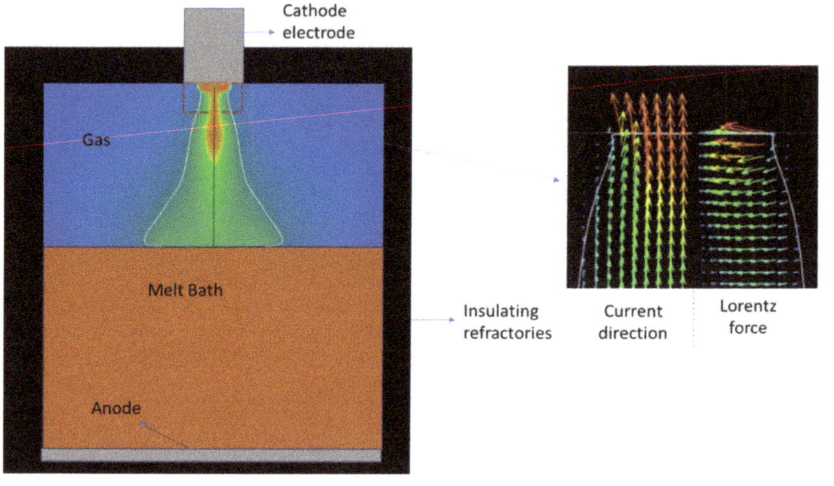

Figure 1. Electric arc furnace schematic.

The arc propagates in this gap providing the thermal and mechanical drive for the furnace.

The arc converts the electric energy supplied by the circuit into thermal and mechanical energy. This arises due to the electrical breakdown of gas as a result of a high potential difference in the arc gap. At high temperatures (5000–7000 K), thermal dissociation is accomplished rendering neutral molecules of gas into conductive ions and electrons [2]. The gas is transformed into thermal plasma that is electrically conductive. The current flows through the arc interacting with its magnetic field. This interaction results in the Lorentz force acting radially inwards along the arc, constricting the arc diameter. At the cathode spot, the Lorentz force acts in an inclined direction toward the cathode spot thus sucking the gas inside, as shown in Figure 1. A powerful jet is created; it is restrained near the cathode spot and then expands gradually as it propagates downwards. Experimental measurements [3] predicted that the speed of the gas inside the jet can reach the orders of km/s. This is capable of creating a thrust that disturbs and penetrates through the surface of the slag and liquid metal, creating a cavity and causing significant splashing that alters the nature and behaviour of the arc. As the electric current flows through the gas arc gap, a high amount of thermal energy is released due to resistive heating that prevents the extinction of the arc as simultaneous heat is released through radiation and convection into the metal bath. This ensures the delivery of the high thermal energy into the metal avoiding any undesired solidification. The impingement also creates a mechanical drive to recirculate and stir the liquid metal.

Historically, the development of arc furnace technology was purely experimental (trial and error). The first mathematical studies were performed on the arc region by Ushio et al. [4] and Szekely et al. [2]. In their work, they represented the thermal and electromagnetic fields through solving the arc region by turbulent Navier–Stokes equations for flow. Alexis et al. [5] developed a mathematical model that predicted the fluid flow and electromagnetic field in addition to the thermal field effect on the liquid metal bath. Although these studies focused on the arc thermal and electromagnetic fields in detail, the major drawback was the lack of momentum coupling with the liquid metal bath. The effect of the electromagnetic fields on the liquid metal bath was covered Szekely et al. [3]

where the thermal effect of the arc on the metal bath without an electromagnetic force was studied. Ramírez et al. [6] performed a simulation to analyse the heat transfer from the arc and electromagnetic effects on the flow inside the fluid bath. The mixing due to the electro-vortex flow drastically affected the distribution of heat from the arc and the durability of the refractories. Kazak [7] studied the effect of the Lorentz force and thermal stresses on the walls of the furnace.

Reynolds [8] analysed the effect of the arc impingement on the liquid bath in a multiphase simulation considering the arc as a gas jet and neglecting the electromagnetic nature and forces. Klementyeva et al. [9] conducted experimental measurements to examine the effect of the arc on the free surface of the liquid metal. A study conducted by Reynolds [10] simulated a lab scale arc impinging the liquid bath of two immiscible phases for different values of the current and arc gap. Adib et al. [11] studied the effect of a high-speed air jet impinging the liquid surface using a VOF formulation and measured the interface deformation and cavity depth. Most of the works reported in the literature have studied one aspect of the DC-EAF but very few of them have introduced the full coupling of the plasma arc and liquid metal interaction. This may be due to the limitation in the computational power and the complexity of the realisation of the arc–liquid interaction.

This paper introduces a study of a multiphase arc simulation. The thermal and electromagnetic fields of the arc are predicted through solving the momentum and induction equations. The liquid metal bath is fully considered with its thermal and electrical properties. This two-way coupling of the arc and free surface of the liquid enable us to capture the arc impingement and the thermal and electromagnetic effect on both the arc and molten metal in a DC-EAF on an industrial scale. The paper relies on several data and parameters from the literature, which was necessary to build upon—instead of spending effort in exploring—what had already been previously achieved. This paper introduces several new advancements to boost the arc modelling especially coupled with liquid bath beneath. Previous studies have mostly focused on the simulation of a single-phase electric arc with a solid surface of a metal bath. A few studies have included the consideration of a liquid metal bath, such as the heat transfer effect, but there has only been one with full coupling and an interaction between the arc and the liquid bath [10] with much smaller current values. This paper assesses the effect of the gas density value and how this assumption helps to reduce the required computational power to achieve the same real-time simulation without greatly affecting the accuracy. The study also aims to explore the major parameters that affect the impingement depth and size.

2. Modelling

The physical arrangement considered in this study covered the micro-environment around the arc rather than the whole furnace geometry for the sake of practical reasons. The configuration consisted of a 2D axisymmetric geometry covering a part of the electrode, the arc gap, and the liquid metal bath. A two-phase flow situation was considered: the gas phase and the liquid metal. The slag layer was not considered. The setup is shown in Figure 2, manifesting the uniform mesh generated for the domain. The numbers in Figure 2 from 1 to 6 indicate the boundary conditions: (1) cathode spot; (2) electrode side; (3) gas top boundary; (4) domain side; (5) metal bath bottom (anode); and (6) the axis of symmetry.

2.1. Governing Equations

The flow was considered to be multiphase where the two fluids were the gas and the liquid metal. The Volume of Fluid (VoF) multiphase model was adopted to simulate the flow in the domain. The continuity, momentum, energy, and induction equations were solved.

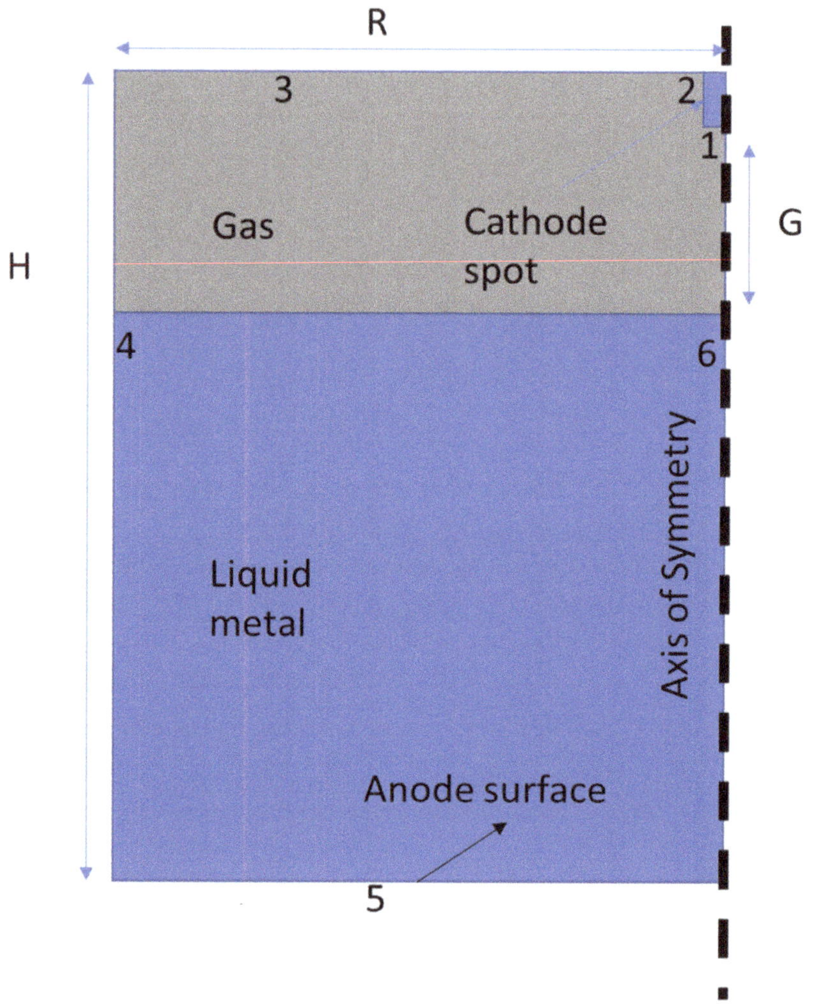

Figure 2. Geometry of the domain: $R = 0.8$, $H = 1$ m. The electrode length was 0.04 and the radius was 0.008; the arc gap (G) was 0.25, and the metal bath depth was initially 0.71 m.

The case was simulated using ANSYS FLUENT® version 14.5 [12]. The equations are described below. The electromagnetic variables were calculated through coupling user-defined functions (UDFs) with fluent solvers.

Continuity:

$$\frac{\partial \rho}{\partial t} + \nabla\left(\rho \vec{U}\right) = 0 \tag{1}$$

Here, ρ is the density and U is the velocity vector field.

Momentum:

$$\frac{\partial\left(\rho \vec{U}\right)}{\partial t} + \nabla \cdot \left(\rho \vec{U}\vec{U}\right) = -\nabla p + \nabla \cdot \left(\mu \nabla \vec{U}\right) + \rho g + F + \frac{1}{\mu_0}\left(\vec{J} \times \vec{B}\right) \tag{2}$$

where p, μ, g, F, μ_0, J, and B stand for the pressure, viscosity, gravitational acceleration, external forces, vacuum permeability, current density, and magnetic field, respectively.

Enthalpy:

$$\frac{\partial}{\partial t}(\rho c_p T) + \nabla\cdot(\rho c_p \vec{U} T) = \nabla\cdot(k\nabla T) + J_{heat} + Rad_{loss} \quad (3)$$

where c_p is the specific heat, T is the temperature, and K is the thermal conductivity. J_{heat} is the joule heating $\frac{\|J^2\|}{\sigma}$, J is the current density, and σ is the electrical conductivity. $Rad_{lo.}$ is the radiation loss. The radiation heat loss was based on experimental measurements [12] for air at 1 atm pressure.

Induction equation:

$$\frac{\partial B_\theta}{\partial t} + \nabla\cdot(\vec{u}B_\theta) = \nabla\left(\frac{1}{\sigma\mu_0}\nabla B_\theta\right) + \frac{\partial}{\partial r}\left(\frac{1}{\sigma\mu_0 r}\right)B_\theta \quad (4)$$

In order to simulate the arc and its interaction with the liquid bath, certain assumptions were adopted due to the high complexities. The flow inside the arc was considered to be always in a local thermal equilibrium. This enabled the use of electrical conductivity solely dependent on the temperature. Electrical conductivity is very low (almost zero) at an atmospheric temperature. It increased to a relatively low order of 1 S/m at a metal bath temperature of 1800 K and increased to reach a high conductivity of 3000 S/m inside the arc where the temperature exceeded 10,000 K [13]. The electrical conductivity of the liquid metal was constant and significantly larger than 80,000 S/m. This created a huge gradient at the interface especially away from the high temperature regions. A small volume fraction of liquid metals can induce a large increase in electrical conductivity. To solve this sudden increase numerically, a harmonic function was implemented:

$$\sigma_{\text{int}} = \frac{\sigma_g\sigma_m}{\alpha_g\sigma_g + (1-\alpha_g)\sigma_m} \quad (5)$$

where σ and α stand for the electrical conductivity and volume fraction, respectively, and g denotes gas and m denotes metal.

Metal properties are assumed to be constant. Gas properties, except density, are temperature-dependent [13]. The dependency of the density induces strong numerical difficulties. This is why we considered the density to be constant and independent of the temperature. This study aimed to predict the overall arc behaviour as a result of an impingement rather than the exact detailed behaviour of the arc. To validate this assumption, a parametric study on the density of the gas was carried out. The assumption of a constant density led to a better stability of the simulation and thus the calculation of the time step could be increased. Other properties, such as the specific heat and dynamic viscosity of the gas, are temperature-dependent, based on Boulos [13]. The implementation of a radiation model was added to the energy equation that considered the liquid metal bulk and solid surfaces to be opaque while the liquid surface had a temperature of 1800 K. The radiation inside the domain was calculated in each cell based on this assumption. The properties of the liquid metal are shown in Table 1.

Table 1. Liquid metal properties.

Density	Specific Heat	Thermal Conductivity	Viscosity	Electrical Conductivity
7000 kg/m^3	800 J/kg	40 W/mK	0.006 Pa.s	800,000 S

2.2. Simulation Settings

The top boundary of the domain was considered to be cooled by water, the surface temperature was 500 K and the temperature of the water was 300 K. All other boundaries were considered to be thermally adiabatic walls. The temperature of the cathode spot was set to be constant at 4000 K [14] and the metal bath was considered to be initially in a liquid phase with a uniform temperature of 1800 K. The total current was supplied at the top of the electrode and the cathode spot had a constant current density of 4.4×10^7 A/m^2 [15]. The cathode spot area was calculated by dividing the total current by the current density and the electrode radius was set to be equal to the cathode spot for simplicity. No external magnetic field was considered.

The initial temperature distribution inside the arc environment was based on the thermal distribution of an arc with a 0.25 cm arc gap [6]. It is important to point out that the arc gap defined the vertical distance between the cathode spot and the initial flat surface of the conducting liquid. This was different from the actual arc length presented at the end of this paper. The actual arc length was the length of the arc connecting the cathode spot and the anode spot at the impinged liquid surface. Figure 3 shows the difference between actual arc length and arc gap. The lower part was liquid metal at 1800 K. The boundary conditions for the induction equation are presented in Table 2.

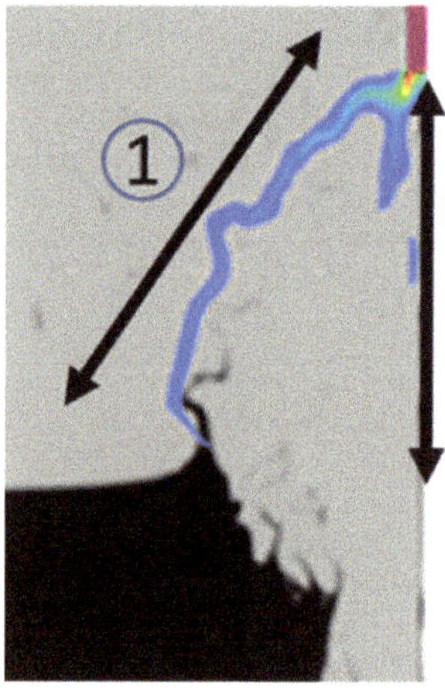

Figure 3. Electric arc during the impingement of the liquid metal: (1) actual arc length; (2) arc gap.

Table 2. Induction equation boundary conditions for the domain.

Boundary	Flow Boundary Conditions	Thermal Boundary Conditions	Induction Equation Boundary Conditions
1. Electrode Bottom	$\vec{U} = 0$	$T = 4000$ K	$\frac{\partial B_\theta}{\partial z} = 0$
2. Electrode Side	$\vec{U} = 0$	$\frac{\partial T}{\partial r} = 0$	$B_\theta = \frac{I_0 \mu_0 r}{2\pi R_c{}^2}$
3. Gas Top	$\frac{\partial U_r}{\partial z} = 0, U_z = 0$	$T = 500$ K	$B_\theta = \frac{I_0 \mu_0}{2\pi r}$
4. Gas and Melt Side	$\vec{U} = 0$	$\frac{\partial T}{\partial r} = 0$	$B_\theta = \frac{I_0 \mu_0}{2\pi r}$
5. Metal Bath Bottom (Anode)	$\vec{U} = 0$	$\frac{\partial T}{\partial z} = 0$	$\frac{\partial B_\theta}{\partial z} = 0$
6. Axis of Symmetry	$\frac{\partial U_z}{\partial r} = 0, U_r = o$	$\frac{\partial T}{\partial r} = 0$	$B_\theta = 0$

The equations were solved using the finite volume method. The simulations were completed in a transient mode for a relatively significant real flow time to achieve the arc impingement effects. As indicated, several simulations were performed to assess the effects of the different parameters on the arc impingement and the stability of the arc. The main variables covered in this study were the arc gap distance, the density of the plasma, and the global current.

The mesh used in the study was a quadratic uniform mesh composed of 200,000 elements over the domain. The cell face size was 2 mm. A mesh dependency test was performed to ensure that the results were not dependent on the mesh size. A total of 4 different meshes were tested with cell sides of 4 mm, 2.66 mm, 2 mm, and 1 mm, which resulted in mesh sizes of 50,000, 125,000, 200,000, and 400,000, respectively. The maximum velocity of the arc before hitting the liquid surface for the different mesh sizes is shown in Figure 4. The depth of the first impingement wave was also plotted, which showed almost no difference over the 4 meshes (Figure 5). This showed that for mesh sizes larger than 200,000 elements, the results were not dependent on the mesh size.

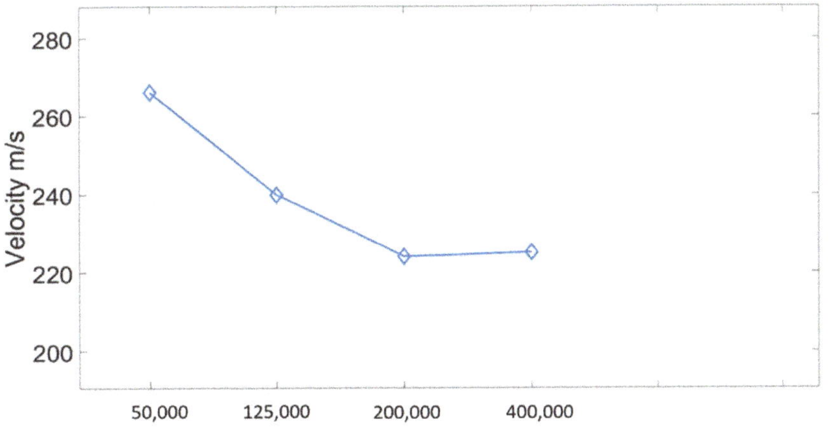

Figure 4. Velocity magnitude as a function of the mesh size.

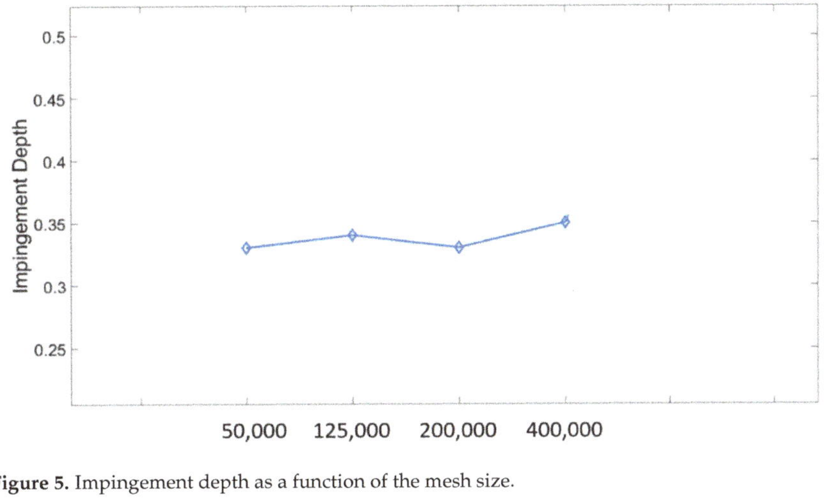

Figure 5. Impingement depth as a function of the mesh size.

3. Results and Discussion

3.1. Reference Case

For an EAF, the arc gap is often set between 0.1 m and 0.5 m [16]; here, the arc gap of choice was 0.25 m as the reference configuration (control case). For the total current, the initial current was equal to 10 kA and the density was constant and equal to 1 kg/m^3. The time step was 10^{-5} s.

Figure 6 depicts the initial state of the domain. The jet emerging from the cathode spot shows the initial current distribution inside the gas before diffusing into the flat surface of the conducting liquid. The yellow contour shows the zoomed area shown in the arc impingement simulation for the gap and density parametric study.

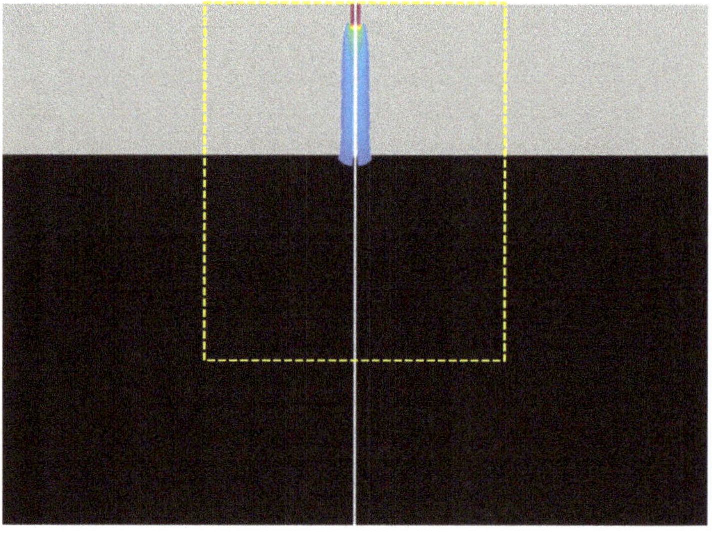

Figure 6. Simulation domain. The yellow frame shows the zoomed domain for a better visual depiction.

Once the arc jet reached the liquid surface, an impingement wave formed pushing the liquid metal downwards and displacing it away from the jet path (Figure 7). Figure 7 illustrates the arc impingement for the reference model during different time snaps. The left half shows the electric current lines (J is the current density) and the right half shows the velocity magnitude of the gas (U is the velocity magnitude). The liquid metal is shown in black. This applies for all the following figures that display the arc impingement. The wave initially propagated vertically and then created a wide cavity. The arc jet pushed the liquid metal droplets upwards, as shown in the 0.286 s image in Figure 7. The cavity expanded and shrank continuously and a wave travelled along the surface. The splashing of droplets appeared to be continuous and the droplets were transported far away from the arc impingement location. A few of the droplets recirculated inside the arc region. The motion of the arc was instantaneously compared with the aerodynamics and motions of the liquid metal. The arc tended to choose the shortest distance in the gas to reach the liquid metal (Figure 7, 0.8 s). When the shortest distance between the electrode and the liquid metal was different from the original path position (near the axis) the arc deviated, as shown in Figure 7, 0.8 s.

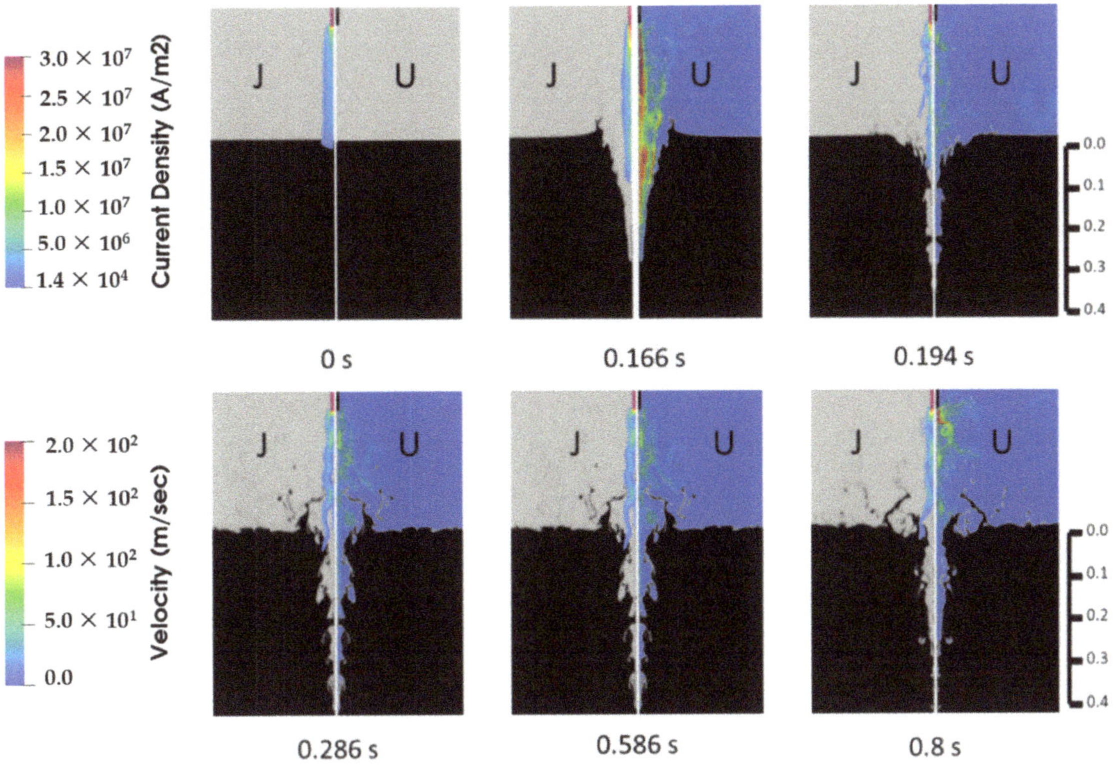

Figure 7. Arc impingement arc gap of 2.5×10^{-1} m.

3.2. *Effect of the Arc Gap*

Apart from the reference case, two other arc gaps of 0.2 m and 0.3 m were also considered. For the gap of 0.2 m, the splashing and impingement width was larger compared with the arc impingement of the 0.25 m gap, as shown in Figure 8. The snaps of the arc were not synchronised between the different cases due to different dynamics. The different snaps in time enabled us to capture the impingement details for each case without the same time constraint. The arc jet was trapped inside the liquid metal. A wave travelled horizontally and it disturbed the surface more than the wave for the 0.25 m reference case. Another important effect was when the liquid metal entered the arc domain. The Lorentz force drifted the liquid metal upward toward the cathode (Figure 8, 0.358 s). This effect led to more violent splashing afterward and a column of liquid metal was pushed downwards with a high speed. A large cavity was created (Figure 8, 1.022 s). The initial impingement depth was less than that in the initial arc gap. This could be related to the shape of the impingement because the force was distributed horizontally due to a wider impingement.

When the arc gap increased to 0.3 m, the arc became less stable and more chaotic movements were observed. The current and the thermal jet had to pass through a longer distance, implying the development of more wavy instability along the arc. This effect was observed initially but as time progressed the arc became more stable, as shown in Figure 9. The impingement depth was similar to the depth of the case with the 0.25 m arc gap.

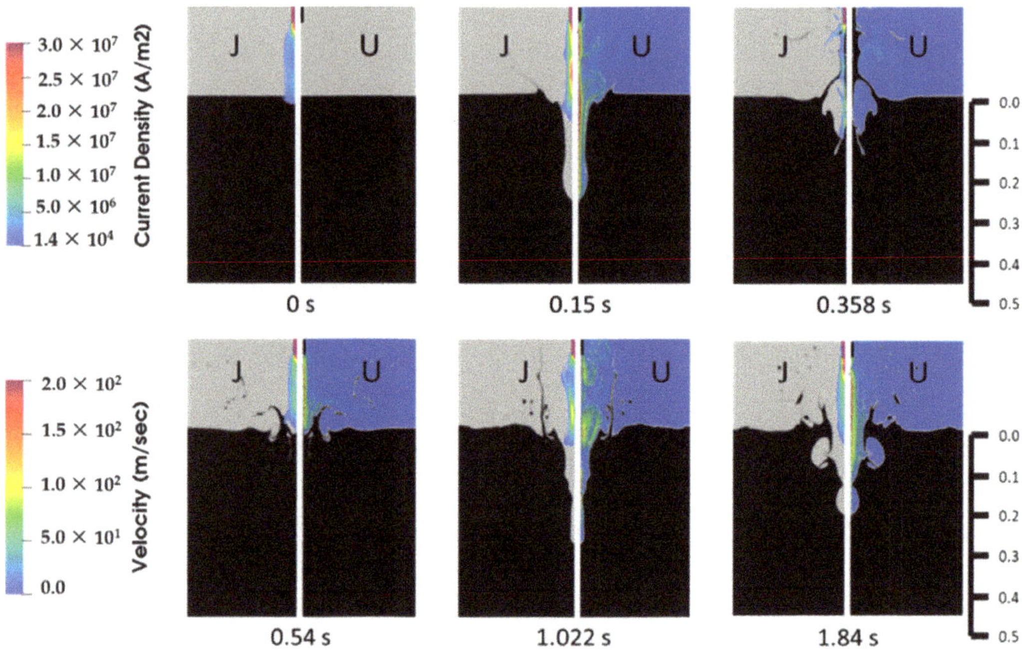

Figure 8. Liquid metal impingement by an arc jet for a 2.0 × 10^{-1} m arc gap.

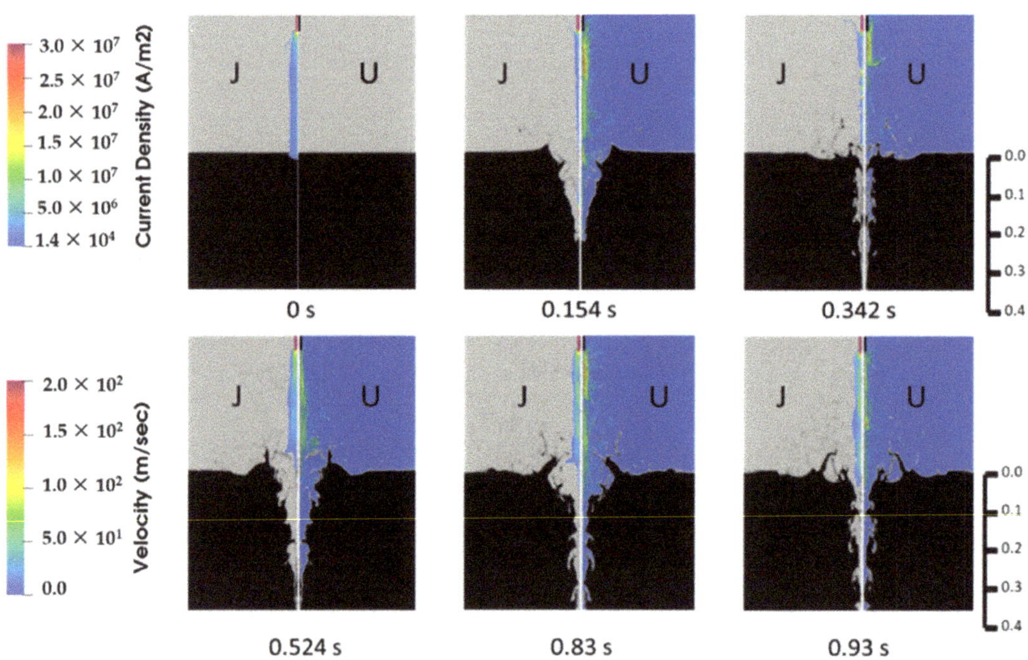

Figure 9. Liquid metal impingement by an arc jet for a 3.0 × 10^{-1} m gap distance.

3.3. Effect of the Gas Density

It is important to note that with a density of unity, a significant drop in the velocity of the gas inside the arc (u = 140 m/s) was noticed whereas the stated velocity in the literatures is in the range of 10^3 m/s. To assess the sensitivity of the variation of the density inside the arc due to a variation in the temperature and the importance of the value of the density, two additional simulations were performed. The only variable changed was the density of the gas, which was taken as 0.1 kg/m^3 and 0.01 kg/m^3.

The results showed that as the density lowered, the velocity of the plasma inside the arc increased significantly. When the density of the gas was equal to 0.01 kg/m^3, the speed reached the speed range in the literature of 10^3 m/s. However, no effect on the overall impingement depth of the arc on the liquid metal was noticed. Figures 10 and 11 show the arc impingements for the different densities of the gas. Although the impingement depth was not significantly changed by the change in the density of the gas, the rate of droplet formation appeared to increase and droplets could reach a higher and further distance despite their smaller diameter. This could be explained by observing the average velocity profile for both cases. The average velocity was 450 m/s for the density of 0.1 kg/m^3 and it was equal to 1.4×10^3 m/s for the density of 0.01 kg/m^3.

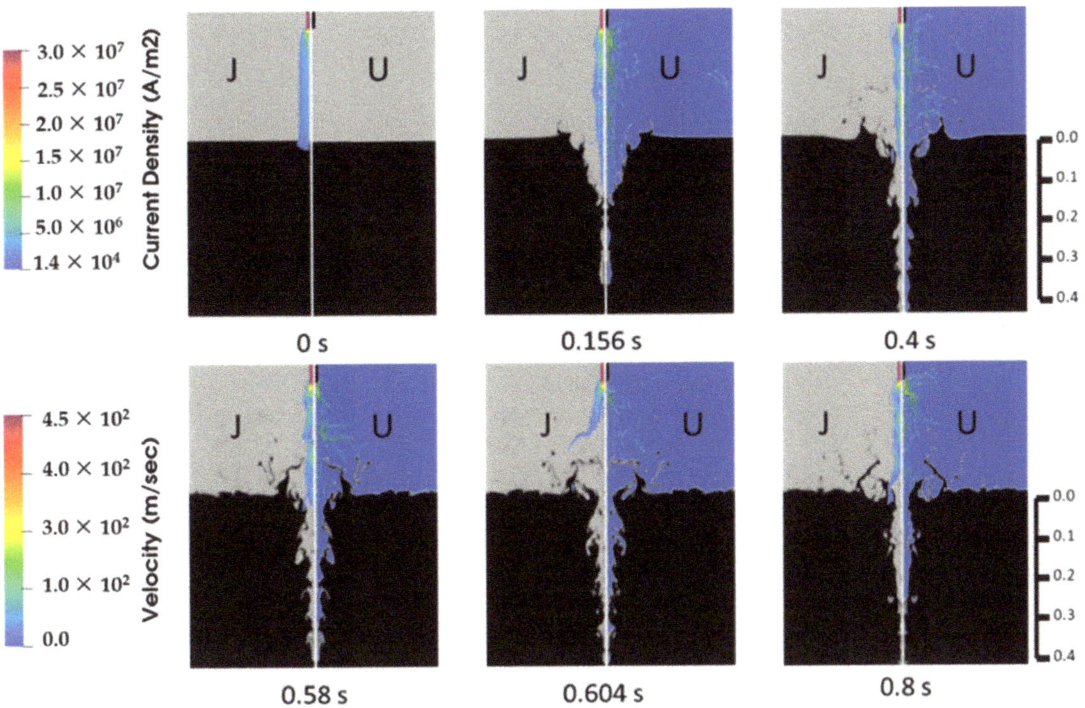

Figure 10. Liquid metal impingement by an arc jet for a gas density of 1.0×10^{-1} kg/m^3.

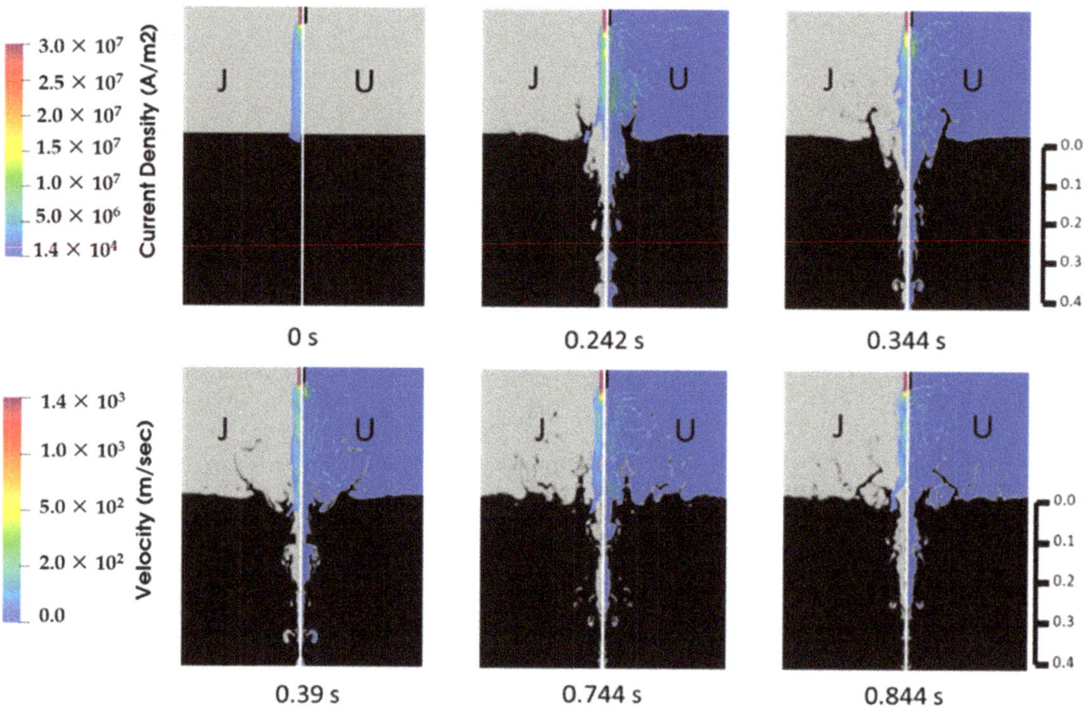

Figure 11. Liquid metal impingement by an arc jet for a gas density of 1×10^{-2} kg/m^3.

3.4. Effect of the Applied Current Magnitude

In addition to the reference case, two Higher current values were considered. A higher applied current led to a deeper arc impingement. Figure 12 shows the extended area shown when a higher current was applied compared with the zoomed area shown in Figure 6. The current values that were applied are 20 kA and 30 kA (Figures 13 and 14, respectively). When a current of 20 kA was applied, a larger arc diameter was observed; this was related to a larger cathode spot of ~12 mm compared with that of the reference case (10 kA) of ~8 mm. When the jet reached the surface, a very large cavity formed, compared with the previous cases. The jet violently pushed the liquid metal in the horizontal and vertical directions. After 0.272 s, the jet reached the bottom of the domain. As the time progressed, the impingement depth was maintained and the droplet splashing reached the top and side boundaries of the domain. Several waves also emerged on the surface and interacted with the droplets falling back into the bath. For the case of 30 kA, similar results appeared but with more powerful effects. Similarly, the arc radius was larger (cathode spot ~15 mm). The cavity was larger than that obtained in the case of 20 kA; thus, much more liquid was displaced. Once again, the jet easily exceeded a cavity depth of 0.71 m and reached the bottom. Due to the powerful jet, several gas bubbles were entrained inside the liquid metal. This led to a Rayleigh–Taylor instability. When these bubbles reached the surface, they caused very violent splashing similar to small explosions due to the very high density difference between the plasma and the liquid metal. The arc behaviour appeared to be very complex although it mainly followed the shortest path to the liquid metal. There appeared to be kink and sausage instabilities in the motion inside the plasma and the arc moved in a helical path. These phenomena were observed through experimental measurements by Reynolds [17].

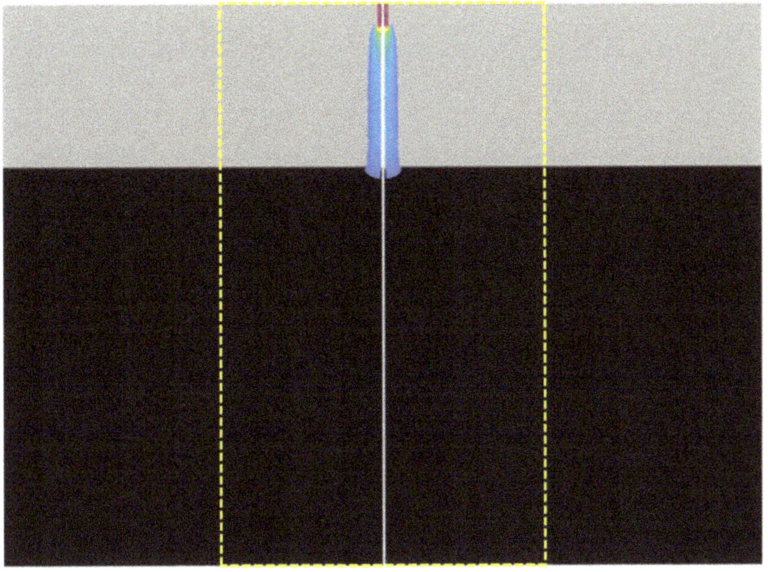

Figure 12. The zoomed domain is enlarged (yellow frame) due to a deeper arc impingement.

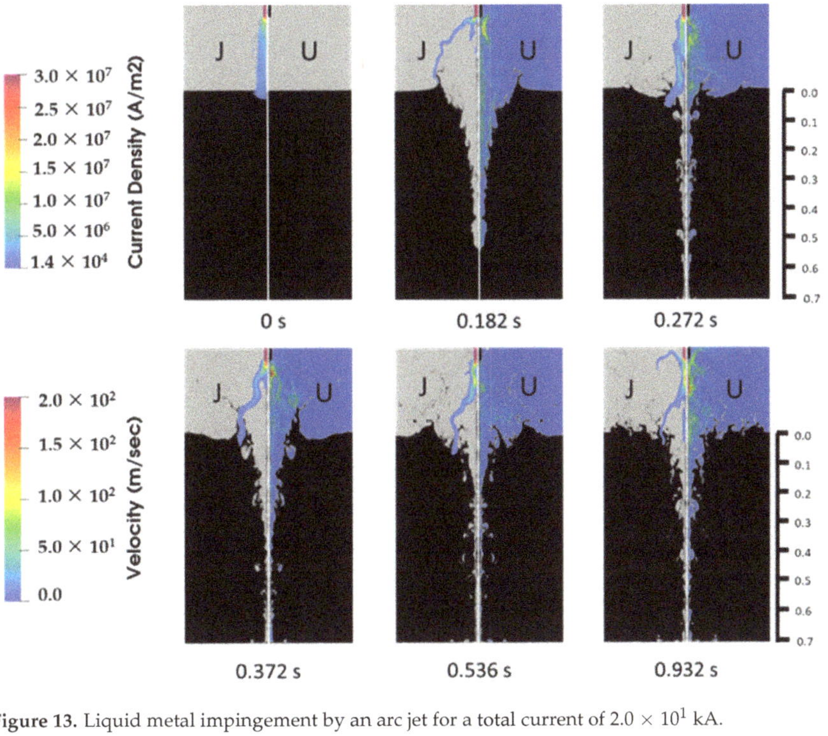

Figure 13. Liquid metal impingement by an arc jet for a total current of 2.0×10^1 kA.

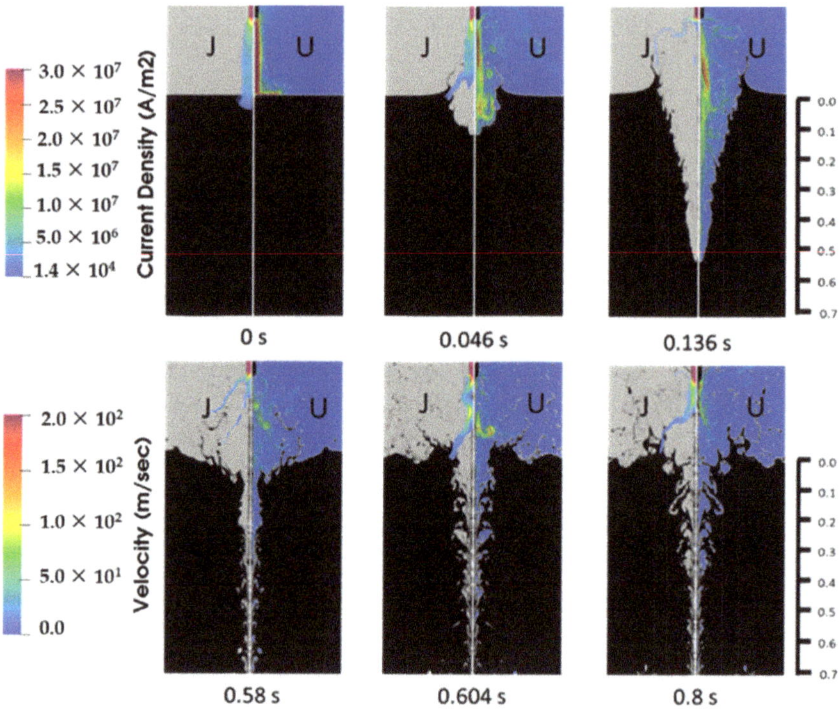

Figure 14. Liquid metal impingement by an arc jet for a total current of 3.0×10^1 kA.

The distance between the cathode and the arc attachment point with the conducting liquid was denoted by the actual arc length after the impingement (Table 3). For the reference case, the actual arc length was equal to 0.28 m. As expected, when the initial arc gap decreased to 0.2 m the actual arc distance shrank to 0.22 m and for an arc gap of 0.3 m, the length increased to 0.3 m. When the applied current was increased to 20 kA, the arc distance increased to 0.3 m and 0.31 m when the applied current was increased to 30 kA. Similar to the arc impingement depth, the applied current was shown to be the most important factor.

Table 3. Parametric arc impingement results.

Current (kA)	Gap (m)	Gas Density (kg/m^3)	Max Arc Jet Velocity (m/s)	Cavity Depth (m)	Actual Arc Length (m)
10	0.25	1	≈200	0.5	0.28
10	0.2	1	≈200	0.4	0.22
10	0.3	1	≈200	0.5	0.3
10	0.25	0.1	≈450	0.5	0.28
10	0.25	0.01	≈1400	0.5	0.28
20	0.25	1	≈200	≥0.7	0.3
30	0.25	1	≈200	≥0.7	0.31

3.5. Discussion of the Thermal Field inside the Arc

The temperature field is vital for the arc existence and propagation. The heat released through joule heating from the arc was the main source of energy in Equation (3). A high temperature ensured a high electrical conductivity for the current to flow. However, the time of simulation was minimal compared with the overall EAF process. Even though this period was enough to understand several important fluid and arc dynamics, it was not significant enough to affect the thermal distribution inside the conducting liquid. A detailed arc simulation was not convenient for studying the heat transfer inside the EAF. Considering the arc as a fixed cylindrical channel with constant boundary conditions including inlets and outlets for the gas and heat transfer on the circumferential area is more efficient and less computationally demanding [18].

4. Conclusions

A 2D axisymmetric two-phase model was developed to study the arc impingement inside an electric arc furnace. The obtained results attested that the proposed model could predict arc impingement phenomena. The assumptions made in this model could be justified as this model aimed to assess the effect of the arc jet on liquid metal and understand how the impingement affected the behaviour and stability of the arc. The main insights from this study are stated as follows.

The arc gap played an important role in the stability and the initial profile of the arc. A small gap induced a powerful splashing and drifted the metal inside the arc region, resulting in greater splashing as time progressed. For a larger gap, the stability of the arc reached a critical level at which the arc was extinguished. The cavity depth was not affected noticeably by the arc gap value.

The most important factor that linked the impingement depth to the mixing of the liquid metal was found to be the total applied current. The increase of the current value from 10 kA to 20 kA increased the penetration depth from ~0.5 m to >0.71 m (Table 3) and reached the bottom boundary. Rayleigh instabilities were created due to the entrapment of gas bubbles inside the liquid metal. These bubbles led to small explosions and powerful splashing when they reached the liquid surface due to a high density difference.

The density of the plasma did not affect the level of the impingement depth but the distribution of the droplets and their size were affected. When the density decreased, the splashing rate increased qualitatively; this was due to a higher arc velocity. The velocity increased as the density decreased. However, the rate of velocity change was not linearly proportional to 1 over density. The rate validated the conservation of momentum ρu^2 = const. Moreover, the actual arc length (similar to the impingement depth) was not affected by the density value, which again enforced the validity of the constant density assumption. The obtained arc length reconfirmed that the applied current was the chief factor in the arc impingement. This explained the obtained results for the unvarying depth of the impingement and validated the use of a constant density for less computational effort.

Two ideas that could improve the modelling in future are the consideration of the compressibility of the plasma and the freedom of the cathode spot motion although these adjustments impose more technical constraints and the need of a very high computational power as a result of the necessity to drastically decrease the time step. A further step to take the model forward is the development of a 3D model. This step could enhance our understanding of the arc behaviour, giving the arc freedom to be deflected and non-symmetrical.

Author Contributions: Conceptualization, A.K., investigation, M.A.-N., M.W. and H.B.; supervision, A.I., M.G. and C.R.; visualization, M.A.-N. and A.L.; writing—original draft, M.A.-N. and A.K.; writing—review and editing, C.P., G.H. and M.G. All authors have read and agreed to the published version of the manuscript.

Funding: This research was funded by the Austrian Federal Ministry of Economy, Family and Youth and the National Foundation for Research, Technology and Development within the framework of the Christian-Doppler Laboratory for Metallurgical Applications of Magnetohydrodynamics (Abdellah Kharicha).

Institutional Review Board Statement: Not applicable.

Informed Consent Statement: Not applicable.

Data Availability Statement: Data presented in this article is available on request from the corresponding author.

Acknowledgments: Abdellah Kharicha and Mohamad Al-Nasser acknowledges financial support from the Austrian Federal Ministry of Economy, Family and Youth and the National Foundation for Research, Technology and Development within the framework of the Christian-Doppler Laboratory for Metallurgical Applications of Magnetohydrodynamics.

Conflicts of Interest: The authors declare no conflict of interest.

Abbreviations

Symbol	Name	Unit
ρ	Density	Kg/m^3
U	Velocity	m/s
p	Pressure	N/m^2
μ	Kinetic viscosity	m^2/s
g	Gravitational acceleration	Nm^2/kg^2
t	Time	s
T	Temperature	K
cp	Specific heat	$J/k{\cdot}kg$
K	Thermal conductivity	$W/m{\cdot}K$
B	Magnetic field	T
J	Current density	A/m^2
μ_0	Vacuum permeability	H/m

References

1. Jones, R.T.; Reynolds, Q.G.; Curr, T.R.; Sager, D. Some myths about DC arc furnaces. *J. South. Afr. Inst. Min. Metall.* **2011**, *111*, 665–674.
2. Szekely, J.; McKelliget, J.; Choudhary, M. Heat-transfer fluid flow and bath circulation in electric-arc furnaces and DC plasma furnaces. *Ironmak. Steelmak.* **1983**, *10*, 169–179.
3. Bowman, B.; Krüger, K. *Arc Furnace Physics*; Verlag Stahleisen: Düsseldorf, Germany, 2009; pp. 176–190.
4. Ushio, M.; Szekely, J.; Chang, C.W. Mathematical modelling of flow field and heat transfer in high-current arc discharge. *Ironmak. Steelmak.* **1981**, *8*, 279–286.
5. Alexis, J.; Ramirez, M.; Trapaga, G.; Jönsson, P. Modeling of a DC electric arc furnace—heat transfer from the arc. *ISIJ Int.* **2000**, *40*, 1089–1097. [CrossRef]
6. Ramírez, M.; Alexis, J.; Trapaga, G.; Jönsson, P.; Mckelliget, J. Modeling of a DC electric arc furnace—Mixing in the bath. *ISIJ Int.* **2001**, *41*, 1146–1155. [CrossRef]
7. Kazak, O. Numerical modelling of electrovortex and heat flows in dc electric arc furnace with cooling bottom electrode. *Heat Mass Transf.* **2014**, *50*, 685–692. [CrossRef]
8. Reynolds, Q.G. Application of multiphase computational fluid dynamics to the plasma arc impingement problem in DC furnaces. In Proceedings of the 9th South African Conference on Computational and Applied Mechanics, Somerset West, South Africa, 14–16 January 2014; pp. 14–16.
9. Klementyeva, I.B.; Pinchuk, M.E. Parameters of electrical discharges with liquid metal electrode. *J. Phys. Conf. Ser.* **2015**, *653*, 012150. [CrossRef]
10. Reynolds, Q. Computational Modeling of Arc–Slag Interaction in DC Furnaces. *JOM* **2017**, *69*, 351–357. [CrossRef]
11. Adib, M.; Ehteram, M.A.; Tabrizi, H.B. Numerical and experimental study of oscillatory behavior of liquid surface agitated by high-speed gas jet. *Appl. Math. Model.* **2018**, *62*, 510–525. [CrossRef]
12. *Ansys Fluent Software*; Version 14.5; 275 Technology Drive Inc.: Canonsburg, PA, USA, 2011.
13. Morris, J.C.; Bach, G.R.; Krey, R.U.; Liebermann, R.W.; Yos, J.M. Continuum radiated power for high-temperature air and its components. *AIAA J.* **1966**, *4*, 1223–1226. [CrossRef]
14. Boulos, M.I.; Fauchais, P.; Pfender, E. *Thermal Plasmas: Fundamentals and Applications*; Springer: New York, NY, USA, 2013.

15. McKelliget, J.W.; Szekely, J. A mathematical model of the cathode region of a high intensity carbon arc. *J. Phys. D Appl. Phys.* **1983**, *16*, 1007. [CrossRef]
16. Jordan, G.R.; Bowman, B.; Wakelam, D. Electrical and photographic measurements of high-power arcs. *J. Phys. D Appl. Phys.* **1970**, *3*, 1089. [CrossRef]
17. Ramírez, M.A. Mathematical Modeling of DC Electric Arc Furnace Operations. Ph.D. Thesis, Massachusetts Institute of Technology, Cambridge, MA, USA, 2000.
18. Gruber, J.C.; Echterhof, T.; Pfeifer, H. Investigation on the influence of the arc region on heat and mass transport in an EAF freeboard using numerical modeling. *Steel Res. Int.* **2016**, *87*, 15–28. [CrossRef]

metals

MDPI

Article

Application and Evaluation of Mathematical Models for Prediction of the Electric Energy Demand Using Plant Data of Five Industrial-Size EAFs

Alexander Reimann [1,*], Thomas Hay [1], Thomas Echterhof [1], Marcus Kirschen [2] and Herbert Pfeifer [1]

1 Department of Industrial Furnaces and Heat Engineering, RWTH Aachen University, Kopernikusstraße 10, 52074 Aachen, Germany; hay@iob.rwth-aachen.de (T.H.); echterhof@iob.rwth-aachen.de (T.E.); pfeifer@iob.rwth-aachen.de (H.P.)
2 Thermal Process Engineering, University of Bayreuth, Universitätsstraße 30, 95447 Bayreuth, Germany; marcus.kirschen@uni-bayreuth.de
* Correspondence: reimann@iob.rwth-aachen.de

Abstract: The electric arc furnace (EAF) represents the most important process route for recycling of steel and the second most productive steelmaking process overall. Considering the large production quantities, the EAF process is subject to continuous optimization, and even small improvements can lead to a significant reduction in resource consumption and operating cost. A common way to investigate the furnace operation is through the application of mathematical models. In this study the applicability of three different statistical modeling approaches for prediction of the electric energy demand is investigated by using more than 21,000 heats from five industrial-size EAFs. In this context, particular consideration is given to the difference between linear and nonlinear regression models. Detailed information on the treatment of the process data is provided and the applied methods for regression are described in short, including information on the choice of hyperparameters. Subsequently, the results of the models are compared. Gaussian process regression (GPR) was found to yield the best overall accuracy; however, the benefit of applying nonlinear models varied between the investigated furnaces. In this regard, possible reasons for the inconsistent performance of the methods are discussed.

Keywords: electric arc furnace; energy demand; regression; artificial neural network; Gaussian process regression; Köhle formula

Citation: Reimann, A.; Hay, T.; Echterhof, T.; Kirschen, M.; Pfeifer, H. Application and Evaluation of Mathematical Models for Prediction of the Electric Energy Demand Using Plant Data of Five Industrial-Size EAFs. *Metals* **2021**, *11*, 1348. https://doi.org/10.3390/met11091348

Academic Editor: Shi-Hoon Choi

Received: 25 July 2021
Accepted: 24 August 2021
Published: 27 August 2021

1. Introduction

In 2019 the electric arc furnace (EAF) process accounted for approximately 28% of the worldwide crude steel production with the total amount of produced steel reaching an all-time high [1]. Within the European Union the percentage of steel produced in arc furnaces presented as much as 41% of the total production [2]. Benefits of the EAF include its high flexibility regarding raw material input and production volume, making it the most common process for recycling of steel scrap. In view of the current climate targets, the share of steel produced in the EAF is likely to increase while a further reduction of the carbon footprint of the EAF process is pursued [3].

The electrical energy demand represents the most important contribution to EAF conversion costs, besides electrode graphite. Combined with raw materials, the high electrical energy demand accounts for more than 80% of the total operating cost of the EAF [4]. Considering the large production quantities, even small improvements to the specific electric energy demand can generate significant cost savings and reduce the environmental impact of the process. A common way to investigate improvements to operational strategies in the EAF is the application of mathematical models. By employing such models, the effect of proposed changes can be studied without affecting regular production, reducing cost, and eliminating the risk connected with trial campaigns. Furthermore, models can be used to monitor production and detect changes in the process or the input material which could otherwise only be noticed during quality assurance.

A task for which mathematical models are commonly used is estimation of the electric energy demand by analysis of process data. The information gained can be utilized to predict the energy demand of future heats or to identify key factors for overall reduction of the energy demand. However, the flexibility of the EAF process can prove challenging for modelling of the energy demand since the inputs vary over a wide range of materials with variable composition. In addition, due to the nonlinear nature of the process, the impact of individual variables cannot be easily determined.

In general, the applied models can be distinguished into empirical and analytic models. The latter approach considers the furnace based on physical or thermodynamic principles. As such, these models are usually associated with higher development cost, yet allow for use outside of the range of their training data [5]. Extensive analytic process models of the EAF have been published previously by Bekker [6] and Logar [7,8], as well as MacRosty and Swartz [9]. A more comprehensive overview of the published process models is given by Hay et al. [5]. Empirical models, on the other hand, rely on data from observation or experiment. They are often termed "black boxes" as the underlying phenomena are not considered, or are unknown [10]. These models are the focus of this work.

In the past several statistical models of the electric energy demand in the EAF have been discussed, ranging from multiple linear regression (MLR) models to more complex machine learning (ML) algorithms such as artificial neural networks (ANN) [4]. Simple models often lack in accuracy or require detailed process knowledge in the preparation of the data. ML algorithms yield better results; however, the models have a complex structure and are difficult to comprehend. In addition, they require a larger set of data for training of the model parameters. In this paper three types of regression models are implemented in order to predict the electric energy demand of the EAF. The models are used with extensive process data of five different arc furnaces and the results are compared in order to determine the models best suited for application. In this regard, the effect of data quality and treatment on the accuracy of the model results is investigated.

2. Materials and Methods

2.1. Modelling Approach

One of the first widely known empirical models for prediction of the electric energy demand of EAFs was developed by Köhle et al. in the 1990s by statistical analysis of average production values from 14 furnaces. The Köhle model was later improved and extended to post-combustion and alternative ferrous material such as hot briquetted (HBI) or direct reduced iron (DRI) using 5000 single heats from 5 different furnaces. An updated formula for the specific electric energy demand (W_R) published in 2005 is given in Table 1 and Equation (1) [11]. In contrast to later models, the coefficients are not only fitted by linear regression, but in most cases also correspond to values found in thermodynamic analysis of arc furnace process [12,13].

Table 1. Parameters of the Köhle formula.

Parameter	Name	Unit	Parameter	Name	Unit
G_A	Tap weight	t	t_s	Power-on time	min
G_E	Weight of ferrous material	t	t_N	Power-off time	min
$G_{DRI/HBI}$	Weight of DRI	t	M_G	Specific burner gas	m³/t
G_{SHR}	Weight of HBI	t	M_L	Specific lance oxygen	m³/t
G_{HM}	Weight of hot metal	t	M_N	Specific post-combustion oxygen	m³/t
G_Z	Weight of slag formers	t	NV	Furnace specific factor	-
T_A	Tapping temperature	°C	W_V	Energy losses	kWh/t

While the results given by the Köhle model are in good agreement with the average electric energy demand of the furnaces, results from single heats can significantly differ, as will be shown in the results section. The formula is, however, still used for benchmarking of the operation of arc furnaces [14,15]. The Köhle model was also specified for an almost 100% DRI EAF operation at Mittal Steel Lázaro Cárdenas [16]. In order to predict the energy demand of single heats more reliably, a number of models based on more complex algorithms have been developed in the last decade [4,17–20]. In this study, three different kinds of regression models will be utilized for prediction of the electric energy demand of the EAF. However, proper adjustment of the applied models is a wide area of research and description of all possible settings is beyond the scope of this paper. Therefore, in the following the basics of the applied methods are described in short, and reasoning is given concerning the choice of hyperparameters.

$$\begin{aligned} W_R = 375 + 400\left[\tfrac{G_E}{G_A} - 1\right] + 80\tfrac{G_{DRI/HBI}}{G_A} - 50\tfrac{G_{Shr}}{G_A} - 350\tfrac{G_{HM}}{G_A} + 1000\tfrac{G_Z}{G_A} \\ +0.3[T_A - 1600] + t_{T2T} - 8M_G - 4.3M_L - 2.8M_N + NV\left[W_V - \overline{W_V}\right] \end{aligned} \quad (1)$$

For regression, the measured data is first standardized by subtracting the mean value of every predictor and dividing by its standard deviation. Calculation of the so-called z-score is shown in Equation (2). In using standardization variables with varying scales, different units of measurement are brought to the same scale and can contribute equally to the result. This might also increase training speed of the models. On the contrary, standardization gives equal weight to data with comparatively small variance and may thus excessively incorporate noise into the calculation. Furthermore, information on the mean and standard deviation of the explanatory variable is lost.

$$z_i = \frac{x_i - \overline{x}}{S} \quad (2)$$

For optimization of the model parameters, the mean square error (MSE) between the total demand of electric energy and the model prediction is minimized. Calculation of the MSE is shown in Equation (3). The measured electrical energy demand is named y_i. The calculated value of the electrical energy demand is labeled f_i with the number of data points denoted as n. In other works [18,19], the specific electric energy demand per ton of produced steel is used for analysis. In the context of this study, application of the models for the prediction of future heats shall be investigated. The mass of tapped steel is, however, unknown prior to tapping. Hence, tuning of the model parameters is performed using the absolute demand of electric energy for this study.

$$MSE = \frac{\sum (f_i - y_i)^2}{n} \quad (3)$$

Although Köhle performs a nonlinear transformation on some variables, for example by dividing by the mass of tapped steel, the base model remains a multiple linear regression (MLR). Furthermore, Köhle did only use data available for all furnaces and abstained from standardization of the data. MLR is one of the earliest and most basic methods for supervised learning, which is mapping of input to an output based on a set of training examples. In MLR, the predicted response is calculated by linear combination of the explanatory variables as stated in Equation (4). In doing so, it is assumed that the relationship between true response and explanatory variables is linear and explanatory variables are not correlated. In the case of the arc furnace, both assumptions are, however, violated. Thermal radiation is increasing with the fourth power of the melt's temperature, and energy loss through cooling of the furnace therefore increases at later stages of the process when the melts temperature is higher, and the furnace walls are not shielded by scrap. Other mechanisms such as slag foaming can further impact the overall energy demand nonlinearly [21]. Nevertheless, due to their simplicity and low computational demand,

MLRs are still commonly used. Within this work, MLR will be used as a benchmark for the nonlinear model types.

$$f_i = \beta_0 + \sum \beta_i x_i \tag{4}$$

Limitations of linear models, such as their inability to account for interactions between the input variables, gave rise to the popularity of ANNs for estimation of the electric energy demand of EAFs [17–19]. A network in which information only moves forward through the layers without feedback is called feedforward network or multilayer perceptron. These networks are the quintessential deep learning models [22]. The structure of a simple feedforward network with only one hidden layer is displayed in Figure 1. At each neuron, the values from the previous layer are multiplied with a set of weights and a bias is added. The resulting value is transferred to the subsequent layer through application of an activation function. In the past, tangens hyperbolicus or the logistic function were frequently used as activation function. However, these sigmoid activation functions are only strongly sensitive when the input is close to 0, for high or low values the function quickly saturates, affecting gradient-based learning [22]. In modern applications of ANN's, linear units are often utilized. For prediction of the energy demand an exponential linear unit with $\alpha = 1$ was chosen. The output of this threshold operation is given by Equation (5)

$$f(x) = \begin{cases} x, & x \geq 0 \\ \alpha(exp(x) - 1), & x < 0 \end{cases} \tag{5}$$

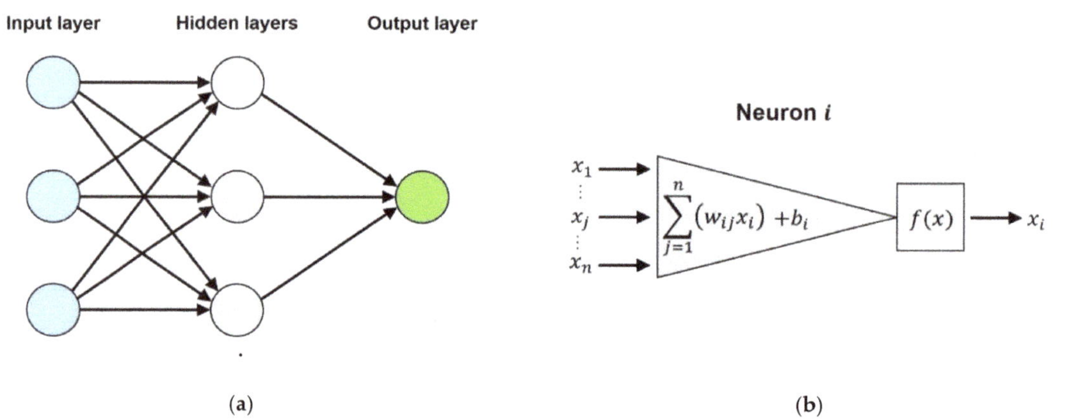

Figure 1. (**a**) structure of feedforward neural net featuring a single hidden layer; (**b**) calculation of the output of a neuron and transfer function.

Within this work the neural network applied for estimation of the energy demand contains 2 hidden layers featuring n and $n/2$ neurons respectively, where n is the number of explanatory variables. During training of the model, the weights and biases of the ANN are tuned by minimizing the loss function (MSE) and backpropagation of the error to each neuron in each layer. For optimization, a stochastic gradient descend with momentum was used with an initial learn rate of 0.01.

Another approach to supervised learning is through application of Gaussian processes. A Gaussian Process is defined as a collection of random variables, every finite collection of which have a multivariate normal distribution. It is a generalization of the Gaussian distribution over functions with a continuous domain and is fully specified by a mean $m(x)$ and covariance function $K(x, x')$ as stated in Equation (6) [23]. In consequence, the Gaussian process is a nonparametric model. Rather than calculating parameters such that a given class of functions (e.g., linear functions) fits the data, the prior distribution contains all functions defined by the chosen mean and covariance function.

$$f(x) \sim GP(m(x),\ K(x, x')) \tag{6}$$

By incorporating the observation from the training data, functions which do not pass the data points (or do not closely pass in case of noisy data) are removed from the infinite set, in order to form the posterior distribution. As a result, the posterior uncertainty in the vicinity of the observations is reduced. This is also called conditioning of the Gaussian prior distribution on the observations. In Figure 2a three samples from the prior distribution are shown. The posterior distribution after observation of five data points is depicted in Figure 2b. The underlying (unknown) function is a polynomial of the third degree. Making a prediction using the Gaussian process ultimately amounts to drawing samples from its posterior distribution.

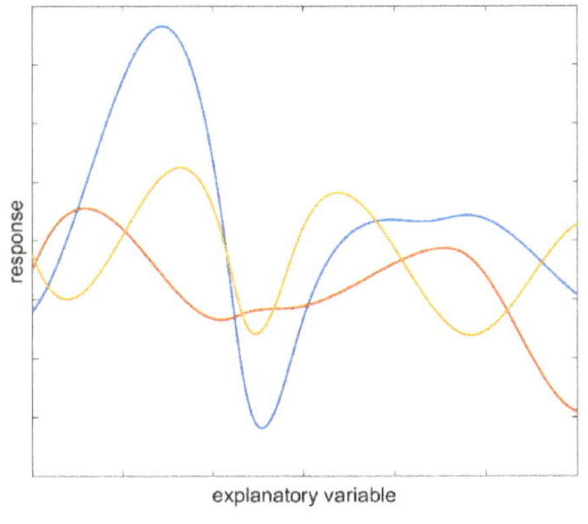

(**a**)

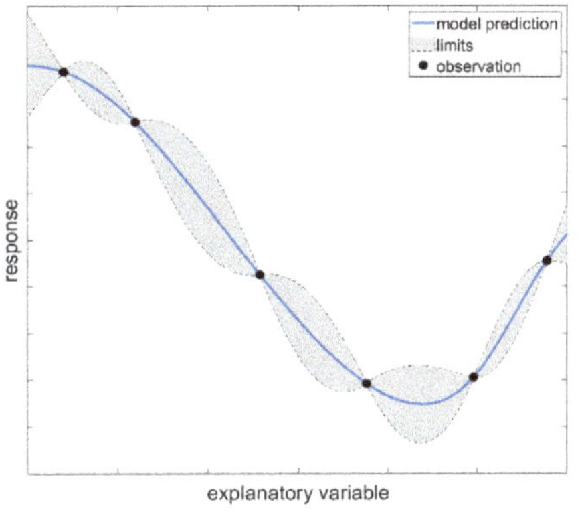

(**b**)

Figure 2. (**a**) three samples of the prior distribution specified by the mean and covariance function; (**b**) posterior distribution after observation of six data points.

That being said, the predictive performance of Gaussian processes depends exclusively on the chosen kernel [24]. For prediction of the electric energy demand the Matérn covariance function with $v = 3/2$ given in Equation (7) was chosen. In contrast to other popular kernels, such as the infinitely differentiable squared exponential kernel displayed in Equation (8) (Gaussian function), its shape is rather rough. However, strong smoothness is argued to be unrealistic for modelling of physical processes [25].

$$k_{v=3/2}(r) = \left(1 + \frac{\sqrt{3}r}{l}\right) exp\left(-\frac{\sqrt{3}r}{l}\right) \tag{7}$$

$$k_{SE}(r) = exp\left(-\frac{r^2}{2l^2}\right) \tag{8}$$

The performance of the models on the validation data is evaluated using the adjusted coefficient of determination $\left(R^2\right)$, as well as the mean absolute error (MAE), standard deviation of the result (SD) and relative standard deviation (RSD). These values are calculated as shown in Equations (9)–(12). The relative standard deviation (coefficient of variation) is utilized in order to illustrate the extend of variability in relation to the average demand of electric energy [26]. The mean values of the measured and calculated electric energy demand is denoted as $\overline{y}$ and $\overline{f}$, respectively. The coefficient of determination ranges from 0 to 1 and is often used as an indicator for the goodness of the fit, with 1 meaning the results perfectly match the measurements.

$$R^2 = 1 - \frac{RSS}{TSS} = 1 - \frac{\sum (f_i - y_i)^2}{\sum (y_i - \overline{y})^2} \tag{9}$$

$$MAE = \frac{\sum [f_i - y_i]}{n} \tag{10}$$

$$SD = \sqrt{\frac{\sum \left(f_i - \overline{f}\right)^2}{n}} \tag{11}$$

$$RSD = \frac{SD}{\overline{y}} \tag{12}$$

2.2. Datasets of EAF Heats Used in This Study

For evaluation of the described models, process data of five electrical arc furnaces for industrial steel production was used. In total, the data sets contain material consumption and furnace operation data of roughly 21,000 heats. However, the investigated furnaces differ considerably regarding their capacity and material input, as well as the measurements taken during operation. The characteristics of the furnaces are summarized in Table 2. With an average tap weight of about 80 t EAF-A has a notably smaller capacity than the remaining furnaces. Likewise, the average tap-to-tap time of EAF-A is shorter. Furnace B, C and D have similar capacities and tap-to-tap times. The highest specific electrical energy demand is found for EAF-C. The different specific electric energy demand of the furnaces can in part be attributed to the differences in the charged ferrous material. For both EAF-B and EAF-C, the input material contains large quantities of DRI or HBI while the remaining furnaces use scrap of varying quality.

Table 2. Specification and key performance indicators of the investigated electric arc furnaces.

Furnace	EAF-A	EAF-B	EAF-C	EAF-D	EAF-E
Average tap weight [t]	81	153	142	142	123
Average tap-to-tap time [min]	45	62	69	61	57
Average specific electric energy demand [kWh/t]	325	467	535	422	345
Ferrous Material	Scrap	DRI	DRI/Scrap Mix	Scrap	Scrap
Number of overall heats	5220	1046	6139	8088	2341
Number of excluded heats	150	32	791	785	163
Percentage of removed heats	2.9	3.1	12.8	9.7	7.0

Not all documented heats can be used for evaluation of the electric energy demand. In the first step data, treatment is performed on each set. The overall goal is the removal of faulty or irregular data originating from erroneous data logging or irregular operation such as trial heats, aborted heats, or equipment malfunctions. Including these heats would otherwise have a negative impact on training of the models for regular heats, which are the main subject of the investigation. Table 2 shows the total amount and the percentage of excluded heats. In the following the applied decision rules for removal of data are described.

When crucial data like electric energy demand or tap weight are missing, the applied regression models cannot accurately predict the electrical energy demand, and therefore the heats in question must be excluded from consideration. In addition, heats are excluded if the measurements are unreasonable. This is, for example, the case if the tap weight exceeds the maximum capacity of the furnace, or the recorded tap-to-tap time is lower than the power on time of the heat. Significant outliers were also removed from the data sets. These include heats with an abnormal tap-to-tap time since they are likely to contain long power off times as a result of production delays by unscheduled events or regular maintenance stops. Likewise, heats are removed if the number of buckets differs from the rest of the batches. Finally, heats are removed if their ratio between charged ferrous material and tapped steel is below 0.75 or above 1.05, respectively. This is due to the mass of the hot heel not being measured for most of the furnaces. In keeping part of the molten steel inside the furnace after tapping, the melting rate of the subsequent heat can be increased. This results in lower thermal losses and a lower overall energy demand [27]. Moreover, in DC furnaces a hot heel is necessary for operation as it is covering the electrode in the bottom of the vessel and closes the circuit. However, when the furnace is completely emptied the amount of energy needed for initial melting of the hot heel is unaccounted for, while the energy demand of the next heat is higher compared to regular heats. In consequence, these heats, usually occurring before and after maintenance periods, are removed from consideration. In total, between 3% and 13% of the heats were removed. EAF-C, EAF-D and EAF-E show a notable larger percentage of the excluded data compared to EAF-A and EAF-B. This can mostly be attributed to unusually long heats, i.e., frequent production interruptions (270 for EAF-C and 160 for EAF-E) and missing measurements or recorded data. For EAF-D 775 out of 785 removed heats are missing the mass of charged material and about 270 heats from EAF-C are lacking temperature measurement from the molten steel.

Apart from the data quality and overall differences in the operation of the furnaces, the amount of data recorded during operation also differs significantly. Table 3 shows an overview of the available measurements. The electric energy demand, as well as the mass of charged ferrous material, coal and slag formers are measured for all furnaces. For the first three furnaces only a basic breakdown into scrap, DRI and alloying metals is given while EAF-D and EAF-E have a detailed record of the charged scrap grades. The exact chemical composition of the input materials is, however, unknown and is likely to differ between plants and even between heats. Furthermore, the mass of charged slag former in EAF-B and EAF-C is provided with an accuracy of 0.5 tons. This suggests that the stated mass is estimated or measured with limited accuracy only. Although this was the only obvious case it must be noted that all measurements are associated with a degree of uncertainty since no information on the methods and accuracy of the measurements was given.

Table 3. Overview of the available measurements at the investigated EAFs.

Furnace	EAF-A	EAF-B	EAF-C	EAF-D	EAF-E
Electric energy demand [kWh]	x	x	x	x	x
Mass of charged scrap grades [t]	x	x	x	x^1	x^1
Mass of slag formers [t]	x	x	x	x^1	x^1
Charged or injected Coal [t]	x	x	x	x	x
Bath height [m]	-	-	-	-	x
Oxygen consumption [m3]	(x)	(x)	(x)	x	x
Natural gas consumption [m3]	x^2	-	x	x	x
Power-on time [min]	x	x	x	x	x
Tap-to-tap time [min]	x	x	x	x	x
Sub-process times [min]	x	-	-	-	x
Tap weight [t]	x	x	x	x	x
Melt temperature [°C]	x	x	x	x	x
Mass of hot heel [t]	-	x	-	-	x
Steel composition [kg/kg]	-	-	-	x	(x)
Slag composition [kg/kg]	-	-	-	(x)	-

(x): only limited information available; x^1: detailed record of scrap grades; x^2: chemical heat input provided in kWh.

During operation, the injected oxygen and carbon mass was measured along with the consumption of natural gas. EAF-B is the only furnace without operation of natural gas burners. Moreover, in the records from EAF-D and EAF-E, oxygen input is further separated into different applications within the furnace such as oxygen for burners, lances, or post-combustion. In the remaining data sets oxygen input is separated only for the purpose of post-combustion with all other flows combined into a single measurement. Power-on time as well as tap-to-tap time are measured for all furnaces, yet only EAF-A and EAF-E have a detailed breakdown of sub-process times such as charging, melting, and tapping provided. Energy losses can vary considerably throughout the different stages of the melting process. Therefore, by providing information on the length of the sub-processes the quality of the prediction can be improved. Furthermore, the weight and temperature of the tapped steel are available. Temperature measurement is carried out shortly before tapping in order to ensure the target temperature was reached. The temperature and mass of the tapped steel are directly related to the energy demand for melting. As stated before, the mass of the hot heel which remains inside the vessel after tapping is however only measured at two of the furnaces with the method and accuracy of the measurement unknown. Lastly, at EAF-D the composition of both steel and slag is analyzed after each heat, while EAF-E has steel composition measured in regular intervals for 98 heats in total. As can be seen from the overview of furnace characteristics in Table 2 and measurements in Table 3 a single regression model cannot be applied for all furnace without the need to drastically reduce the data sets in order to form a common denominator. Even in doing so, the measurements are performed with different precision and in the case of scrap grades and slag formers classification is not necessarily uniform. In consequence the furnaces must be considered separately, while the general design of the investigated models is maintained.

For each furnace, heats are divided into a training and validation set. The training set contains 70% of the data and is drawn at random. Subsequent validation is performed on the remaining 30% of heats. All applied models are trained on the same data set. However, since selection of the training data influences the model accuracy, training and validation are performed on 5 separate training samples and the median results of the regressions are discussed. Beyond that, process data can be divided into two groups: measurements available before and only after the heat is finished. In the literature the entire dataset is often used for modeling of the electric energy demand [11,17,20,28]. While those models yield better results in terms of accuracy, they cannot be applied to predict the energy demand of a future heat. Subsequently, the investigated regression models will be used on the entire and limited dataset and the results of both approaches will be compared.

3. Results

At first the results of the Köhle formula were calculated for each EAF. In place of the furnace-specific parameter NV, a bias was added to the results, such that the average deviation for each furnace assumes the value of 0. The MAE, SD, and RSD for prediction of the electric energy demand of single heats is presented in Table 4. The results show that the accuracy of prediction significantly differs between the furnaces, with the best result obtained for EAF-B. This is likely due to EAF-B only having DRI and hot metal charged. Both input materials are represented in the Köhle formula while different scrap grades are not considered, resulting in a large deviation of the calculated energy demand.

Table 4. Results of the Köhle formula on the investigated furnaces.

Model Performance	EAF-A	EAF-B	EAF-C	EAF-D	EAF-E
MAE [kWh/t]	25.1	14.8	31.6	21.1	40.0
SD [kWh/t]	36.0	19.0	41.6	27.3	53.1
RSD [%]	11.1	4.1	7.8	6.5	15.4

Subsequently, the previously described regression models were applied to the process data of the five EAFs. After parameter optimization on the training data was finished, the regression models were used to estimate the electric energy demand of the heats within the test set. The median results of the regression models on the entire dataset are summarized in Table 5. For the sake of comparability, the mean absolute error and standard deviation are calculated using the specific electric energy demand rather than the absolute electric energy demand. The coefficient of determination is calculated on the absolute electric energy demand per heat. It can be seen from the table that the Gaussian process regression shows the best overall accuracy with regards to the mean absolute and standard deviation as well as the coefficient of determination, ranging from 0.651 to 0.941 for the investigated furnaces. That being said, by applying a multiple linear regression the quality of prediction can still be significantly improved compared to the results of the Köhle formula on single heats.

Table 5. Median result of the applied regression algorithms on the entire data set for validation.

Furnace	Model Performance	Linear Regression	ANN	GPR
EAF-A	R^2	0.842	0.847	0.859
	MAE [kWh/t]	7.9	7.6	7.1
	SD [kWh/t]	11.2	10.5	10.6
	RSD [%]	3.4	3.2	3.3
EAF-B	R^2	0.899	0.871	0.943
	MAE [kWh/t]	11.7	12.7	8.3
	SD [kWh/t]	15.4	16.4	11.1
	RSD [%]	3.3	3.5	2.4
EAF-C	R^2	0.881	0.923	0.941
	MAE [kWh/t]	14.9	12.3	10.3
	SD [kWh/t]	20.8	16.4	14.4
	RSD [%]	3.9	3.1	2.7
EAF-D	R^2	0.754	0.755	0.769
	MAE [kWh/t]	9.9	9.9	9.6
	SD [kWh/t]	12.8	12.8	12.4
	RSD [%]	3.0	3.0	2.9
EAF-E	R^2	0.519	0.587	0.651
	MAE [kWh/t]	6.3	5.8	5.3
	SD [kWh/t]	8.4	7.8	7.2
	RSD [%]	2.4	2.3	2.1

By utilizing a Gaussian process regression for EAF-B and EAF-C the mean absolute deviation, as well as the relative standard deviation of the results can be decreased by approximately 30% when compared to the results of the linear regression. In contrast with the remaining furnaces, the benefit of applying nonlinear models is notably lower, with EAF-A and EAF-D hardly showing any differences between the investigated models. A possible reason might be the use of DRI in both EAF-B and EAF-C instead of the various scrap grades. Although a larger amount of energy is required for the melting of DRI [12], the variance in chemical composition of the material is lower than that of the scrap mix. The charged scrap can have various contaminants which affect the process and energy requirement for melting. At the same time, at the remaining furnaces the number of heats including individual scrap grades is significantly lower when compared to the number of heats containing DRI at EAF-B and EAF-C. The smaller effective sample size could have a negative impact on training of the ANN in particular. In Figure 3 the electric energy demand of EAF-C is displayed in relation to the percentage of charged DRI and the number of baskets. As can be seen in the diagram, a large number of samples is available for each category. By applying an ANN or GPR, the nonlinear relationship between these process parameters can be modelled and its large-scale effect on the electric energy demand is estimated more accurately than by using a linear regression on the raw data. For the remaining furnaces, the smaller sample size for the input of individual scrap grades and higher variance within grades results in an equal level of accuracy across the model types, even when considering possible nonlinear interaction.

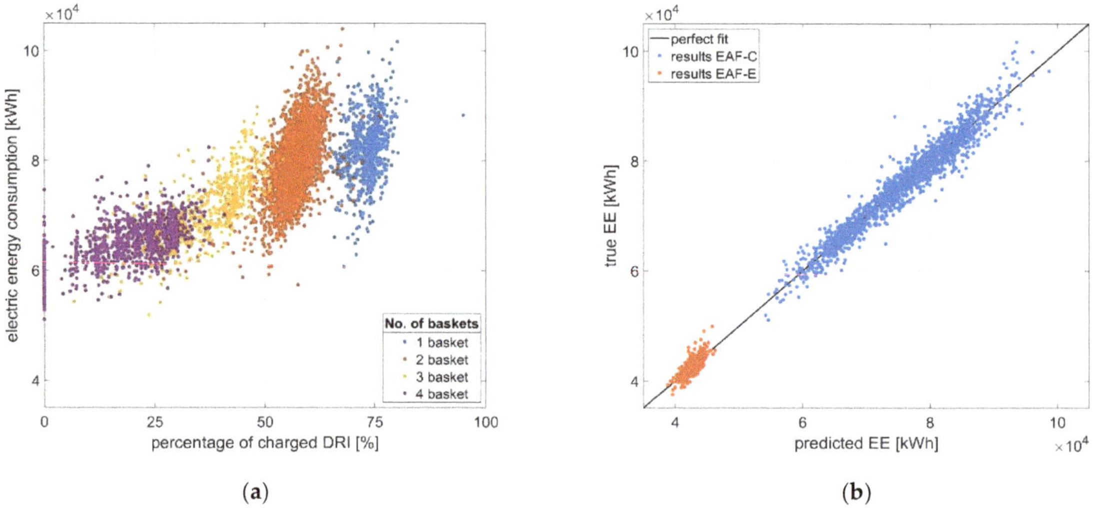

Figure 3. (**a**) Percentage of DRI charged in EAF-C with varying number of baskets; (**b**) Comparison of the estimated electric energy demand for EAF-C and EAF-E.

Furthermore, although EAF-E exhibits the smallest coefficient of determination (implying a larger deviation between the model results and measurements) its mean absolute error and standard deviation are in fact the smallest among the investigated furnaces. On average the regression models deviate from the true electric energy demand by 721 kWh/heat for EAF-E and 1789 kWh/heat for EAF-C, i.e., 5.9 kWh/t for EAF-E and 12.6 kWh/t for EAF-C. The difference in average tap weight displayed in Table 2 cannot explain the large deviation. This is also shown in by the relative standard deviations for both furnaces. As is stated in Equation (7), the coefficient of determination represents the ratio between the deviation of the calculated electric energy demand from the average measured demand and the measured deviation from its mean value. It is often interpreted as the proportion of energy demand which is explained by the regression model [29]. As depicted in Figure 3 the electric energy demand of EAF-C spans between roughly 40 MWh and 100 MWh per heat in contrast to the much narrower production parameters of EAF-E. As a result, the calculated ratio is smaller for EAF-C, although the electric energy demand is more accurately described for EAF-E.

In consequence, when discussing the accuracy of regression models for multiple furnaces, the coefficient of determination on its own is not suited for evaluation. In this context, the values for R^2 given in the literature must also be examined critically as the investigated furnaces most likely differ as well. The same applies to depictions of normalized results. In this regard, in Figure 4 the normalized estimated energy demand of EAF-C and EAF-E are compared. For EAF-E the model appears to predict the measurement more accurately. However, in terms of absolute values, the residuals for EAF-C are on average about twice that to the results of EAF-E as stated in Table 5.

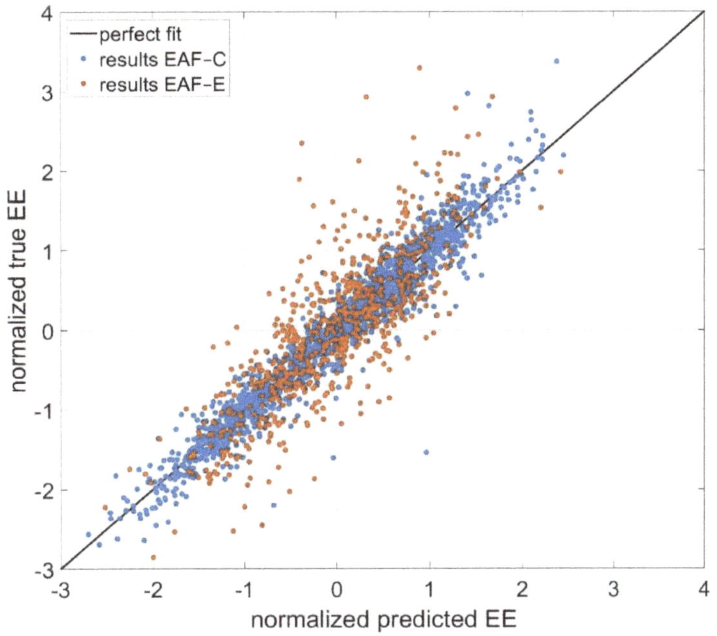

Figure 4. Comparison of normalized estimated electric energy demand of EAF-C and EAF-E.

Another problem arises from measurements being unavailable until the heat is finished. Naturally, this involves for example the tap weight and steel temperature as well as the consumption of natural gas and oxygen. As mentioned before, training of the regression models is therefore repeated for a reduced data set, containing only the mass of material charged at the start of each heat. The aim is to evaluate the applicability of the regression models for prediction of the electric energy demand of future heats. In Table 6 the results of the applied regression algorithms on the limited data set are shown. In comparison to the previous results in Table 5 a significant reduction in the quality of the model results can be observed. Application of the GPR still yields the best overall results; nevertheless, the mean absolute error calculated on the validation data is increased by up to 10 kWh/t. The standard deviation rises on an equal scale. Similar to the previous case, the largest difference between the models can be found for EAF-B and EAF-C. This suggests that the difference in charged input materials is indeed responsible for the inconsistent performance of the applied nonlinear regressions. On the other hand, the drastically reduced prediction quality illustrates the information lost by removing the a-posteriori measurements. Considering arc furnaces are usually operated on distinct power levels, power-on time of the arc, for example, is closely correlated to the electric energy demand. Including the power-on time therefore naturally increases the accuracy of the model. However, process times might also be an indicator for the quality of the input material. Likewise, injection of coal and consumption of natural gas are directly reducing the demand of electric energy by supplying chemical energy. Yet excessive consumption of natural gas and coal can also indicate poor operation, resulting in long tap-to-tap as well as power-on time and ultimately a high electric energy demand. When predicting the energy demand of future heats, these in-process measurements are, however, unavailable. The difference in the quality of results is particularly high when a large number of different scrap types is used, as can be seen from the results of EAF-A and EAF-D in Tables 5 and 6. Another reason for the differences in the results is the use of natural gas, carbon, oxygen, and other additives, which can vary considerably between single heats, corresponding to irregularities in the operation of the furnace or contaminants within the ferrous material.

Table 6. Median results for regression on the data available before finishing of the heat.

Furnace	Model Performance	Linear Regression	ANN	GPR
EAF-A	R^2	0.185	0.183	0.227
	MAE [kWh/t]	17.4	17.4	17
	SD [kWh/t]	24.4	24.5	23.7
	RSD [%]	7.5	7.5	7.3
EAF-B	R^2	0.71	0.748	0.821
	MAE [kWh/t]	20.3	17.0	13.4
	SD [kWh/t]	24.7	23.0	19.4
	RSD [%]	5.3	4.9	4.2
EAF-C	R^2	0.742	0.755	0.853
	MAE [kWh/t]	24.3	21.9	17.0
	SD [kWh/t]	30.6	29.8	23.1
	RSD [%]	5.7	5.6	4.3
EAF-D	R^2	0.244	0.242	0.228
	MAE [kWh/t]	17.3	17.3	17.2
	SD [kWh/t]	22.5	22.5	22.7
	RSD [%]	5.3	5.3	5.4
EAF-E	R^2	0.245	0.299	0.360
	MAE [kWh/t]	8.0	7.5	7.3
	SD [kWh/t]	10.5	10.1	9.7
	RSD [%]	3.0	2.9	2.8

On a side note, the correlation between the consumption of coal, natural gas, and oxygen with the demand of electric energy in the EAF can result in positive parameter values for these explanatory variables. Interpretation of the parameter values would imply an increase in energy demand through the use of coal for example. However, a regression model cannot provide direct information on causality, which has to be kept in mind when interpreting the results [29]

In an attempt to utilize all available information, training of model parameter was carried out using the entire data set, while applying only the limited data for validation of the results on the test data. Missing values, such as consumption of natural gas and oxygen, were replaced by mean values of previous heats. However, this approach yielded very similar results as displayed in Table 6. In this regard, investigation of the consumption of natural gas, carbon, oxygen, and other additives in relationship to the material input and produced steel grade is required. By further classification of the heats, the variation of process parameters within the subsets can possibly be limited and prediction accuracy on future heats can be improved. In this context, an analysis of the quantity of contaminants within single scrap types would be beneficial.

4. Discussion

Within this study, the applicability of three different approaches for regression were examined in order to estimate the demand of electric energy in the operation of an electric arc furnace. To this end, the examined methods were tested on process data containing over 21,000 heats originating from five industrial-size EAFs.

Application of Gaussian process regression yielded the best overall results in terms of prediction accuracy. In some cases, the mean error, as well as the standard deviation, could be reduced by up to 30% compared to the linear regression. However, large differences were found across the investigated furnaces. The quality of the measured data was identified as one of the main reasons for the inconsistent behavior. This includes the categorization of charged scrap grades and slag formers. In general, application of a wide range of materials resulted in a lower accuracy of the implemented models as opposed to the predominant use of single grades with limited variance such as DRI. Even in utilizing non-linear methods, during training the models are unable to appropriately tune the weights or parameters due to, for example, various contaminants affecting the chemical composition. In consequence, the benefit of applying nonlinear models over linear regression is heavily dependent on the process parameters and measurement quality. In this regard, the crucial role of in-process measurements on the model precision was highlighted. However, when predicting the energy demand of future heats, this information cannot be used. Careful classification of the charged scrap types and slag formers is therefore particularly important in order to increase the model accuracy, and including further information on the properties of the charged material is recommended, if possible.

Lastly, by comparison of the achieved results, it was shown that the often-reported coefficient of determination is not sufficient for evaluation of a model's predictive quality since the metric is heavily influenced by the observed variation in the target values. The same argument was given for evaluation of normalized results.

Author Contributions: Conceptualization, A.R.; methodology, A.R.; software, A.R.; validation, A.R.; formal analysis, A.R.; investigation, A.R.; resources, A.R., T.E., H.P.; data curation, A.R.; writing—original draft preparation, A.R.; writing—review and editing, A.R., T.H., T.E. and M.K.; visualization, A.R.; supervision, T.E., H.P.; project administration, T.E., H.P. All authors have read and agreed to the published version of the manuscript.

Funding: This research received no external funding.

Institutional Review Board Statement: Not applicable.

Informed Consent Statement: Not applicable.

Data Availability Statement: Data sharing not applicable.

Conflicts of Interest: The authors declare no conflict of interest.

References

1. World Steel Association. *Steel Statistical Yearbook 2019: Concise Version*; World Steel Association: Brussels, Belgium, 2019.
2. Wirtschaftsvereinigung Stahl. *Statistisches Jahrbuch der Stahlindustrie: 2020 | 2021*; Wirtschaftsvereinigung Stahl: Berlin, Germany, 2021.
3. European Commission. Climate Strategies & Targets. Available online: https://ec.europa.eu/clima/policies/strategies_en (accessed on 29 June 2021).
4. Carlsson, L.S.; Samuelsson, P.B.; Jönsson, P.G. Predicting the Electrical Energy Consumption of Electric Arc Furnaces Using Statistical Modeling. *Metals* **2019**, *9*, 959. [CrossRef]
5. Hay, T.; Visuri, V.-V.; Aula, M.; Echterhof, T. A Review of Mathematical Process Models for the Electric Arc Furnace Process. *Steel Res. Int.* **2021**, *92*, 2000395. [CrossRef]
6. Bekker, J.G.; Craig, I.K.; Pistorius, P. Modeling and Simulation of an Electric Arc Furnace Process. *ISIJ Int.* **1999**, *39*, 23–32. [CrossRef]
7. Logar, V.; Dovžan, D.; Škrjanc, I. Modeling and Validation of an Electric Arc Furnace: Part 1, Heat and Mass Transfer. *ISIJ Int.* **2012**, *52*, 402–412. [CrossRef]
8. Logar, V.; Dovzan, D.; Škrjanc, I. Modeling and Validation of an Electric Arc Furnace: Part 2, Thermo-chemistry. *ISIJ Int.* **2012**, *52*, 413–423. [CrossRef]
9. MacRosty, R.D.M.; Swartz, C.L.E. Dynamic Modeling of an Industrial Electric Arc Furnace. *Ind. Eng. Chem. Res.* **2005**, *44*, 8067–8083. [CrossRef]
10. Cameron, I.T.; Hango, K. *Process Modelling and Model Analysis*; Academic Press: London, UK, 2001; ISBN 978-0121569310.
11. Kleimt, B.; Köhle, S.; Kühn, R.; Zisser, S. Application of models for electrical energy consumption to improve EAF operation and dynamic control. In Proceedings of the 8th European Electric Steelmaking Conference, proceedings. European Electric Steelmaking Conference, Birmingham, UK, 9–12 May 2005.
12. Pfeifer, H.; Kirschen, M.; Simoes, J.-P. Thermodynamic analysis of electrical energy demand. In Proceedings of the 7th European Electric Steelmaking Conference, Venice, Italy, 26–29 May 2002; pp. 1.413–1.428, ISBN 88-85298-44-3.
13. Kirschen, M.; Zettl, K.-M.; Echterhof, T.; Pfeifer, H.; Models for EAF Energy Efficiency. Steel Times International [Online], 6 April 2017. Available online: https://www.steeltimesint.com/features/models-for-eaf-energy-efficiency (accessed on 25 July 2021).

14. Kleimt, B.; Pierre, R.; Kordel, T.; Rekersdrees, T.; Schlinge, L.; Hellermann, O.; Elsabagh, S.; Haverkamp, V.; Gogolin, S. *Adaptive EAF Online Control Based on Innovative Sensors and Comprehensive Models for Improved Yield and Energy Efficiency*; European Commission: Brussels, Belgium, 2019; ISBN 978-92-79-98359-7.
15. Malfa, E.; Nyssen, P.; Filippini, E.; Dettmer, B.; Unamuno, I.; Gustafsson, A.; Sandberg, E.; Kleimt, B. Cost and Energy Effective Management of EAF with Flexible Charge Material Mix. *BHM Berg-und Hüttenmännische Mon.* **2013**, *158*, 3–12. [CrossRef]
16. Conejo, A.N.; Cárdenas, J. Energy Consumption in the EAF with 100% DRI. In Proceedings of the AISTech 2006 Conference, Cleveland, OH, USA, 1–4 May 2006; pp. 529–535.
17. Carlsson, L.S.; Samuelsson, P.B.; Jönsson, P.G. Using Statistical Modeling to Predict the Electrical Energy Consumption of an Electric Arc Furnace Producing Stainless Steel. *Metals* **2019**, *10*, 36. [CrossRef]
18. Gajic, D.; Gajic, I.S.; Savic, I.; Georgieva, O.; Di Gennaro, S. Modelling of electrical energy consumption in an electric arc furnace using artificial neural networks. *Energy* **2016**, *108*, 132–139. [CrossRef]
19. Baumert, J.-C.; Engel, R.; Weiler, C. Dynamic modelling of the electric arc furnace process using artificial neural networks. *Revue de Métallurgie* **2002**, *99*, 839–849. [CrossRef]
20. Chen, C.; Liu, Y.; Kumar, M.; Qin, J. Energy Consumption Modelling Using Deep Learning Technique—A Case Study of EAF. *Procedia CIRP* **2018**, *72*, 1063–1068. [CrossRef]
21. Bowman, B.; Krüger, K. *Arc Furnace Physics*; Verlag Stahleisen: Düsseldorf, Germany, 2009; ISBN 9783514007680.
22. Goodfellow, I.; Bengio, Y.; Courville, A. *Deep Learning*; The MIT Press: Cambridge, MA, USA, 2016; ISBN 0262035618.
23. Rasmussen, C.E.; Williams, C.K.I. *Gaussian Processes for Machine Learning*, 1st ed.; The MIT Press: Cambridge, MA, USA, 2016; pp. 33–77.
24. Murphy, K.P. *Machine Learning: A Probabilistic Perspective*; MIT Press: Cambridge, MA, USA, 2013; ISBN 978-0262018029.
25. Stein, M.L. *Interpolation of Spatial Data: Some Theory for Kriging*, 1st ed.; Springer: New York, NY, USA, 1999; ISBN 978-0-387-98629-6.
26. Hartung, J.; Elpelt, B.; Klösener, K.-H. *Statistik: Lehrund Handbuch der Angewandten Statistik*; Oldenbourg Wissenschaftsverlag: Munich, Germany, 2012; ISBN 978-3-486-71054-0.
27. Toulouevski, Y.; Zinurov, I. *Innovation in Electric Arc Furnaces: Scientific Basis for Selection*; Springer: Berlin, Germany, 2010; ISBN 978-3-642-03800-6.
28. Kovačič, M.; Stopar, K.; Vertnik, R.; Šarler, B. Comprehensive Electric Arc Furnace Electric Energy Consumption Modeling: A Pilot Study. *Energies* **2019**, *12*, 2142. [CrossRef]
29. Wooldridge, J.M. *Introductory Econometrics: A Modern Approach*, 5th ed.; South Western Educ Pub: Mason, OH, USA, 2012; ISBN 978-1-111-53104-1.

metals

Article

Development of a Fast Modeling Approach for the Prediction of Scrap Preheating in Continuously Charged Metallurgical Recycling Processes

Christian Schubert *, Dominik Büschgens, Moritz Eickhoff, Thomas Echterhof and Herbert Pfeifer

Department for Industrial Furnaces and Heat Engineering, RWTH Aachen University, 52074 Aachen, Germany; bueschgens@iob.rwth-aachen.de (D.B.); eickhoff@iob.rwth-aachen.de (M.E.); echterhof@iob.rwth-aachen.de (T.E.); pfeifer@iob.rwth-aachen.de (H.P.)
* Correspondence: schubert@iob.rwth-aachen.de; Tel.: +49-24-1802-5959

Abstract: Improving the overall energy efficiency of processes is necessary to reduce costs, lower the specific energy consumption and thereby reduce the direct or indirect emission of gases that contribute to climate change. In many metallurgical processes, a large amount of energy is lost with the off-gas. In metallurgical recycling processes, off-gas often can be used to preheat the to-be-recycled metal scrap, leading to significantly higher energy efficiency. However, the application of preheating has the disadvantage that it often requires more precise planning in the design and better control of the process. In this paper, a simplified look at a continuously charged scrap preheating aggregate for the widely used electric arc furnace (EAF) in the steel processing industry is used as illustration. Continuous scrap charging in EAF-type furnaces in general has much higher demands on process control and general process knowledge, which is why they are found only very rarely. General issues and basic modeling approaches to mitigate such issues allowing a better process control will be described. In particular, a fast, one-dimensional modeling approach for the determination of the temperature distribution inside a constantly moving scrap bulk, with hot air (or exhaust gases) flowing through it, will be described. Possible modeling applications, assumptions, possible enhancements and limitations are shown. The first results indicate that this approach can be used as a solid basis for the modeling of scrap bulks with thermally thin parts, consisting of materials with similar thermodynamic material properties. Therefore, as a basis, this approach may help improve the design and control of future or existing preheating devices in metal recycling processes.

Keywords: scrap preheating; electric arc furnace; continuous charging

Citation: Schubert, C.; Büschgens, D.; Eickhoff, M.; Echterhof, T.; Pfeifer, H. Development of a Fast Modeling Approach for the Prediction of Scrap Preheating in Continuously Charged Metallurgical Recycling Processes. *Metals* **2021**, *11*, 1280. https://doi.org/10.3390/met11081280

Academic Editor: Emin Bayraktar

Received: 14 July 2021
Accepted: 9 August 2021
Published: 12 August 2021

1. Introduction

Making better use of waste energy, usually in the form of heat, is necessary to increase the energy efficiency of nearly every process. A large amount of such waste heat, for example, occurring in recycling metallurgical melting processes, is often lost within the off-gas. Today, recycling processes are very important due to their overall better energy efficiency. To give an example, the steel production from recycled materials saves about 1.5 tons of CO_2 per ton of steel, saving around 945 million tons of CO_2 emissions per year [1], not to mention the significant damages to the environment induced by primary iron ore or coal mining. It is estimated that roughly 630 million tons of steel are produced from recycled material every year, with increasing tendency [1]. The most relevant process for steel recycling is the electric arc furnace process. Furthermore, with CO_2-neutral primary iron production processes such as direct reduced iron (DRI), the importance of the electric arc furnace will continue to increase, as it is also required in this process chain. Therefore, further improving its efficiency is very relevant to decrease the worldwide CO_2 emissions; it is also a good investment in the foreseeable future. In the EAF process, a large amount of the supplied energy (approximately 30% [2]) is lost within the off-gas, whose temperature ranges between 750 and 1200 °C, depending on the process and point of measurement. According to Nardin et al. [2], there are many strategies for making better use of the waste energy especially in this kind of process. One strategy is clearly obvious, namely, to use the energy of the off-gas for preheating of the charged material, which can be assigned under direct heat recovery techniques. Thereby, the enthalpy of the charged material is increased, and the heat required for melting the scrap inside the metal bath is reduced. So why is this not the common practice by now? Of course, there are some drawbacks introduced by this approach which only can be surpassed via good process control. For example, according to Toulouevski and Zinorov [3], a maximum temperature between 1560 and 1580 °C must be held inside the metal bath to prevent excessive refractory wear. This is only around 50 °C above liquidus temperature. Furthermore, the heat transfer between molten metal and scrap is limited due to the low convection speeds inside the melt. Since the scrap cannot be homogeneously charged across the liquid bath and a continuous charging between the hot spot inside the melt, between the electrodes, is difficult from a constructive point of view, the temperature and mass flow rate of scrap that is charged must be strictly controlled. Additionally, the off-gas has to undergo some post combustion process to remove CO and lower NO_X. A sensible approach is to split the off-gas between a scrap preheating device and another off-gas channel and bring them back together later for post combustion. Therefore, it is important that the mixed off-gas still has enough energy in combination with some mixed-in air that post-combustion is still possible. At least, it should still have enough energy, so that only little energy, for example, from a post combustion burner, must be added. Otherwise, scrap preheating would not be very effective. In addition, due to organic residues on the scrap, there are usually increased dioxin emissions when preheating techniques are applied [4,5]. The dioxin amount in the off-gas can be reduced, but to achieve this in a cost and energy efficient way, a relatively high temperature (above 850 °C) has to be maintained in the post-combustion chamber, followed by a quick cooling step between 600 and 200 °C to avoid reformation of dioxins [6]. To keep the described problems and the dynamic interactions between the individual problems under control, a significantly improved process control is required. Here, different modeling techniques can be used to better understand, plan and later better control those process interactions between preheating, off-gas temperature and bath temperature.

There have been several publications of concepts, patents and other materials for continuous preheating concepts in the recent years [7–10]. In the latest literature, CFD models and process models regarding the topic of scrap preheating can be found [11–15], but they either lack a detailed description of the model, the simplification level is too high (for example lacking a prediction of the temperature distribution in the bulk) or these approaches are too computationally intensive to be used to build online control models. Zhang and Oeters [16] described a similar modeling approach as in this paper, with the additional modeling of some chemical reactions, but for a different application.

In this paper a basic, fast and also extensible 1D modeling approach for the preheating of flown through bulk material is described. As an example, for relevant application, a simplified view on a continuously charged EAF is presented. In particular, we refer to the so-called ISMELT® technology, recently developed by Inteco Inc. (Bruck an der Mur, Austria) [10], and use it as an illustrative example. In this paper, the modeling will focus on the prediction of the scrap preheating and the cooling of the off-gas. In the future, the model can be used to be coupled with more general EAF models, for example, as described by Hay et al. [17] to predict the overall modified process behavior. Furthermore, thanks to its simplicity and the implementation with the Julia programming language, with its powerful dispatch model [18,19], it is extensible, fast, versatile and may help to further develop suitable online control models.

Description of the Modeled Process

The illustrative process is shown in Figure 1. The scrap is charged into an inclined shaft with some sort of transportation support and control mechanism, for example, a shaking floor, at the side of the EAF. The off-gas can be controlled to be distributed between the shaft and the usual EAF off-gas exhaust. As already described, this is necessary, since the two off-gas flows will be mixed again for post combustion of organic components and purification.

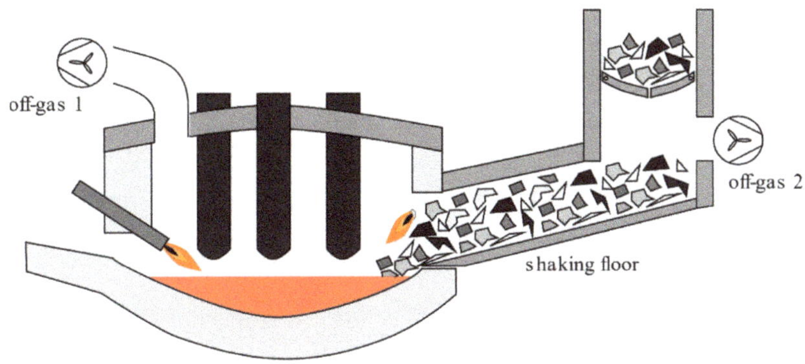

Figure 1. EAF off-gas preheating principle.

Although it is called continuous scrap charging, the EAF process is inherently a batch process with multiple process phases. A continuously charged EAF process should be run in two phases [3]. First, the flat-bath melting phase, and second, the refining phase. During the more or less continuous flat-bath melting phase, the scrap is continuously charged into the residual melt remaining from the previous tapping. During the refining phase, the charging of scrap is stopped and the slag layer above the hot melt is broken up by the now activated oxygen lances, thus, a higher amount of thermal radiation from the melt and electric arcs reaches the scrap in front of the shaft, and since charging is stopped, the scrap bulk is standing still in the charging shaft. Therefore, there is a higher risk that partial melting of scrap parts may occur. This could potentially block scrap charging by clumping or damaging the charging mechanism. To evaluate the possibility of such behavior, simplified modeling approaches for the two phases under the following assumptions are used:

- Flat-bath melting phase model (phase 1): Here, a continuous scrap flow to the melt is modeled. The off-gas temperature and flow rate is assumed to be constant.
- Refining phase (phase 2): No off-gas or scrap flow through the shaking floor tunnel, heat transfer inside the scrap bulk is mainly driven by surface to surface (s2s) radiation effects.

The flat-bath melting phase is assumed to last around 50 min, while the refining phase is assumed to last around 5 min.

2. Materials and Modeling Approach

2.1. Materials

2.1.1. Scrap

The assumption about size distribution and material properties of scrap material has great influence on the following investigations. Therefore, the scrap material is a main interest for this study. There are different implications of the scrap material to the process. The main influences of the scrap material to the process behavior are:

- Scrap composition (steel grades and alloying elements);
- Scrap contamination in terms of adherent organic materials;
- Distribution of the geometrical characteristics of the scrap speaking of the mean scrap part thickness and the surface to volume ratio.

All these unknowns can influence the scrap preheating temperature, the off-gas composition or the pressure loss curve over the length of the preheating shaft. The pressure loss over the length of the shaft is very important, because it must be compensated for to define a controlled volume flow through the shaft. As there is no general information about the consistency of the scrap available, which also may change from charge to charge, some simplifying assumptions will be used for now. Once, it is assumed that the scrap steel uniformly consists of steel 1.0035, with a high Fe amount of 99%, as given by the European steel register [20]. The density is given with a constant value of 7850 kg/m^3; the other properties are shown in Figure 2.

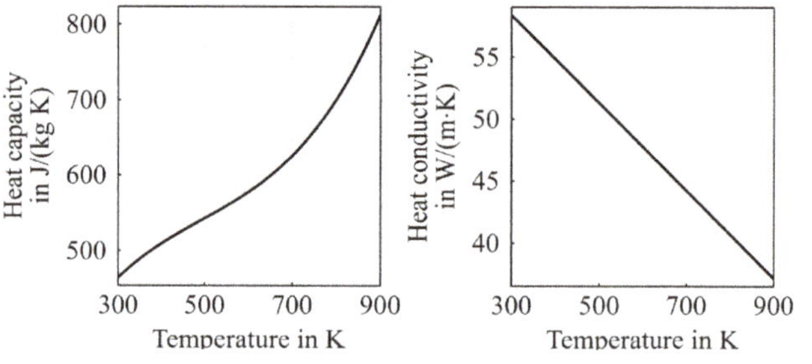

Figure 2. Thermophysical properties of steel 1.0035.

Furthermore, it is assumed that the smallest pieces of scrap will have the most significant influence on the pressure loss over the scrap itself. Therefore, three different characteristic scrap sizes listed in Table 1 will be used for the further investigations.

Table 1. Different scrap classes.

Name	Label	Density in kg/m^3	Dimensions (Length × Width × Thickness) in m	Source
Turning chips	S1	500	0.25 × 0.03 × 0.001	Measurements from CNC scrap of our workshop
Rail crops scrap	S2	1121	1.21 × 0.45 × 0.0127	Datasheet [21]
Plate Scrap	S3	801	0.92 × 0.61 × 0.003175	Datasheet [21]

2.1.2. Off-Gas

As for the scrap, there will be some very significant simplifications used for the off-gas. The composition will be assumed with a (constant) composition of 70 vol.-% N_2, 10 vol.-CO_2 and 20 vol.-% CO, as roughly approximated from the flat-bath melting phase of comparable sized EAF processes. The temperature-dependent material properties were calculated with Cantera [22] using the GRI-Mech 3.0 thermodynamic mechanism set [23]. The properties are shown in Figure 3.

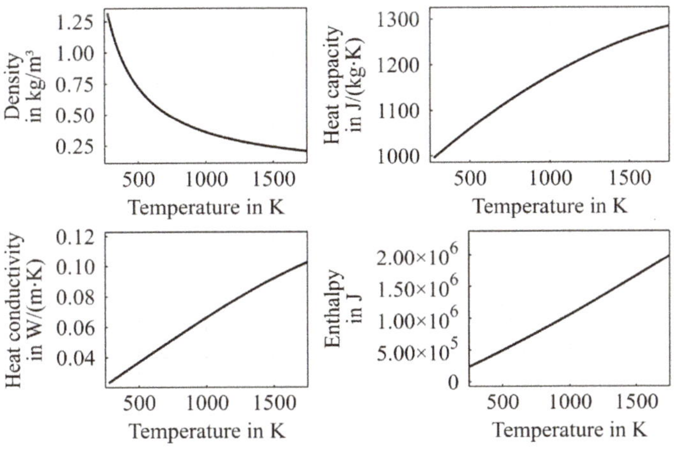

Figure 3. Thermophysical properties of the off-gas (70 vol.-% N_2, 10 vol.-% CO_2 and 20 vol.-% CO).

2.2. Modeling Parameters

A crucial point for the model's predictive power is the appropriate choice of different model parameters, such as characteristic surface areas or characteristic lengths. These parameters are roughly estimated within this work, as can be seen in Table 2, but could also be more exactly defined using more detailed investigations or automated modeling optimizing techniques, when combined with real operation data.

Table 2. Operating data for modeled scrap bulk shaft.

Property	Value	Unit
Input temperature scrap	25	°C
Mass flow scrap	38.6	kg/s
Mass flow off-gas (through scrap bulk)	6.64	kg/s
Floor length	8	m
Floor cross section area	2.9 × 3.2	m^2

2.3. Modeling Approach

The model is essentially developed combining two or three one-dimensional energy transports models, namely, over the fluid phase, over the solid bulk phase and optionally over the individual scrap part's thickness. The implementation itself is carried out using the Julia programming language (ver. 1.6) [18], it can be accessed in the Supplementary File S1. The model, furthermore, is derived using a fixed grid enthalpy-based, energy balance modeling approach, including the capability to include temperature dependent material properties of the off-gas and the scrap material.

Therefore, the shaking floor tunnel is virtually divided into separate cross section layers (bulk layers), perpendicular to flow or scrap movement, as can be seen in Figure 4. Each layer in this discretized bulk volume is then again divided into one off-gas cell and 1 to n_j multiple scrap layer cells, which can be seen in Figure 5. These scrap layer cells, if more than one is used, will model the heat conduction in an exemplary scrap piece of the bulk at the position i. Considering the thickness of the scrap parts may be relevant if they are thermally thick, which usually is the case if the Biot number Bi, Equation (1), is greater than 0.1. Here, α is the heat transfer coefficient, l_{char} is the characteristic thickness of the material and λ_s its thermal conductivity. Here, this approach also allows for some optional approximation of the thermal distribution inside the scrap, under the assumption that the scrap parts can be sufficiently characterized using one characteristic thickness and a representative surface. Therefore, this approach may also be referred to as the 1D(–1D) model.

$$Bi = \frac{\alpha \cdot l_{char}}{\lambda_s} \quad (1)$$

Following the notation in Figure 5, in the following text the cells in the ith direction will be referred to as bulk layer cells and the cells in jth direction as scrap layer cells. Each direction is discretized using homogeneous cell sizes.

Figure 4. Illustration of first discretization step.

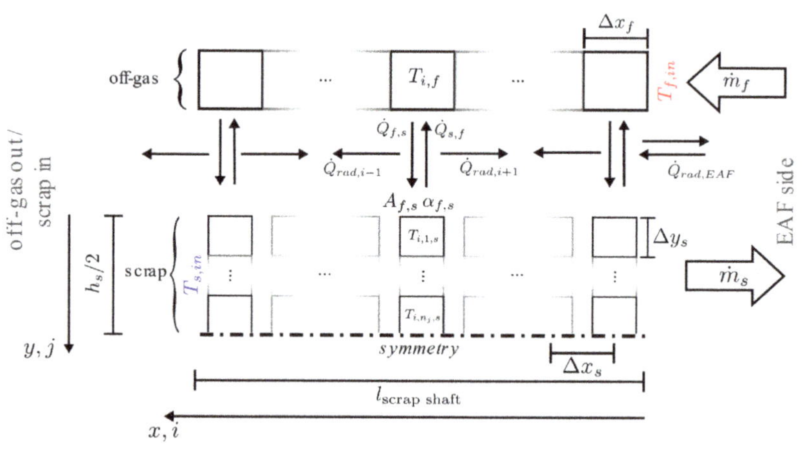

Figure 5. Modeling scheme of the 1D(–1D) model.

In this model mass flow, convective, conductive and radiative heat transfer are expressed as incremental changes in the cell's enthalpy, as isobaric process conditions can be assumed for this case. For the solver implementation, a dT/dt formation is used, which is spatially discretized, to transform the PDE system into an ODE system. The "DifferentialEquations.jl" solver package [24] will be used for the solution of the resulting ODE system. For simplicity, the expressions will be written as change in enthalpy ΔH, which should facilitate the reading of the equations.

For example, the change in enthalpy, during a defined timestep Δt of the off-gas (fluid f) in each cell i, $\Delta H_{i,f}$ can be expressed due to the off-gas mass flow $\dot{m}_f$ using Equation (2). c_p is the true heat capacity.

$$\Delta H_{i,f} = \dot{m}_f \cdot \Delta t \int_{T_{f,i-1}}^{T_{f,i}} c_{p,f}\, \mathrm{dT} \quad (2)$$

In the same way, it is possible to account for the scrap movement; as for the numerical approximation, it is sensible to use upwind approaches according to the movement direction the evaluation direction is flipped for the scrap cells (see Equation (3)).

$$\Delta H_{i,s} = \dot{m}_s \cdot \Delta t \int_{T_{s,i}}^{T_{s,i+1}} c_{p,s}\, \mathrm{dT} \tag{3}$$

Furthermore, the enthalpy change due to heat transfer between all the scrap cells can be expressed within the *i*- and *j*-direction. The different heat transfer mechanisms are heat conduction within the scrap parts, heat conduction within contact areas between the scrap parts, convection between scrap and air and thermal radiation heat transport from the furnace side to the scrap and between individual scrap parts.

2.3.1. Modeling Simplifications

The current implementation of the model comes in hand with some simplifying assumptions; these assumptions are presented for better comprehensibility of the model:

- Scrap consisting of small flat stripes of metal with a characteristic thickness (which should translate to the fact that they have only one direction where thermal conduction will matter the most) and a representative surface area;
- All scrap parts are composed of the same material;
- Scrap pieces are shaped approximately equal and distributed evenly over the scrap bulk;
- The scrap bulk has uniform scrap thicknesses and constant surface to volume ratio;
- There is a uniform scrap movement in direction of the shaking floor;
- Neglection of gas radiation (the usual distance between different parts) is estimated to be in between 1 and 15 cm, and the gray gas emissivity for the given gas composition (over a temperature range from 300 K to 1500 K at 1 atm) varies between 0.01 and 0.08, calculated according to Alberti et al. [25]; therefore, its influence will be neglected for now);
- The emissivity of all surfaces (scrap and furnace walls) will be assumed to be 0.8;
- Assuming temperatures listed in Table 3 for modeling of scrap bulk incident radiation for the refining phase (2);
- Neglecting dissipation and compressible pressure effects in the off-gas flow;
- Constant gas composition over whole process time;
- Constant heat transfer coefficient between scrap and off-gas over the length of the shaft;
- Symmetry assumption over the individual scrap part's thicknesses;
- Neglecting the contact of scrap with the surrounding walls;
- No modeling of thermal conduction between the individual scrap parts in the bulk (convective and/or radiative heating is assumed to dominate the heat distribution of the scrap bulk);
- Homogeneous temperature and mass flow of the off-gas in each bulk layer cell.

2.3.2. Heat Conduction over Characteristic Scrap Thickness

First, the heat transfer through heat conduction in an individual (representative scrap piece), as shown in the lower part of Figure 5, will be modeled. The heat conduction term over the characteristic thickness of the scrap piece can be described using Equation (4). Here, Δy_s is the thickness of each scrap cell (index *s*) in *j*-direction, $A_{y,s}$ is the characteristic cross section surface of a scrap piece, normal to the main heat conduction direction.

$$\Delta H_{i,j,s} = \Delta t \cdot \frac{\lambda_{y,s}(T_{i,j,s})}{\Delta y_s} \cdot A_{y,s} \cdot \left(T_{i,j+1,s} + 2\, T_{i,j,s} - T_{i,j-1,s}\right) \tag{4}$$

2.3.3. Heat Conduction in the Off-Gas

Accordingly, the off-gas conduction in *i* direction can be modeled using Equation (5), which may become relevant for very low off-gas flow rates. Δx_f is the length of the fluid cells (index f) in i-direction; in the current approach, the fluid and solid cells are overlaying each other, and Δx_f is equal to Δx_s.

$$\Delta H_{i,j,f} = \Delta t \cdot \frac{\lambda_{x,f}\left(T_{i,j,f}\right)}{\Delta x_f} \cdot A_f \cdot \left(T_{i+1,f} + 2\, T_{i,f} - T_{i-1,f}\right) \tag{5}$$

However, it its more difficult to come up with a simple model for the conductive heat transfer between the individual pieces and its resulting heat conduction contribution in i direction over the scrap bulk.

2.3.4. Convective Heat Exchange between Scrap and Off-Gas

The energy transfer between the off-gas (fluid) and scrap (solid) models is realized using Equation (6). To also model the heat transfer in the case of no off-gas flow, a minimum value of α_{fs} of 10 W/(m^2 K) should be used. Using Equation (6), the change in the solid cell enthalpy through convective heat transfer is calculated as the product of heat transfer coefficient α_{fs} the scrap surface A_{fs} and the difference between scrap T_s and fluid (off-gas) T_f temperature.

$$\delta Q_{i,fs} = \Delta t \cdot \alpha_{fs} \cdot A_{fs} \cdot \left(T_{i,f} - T_{i,1,s}\right) \tag{6}$$

Here, the interfacial scrap surface A_{fs} can be related to a hypothetical inter-facial area density $SA : V_b$, which represents the ratio of scrap surface m^2 to scrap volume in m^3 in the bulk, and $A_{fs} = SA : V_b \cdot V_{fs}$, $SA : V_b$ can be calculated using the dimensions (distribution) of the scrap pieces, the summed overlapping contact area of the scrap pieces in m^2 per m^3 and the density of the scrap bulk itself.

2.3.5. Balancing

The modeling of the radiation phenomena onto $\delta Q_{rad,i,front}$ and inside $\delta Q_{rad,i,cross}$ the scrap bulk is described in a separate section below.

Based on the summed balance of the different heat Q and enthalpy H changes, generally termed ΔH, the new temperature of each cell can be calculated, using the inverse function $T(H)$ of the enthalpy curve $H(T)$ of the given material, resulting in Equation (7). Here, $H(t)$ is the specific enthalpy of a cell at time t and $m_{i,(j)}$ is the mass of the cell.

$$T_{t+\Delta t} = T\left(H(t) + \frac{\sum \Delta H}{m_{i,(j)}}\right) \tag{7}$$

2.3.6. Radiation Modeling—Bulk Incident Radiation

During the flat-bath melting phase (phase 1), it will be assumed that the off-gas is very dusty and gas radiation dominates, so Equation (8) is used to calculate the radiative heat transfer from the EAF environment to the front of the scrap bulk. Here, $A_{s,front}$ is the surface of the inclined front (Figure 6), with 16.4 m^2 and ϵ_s is the emissivity of the scrap. This equation is then completed by introducing Equation (14), which takes the possible radiation transmittance of the first layer(s) into account.

$$\dot{Q}_{i,rad,EAF} = \sigma \cdot \epsilon_s \cdot A_{s,front} \cdot \left(T_i^4 - T_{\infty,\,EAF}^4\right) \tag{8}$$

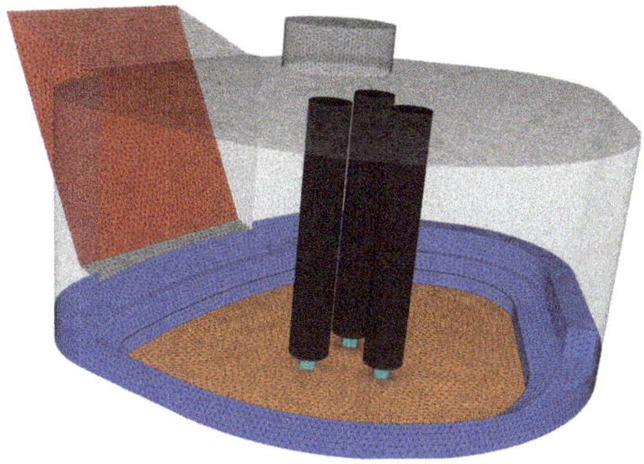

Figure 6. Discretized geometry used for view factor calculation.

Table 3. Temperatures for the calculation of $\dot{Q}_{rad,EAF}$ during phase 2 (geometry is shown in Figure 6).

Color	Red	Blue	Cyan	Orange	Gray
Temperature in °C	Modeled	1500	4500	1650	900

During the refining phase of the exemplary process, thermal surface to surface radiation effects will become the most important mechanisms between the individual scrap parts inside and onto the scrap bulk. Gas radiation effects, on the other hand, will be negligible (in the bulk itself), since the radiation distances between the individual parts are small and the overall heat capacity of the off-gas is low compared to the scrap's heat capacity.

In particular, during the second (refining) phase of the exemplary continuously charged EAF process, where the scrap charging and the off-gas suction through the shaking floor tunnel will be stopped, thermal radiation effects onto and inside the scrap bulk will become highly relevant for the thermal distribution inside the scrap bulk. Therefore, the furnace's scrap bulk temperature in front of the shaking floor may rise a lot more due to the incident thermal radiation.

To describe these radiation effects to the front of the scrap bulk, the surface to surface (s2s) approach is used. According to Hottel or Howel [26,27], the heat flow between thermal radiating surfaces to a special surface i can be computed by Equation (9), which represent, the difference between the in- and outgoing heat fluxes $\dot{q}''$ multiplied with the radiation exchange surface area A_i of a surface i.

$$\dot{Q}_i = \left(\dot{q}''_{out,i} - \dot{q}''_{in,i}\right) \cdot A_i \tag{9}$$

The outgoing heat flux can be evaluated solving the following linear equation system (10), if the geometry specific view factors F_{ik} of the N surfaces are known.

$$\left[\sum_{k}^{N}(\epsilon - 1) \cdot F_{ik} + \delta_{ik}\right] \cdot \left[\dot{q}''_{out,i}\right] = \left[\epsilon_i \, \sigma \, T_i^4\right] \tag{10}$$

To then calculate the surface incident heat flux $\dot{q}''_{in,i}$, Equation (11) can be used.

$$\dot{q}''_{in,i} = \sum_{k=1}^{N} \dot{q}''_{out,k} \cdot F_{ik} \tag{11}$$

Using this approach, the heat flow $\dot{Q}_{rad,EAF}$ (Figure 4) to an area representing the front surface of the scrap bulk is calculated to include the furnace incident radiation effects inside the model. Therefore, the view factor matrix was calculated using an in-house radiation model. The geometry for the view factor calculation is shown in Figure 6.

For the differently colored faces, the temperatures given in Table 3 were assumed. Here, especially the assumed arc temperature (cyan) can vary a lot between different models (<2000–30,000 K) [28–30]. With a view factor of <1%, the impact of this temperature to the scrap front is not very important. Here, the shown values must be seen as a rough approximation that have no particular significance or are supported by measurement data. Nevertheless, to model an existing preheating process, those temperature assumptions must be optimized against existing process data, as these are highly relevant for the heat input over the refining phase 2.

2.3.7. Radiation Modeling—Inside of the Scrap Bulk

As some of the thermal radiation between the scrap cells may penetrate though a single cell layer in i direction, due to its porosity, an approach to model this penetration behavior is necessary. Therefore, a scrap bulk thermal radiation simulation approach, to add additional radiation heat changes ΔQ_{rad} to the evaluation of Equation (7), has been developed. This is somehow difficult, as changes in the modeling approach, according to the chosen cell size, must be made, because the radiation between the scrap pieces is not a continuous phenomenon. This is illustrated in Figure 7a, where you can see that if the cell size is chosen to large, radiation phenomena will probably be underestimated due to its highly nonlinear nature, and if it is chosen to be a lot smaller than the individual scrap pieces (Figure 7c), the individual heat transfer in each scrap piece may be underestimated, as well (especially if those have a high thermal conductivity and are not too elongated), as a single temperature (distribution) for each solid cell i is being used. Therefore, choosing a cell size similar to the dimensions of an individual scrap piece seems reasonable (Figure 7b). This may not always be possible if the individual dimensions of the piece differ greatly from each other or the pieces itself are very large, so that a corresponding large cell size would lead to significant numerical errors.

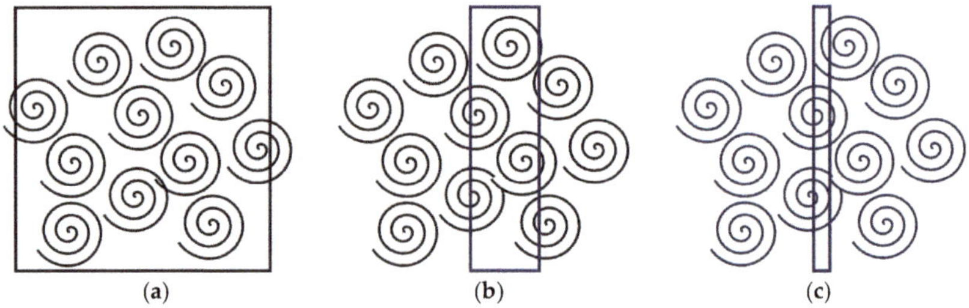

Figure 7. Illustrative sketch of differently used cell sizes. (**a**) cell size >> scrap parts, (**b**) cell size ≈ scrap parts, (**c**) cell size << scrap parts.

For the calculation of radiation passing through the cells, some assumptions will be established. Furthermore, a simple model to calculate a cumulative blocking factor between the cell layer i and its "seen" neighbor cells will be used. Therefore, it is assumed that there is an exposed surface in each cell, which contributes a certain amount of surface $A_{s,rad}$, which may be calculated using Equation (12) and represents the scrap surface, which can be "seen" by its adjacent cell at maximum. Within this equation, it must be considered that in theory, at maximum, some proportion less than half of the size of this surface area can be "seen" from each side of the cell, which leaves a specific unknown factor c somewhere between zero and one, depending on the scrap geometry and distribution. This factor will cover for some side radiation losses, which cannot be represented in the 1D model, and further represent some kind of the only "self-seen amount" of the scrap parts in each cell. If the losses are ignored, which becomes increasingly true, the more cells are used (the smaller the cells become in respect to the shaft dimensions); this factor boils down to represent the not self-seen proportion of the surfaces. Furthermore, c may also cover for the fact that through the structure of the radiation surfaces of the individual scrap pieces in the bulk, ϵ_{eff} instead of the ideal ϵ must be used, because of the cavity formation of the scrap pieces in the individual cells. Depending on the size and shape distribution of the individual scrap parts, c will not be independent of the chosen cell size. Since there is no real knowledge of the quantity of this factor at the moment, a factor c of 0.5 will be used.

$$A_{rad,layer\ i} = c \cdot 0.5 \cdot SA : V_b \cdot V_{fs} \tag{12}$$

In this paper there, is no deep dive into the topic of how one could practically determine the factor c. For now, this factor should be looked at as a kind of optimization factor, which later could be adapted using real process data or adapted within sensitivity studies. If there are different scrap parts with different c values inside the bulk, this model could in general be extended to a 2D or 3D model type, where influences of the composition could also be investigated.

Having the potentially radiation-blocking surface of each cell layer i, which is the same for each cell, a constant surface to volume area ratio and a constant cell size is used; there is a need for a model to estimate the penetration of thermal radiation through the individual cell layers. Therefore, a sub model, where the blocking surfaces of the individual scrap pieces in each cell over the cross-section area of the shaft are kind of unrolled into one dimension (see Figure 8a) has been developed. Furthermore, it is assumed that the scrap pieces are equally distributed over the cross-section area, but the distance between the individual pieces is allowed to vary randomly in between those equally distributed positions. If completely random positions of the individual pieces would be possible, dysfunctional scrap bulks would be part of consideration; for example, convectional heating would not work as intended. Due to the symmetry effects in such a system, this is similar to modeling two randomly positioned accumulated scrap piece areas against each other (Figure 8b). Then, the intersection areas (green lines in Figure 8) between the two layers for many (10,000 and more) randomly varied positions can be calculated.

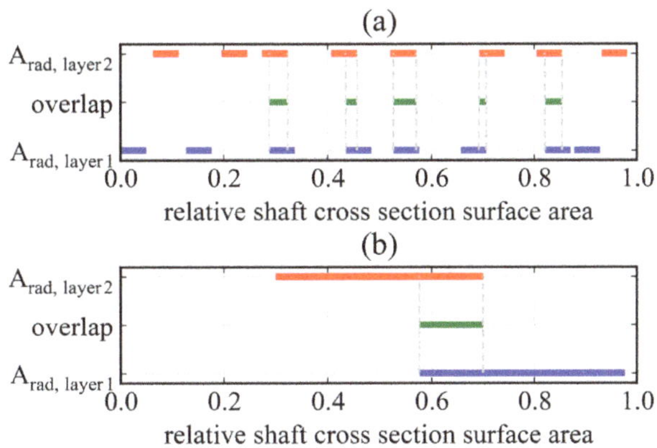

Figure 8. Exemplary representations for the randomized overlap calculation (**a**) using eight individual scrap pieces, (**b**) using one.

Using this method, the overlapping factor o_{fac} for each occupation fraction can be calculated, which is the fraction of the occupied shaft cross-section area by A_{rad}, shown in Figure 9a. Due to the fact that no completely random movement for the individual pieces has been set, the function deviates from a line with a slope of 1, where the average overlapping factor would be equal to the average occupation fraction.

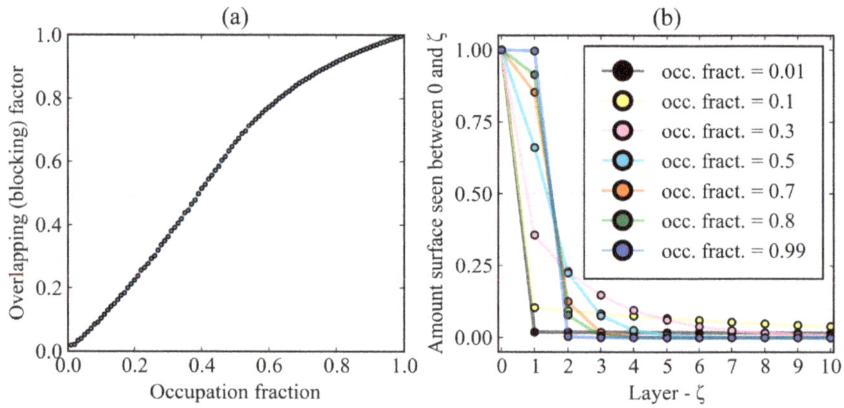

Figure 9. Calculated occupation fractions (**a**) and amount of "seen" surface between a cell i and its neighboring cell ζ (**b**).

Now, with this overlapping factor, the mean amount of surface, which can be seen between different cell layers ζ, can be calculated using the Equation (13), in the following referred to as b_ζ. Here, b_ζ is used as an array of values indexed according to ζ. ζ is a relative coordinate system between a cell i and its neighboring cell(s). This means that $\zeta = 0$ refers to a cell itself and $\zeta = 1$ refers to the exchange with the closest neighboring cell and so on.

$$\begin{aligned} \zeta = 0 &: 1 \\ \zeta = 1 &: o_{fac} \\ \zeta > 1 &: \left(1 - o_{fac}\right)^{\zeta} \cdot o_{fac} \end{aligned} \tag{13}$$

The results are shown in Figure 9b. From this figure, it is seen that different occupation fractions may require a different amount of considered neighboring cells to accurately calculate the thermal radiation exchange between the individual cell layers. In this modeling approach, the exchange between an amount of k neighboring layer cells is considered, where the amount of relative seen surfaces between i and $i \pm k$ is greater than 0.1%. Of course, this is a gross simplification, which also worsens with increasing Δx, because the calculated overlapping only directly correlates to radiation exchange for infinitely small distances, but it also is a fast and easy-to-calculate approach, which in the future also could be extended with additional optimization parameters to better fit reality.

Using the calculated b_ζ values, a modified version b^*_ζ were the entry for the layer index $\zeta = 1$ of b_ζ is removed. Then, the restructured b^*_ζ must be indexed according to $\zeta = i - 1$. The surface-to-surface radiation heat flow for the first k cells of the bulk, going from the EAF side, can be calculated using the Equations (9)–(11). Using the aforementioned modification, the result according to b^*_ζ (see Equation (14)) can be evaluated. This procedure is physically not completely correct, as each cell layer i of the first k cells is simplified as being on the front layer of the bulk (Figure 6), including its corresponding temperature T_i, but it should be sufficiently accurate.

$$\begin{aligned} 0 \le i < k &: \dot{Q}^*_{i,rad,EAF} = b^*_i \cdot \dot{Q}_{i,rad,EAF}(T_i) \\ i \ge k &: \dot{Q}^*_{i,rad,EAF} = 0 \end{aligned} \tag{14}$$

To obtain the overall radiation heat transfer between a cell i and its k surrounding cells, an energy balance considering all cells within a certain range index range k must be established, which again marks the range after which all outgoing radiation from a radiating cell should be absorbed. Therefore, Equation (15) has been derived. Here, the c factor is considered within A_{rad}, which, according to the assumptions, is the same for each cell i. Furthermore, a modified equation for the heat transfer between two parallel equally sized surfaces, modified with the factor b_ζ, which was introduced in the previous sections, is used with $\zeta = |i - \gamma|$.

$$\dot{Q}_{i,rad,cross} = \sigma \cdot A_{rad} \cdot \left(\sum_{\gamma=i-k}^{i-1} b_\zeta \cdot \frac{T_\gamma^4 - T_i^4}{\frac{1}{\epsilon_{s,i}} + \frac{1}{\epsilon_{s,\gamma}} - 1} + \sum_{\gamma=i+1}^{i+k} b_\zeta \cdot \frac{T_\gamma^4 - T_i^4}{\frac{1}{\epsilon_{s,i}} + \frac{1}{\epsilon_{s,\gamma}} - 1} \right) \quad (15)$$

Altogether, these equations should allow a decent modeling of the radiative transported heat inside the scrap bulk for higher temperature ranges. Unfortunately, one must estimate the mesh specific coefficient c in a certain range.

3. Results

3.1. Model Sensitivity to Different Scrap Types

First, some results for three hypothetical scrap bulks consisting of different classes of scrap (see Table 1) are shown. In Figure 10, the predicted preheating temperature of the scrap (Figure 10a) and off-gas outlet temperature (Figure 10b) development over time can be seen. In Figure 10a, the resulting scrap temperature does not differ very much (deviation around 20 K), although the scrap characteristics defer in the range of usual scrap parts. Furthermore, as the outlet temperature of the off-gas is close to the charged scrap temperature (Figure 10b), nearly all energy is transferred from the off-gas to the scrap.

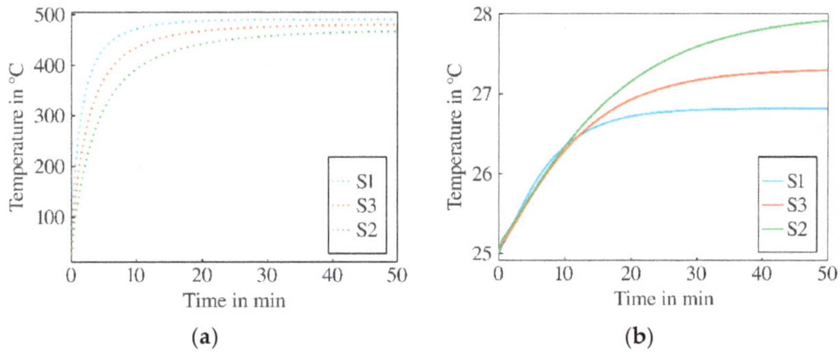

Figure 10. Simulated preheating temperature of scrap (**a**) and off-gas outlet temperature (**b**) over process time of phase 1 for the different characteristic scrap parts S1, S2 and S3 (see Table 1).

As it can be seen from Figure 10a, the scrap preheating temperature difference reduces over time of phase 1 under constant process conditions, which also applies to the spatial temperature distribution inside the bulk (see Figure 11a,b). From here on, the scrap type S3 is used for further investigations.

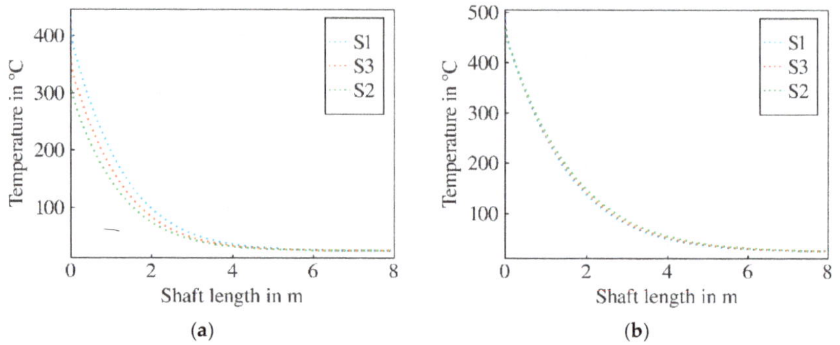

Figure 11. Simulated temperature distribution over the scrap bulk after (**a**) 5 and (**b**) 50 min.

3.2. Mesh Size Implications on the Modeled Process

In Figure 12, the effect of the chosen mesh size on the temperature distribution over the shaft length is shown to be very pronounced. There are mainly two reasons for that. The first is the numerical convergence of the model regarding mesh size (convection, gas conduction) and the second is that the radiative heat transport is physically influenced by the mesh size, which basically is due to the geometry reduction to 1D. The numerical error of the convection modeling is amplified by the fact that there is no heat conduction model implemented over the length of the scrap for the scrap parts, therefore, the heat conduction in x direction is theoretical infinite in each scrap cell and 0 between the scrap cells.

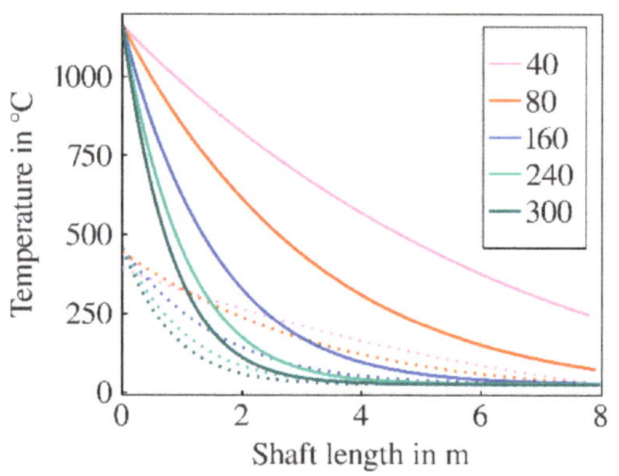

Figure 12. Simulated scrap bulk (S3) temperature and off-gas temperature over the length of the shaft for different cell sizes after 50 min; dotted = scrap temperature, solid = off-gas temperature.

As the heat conduction in and between the scrap parts should not be the main transport mechanism, at least for the exemplary modeled process, the modeling error in this regard should be acceptable, if enough cells to significantly reduce the numerical error of the convective modeling are used. In Figure 13, the difference between the model using radiation modeling and no radiation modeling used in the model is shown at the end of the two different phases. As expected, the influence of thermal radiation, especially in the front of the bulk, are non-negligible in phase 2, while for phase 1, the influence especially in the main part of the bulk is very small.

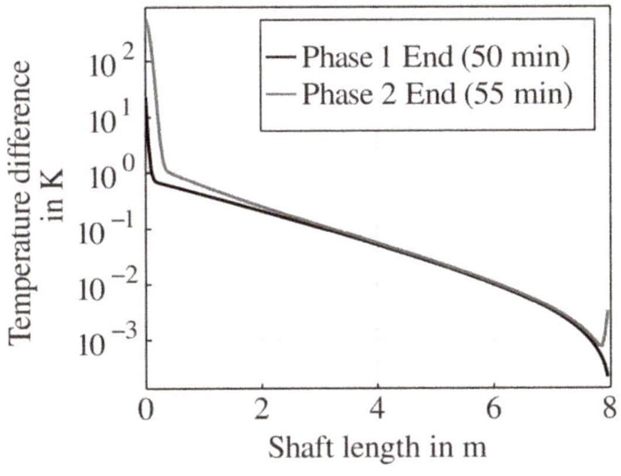

Figure 13. Comparison of resulting temperature differences radiation modeling in the bulk vs. no radiation modeling in the bulk at the ends of phase 1 and 2.

Furthermore, looking at two main target properties, namely, the scrap preheating temperature and the off-gas temperature after the shaft, it can be seen from Figure 14a,b that these properties are converging for cell amount larger than ≈100 cells rather quickly when they approach a steady state condition.

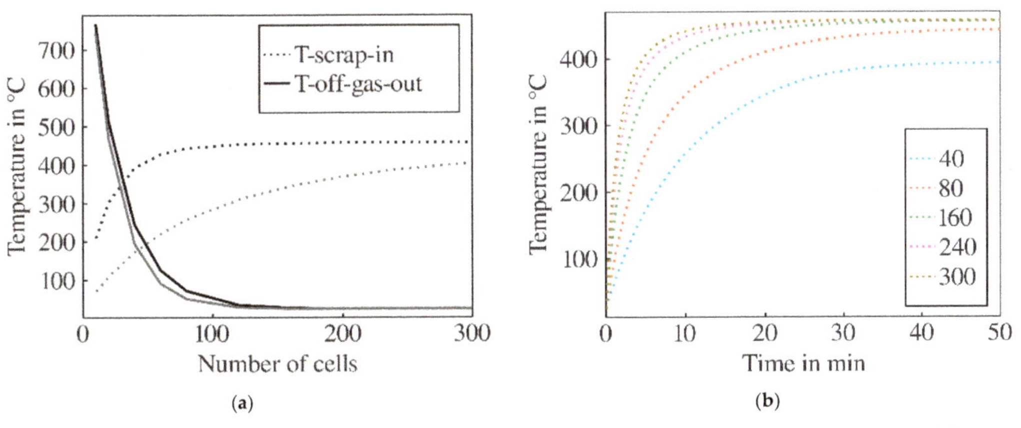

Figure 14. Scrap (S3) preheating and off-gas outlet temperature after 5 (gray lines) and 50 min (black lines) for different cell numbers (**a**); scrap (S3) preheating temperature development for cell numbers (**b**).

3.3. Influence of Convective Heat Transfer Assumptions

The influence of the used heat transfer coefficient is shown in Figure 15a,b. The simulation is very sensitive to the heat transfer coefficient in between 10 and 40 W/(m^2 K).

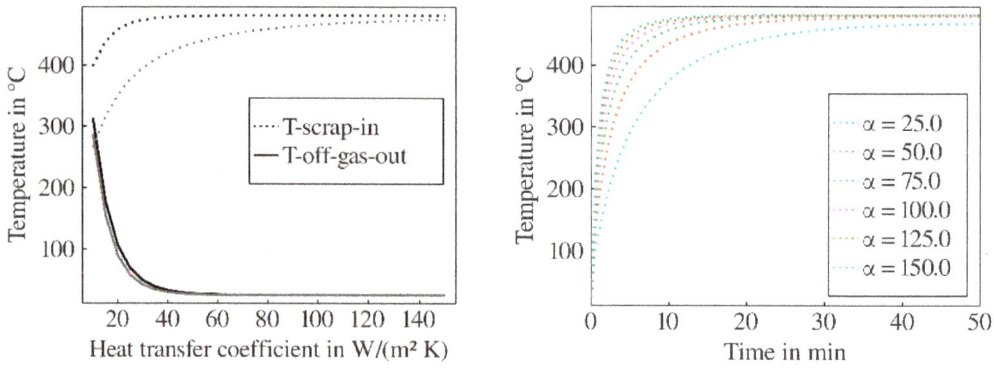

(a) (b)

Figure 15. Scrap (S3) preheating and off-gas outlet temperature after 5 (gray lines) and 50 min (black lines) for different heat transfer coefficients a_{fs} $(= \alpha)$ (**a**); scrap (S3) preheating temperature development for different heat transfer coefficients a_{fs} $(= \alpha)$ (**b**).

3.4. Influence of Factor c on Modeled Radiative Heat Transfer

The importance of the factor *c* does not seems to be of the most importance for the refining phase (phase 2) if the scrap temperatures inside the bulk are rising above 800 °C (Figure 16). During this phase (Figure 16b), deviations of around 50 °C can be observed between the predicted heating of the first scrap layer inside the scrap within varying *c* between 0.3 and 0.9. As the flat-bath melting phase (Figure 16a) is clearly dominated by convective effects, differences are minor and around 10 °C.

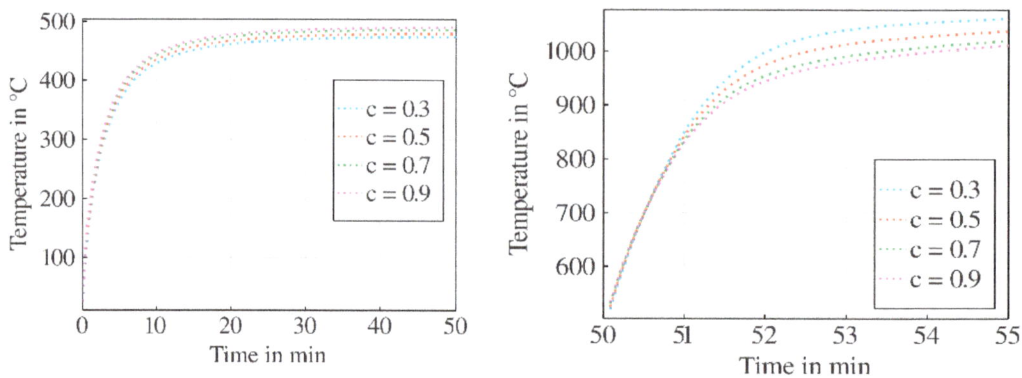

(a) (b)

Figure 16. Scrap (S3) preheating temperature development for different *c* values over the time of (**a**) phase 1 and (**b**) phase 2.

Of course, there are other influences on the model, such as mass flow rates, material properties, etc., but these are rather well known. Therefore, they will not be described further.

3.5. Verification of the Model's Heat Balance

To give some exemplary verification of the correctness of the model's implementation, energy in- and outputs to the shaft (scrap and off-gas), using the model without radiation transport model enabled, for the modeling of phase 1, are shown in Figure 17a.

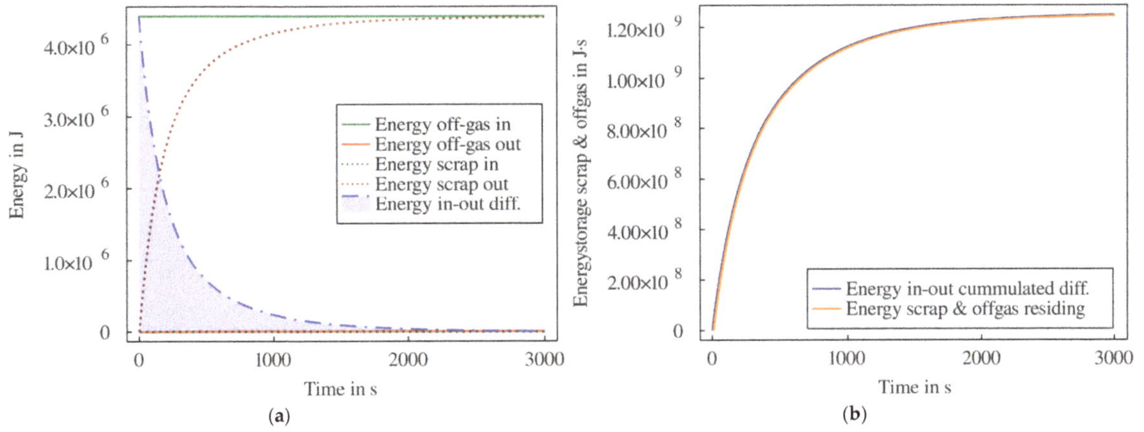

Figure 17. Energy in-, outputs and difference over time (**a**); cumulated energy difference and energy stored inside the bulk (scrap and off-gas) (**b**).

In Figure 17b, the integral (blue area in Figure 17a) is shown as the blue curve. As the temperature distribution inside the bulk reaches a stationary state, the differences between the in- and output energies will approach zero; the energy difference accumulated up to this point in time must reside inside the scrap bulk. This is shown with the orange curve in Figure 17b, which depicts the stored energy inside the bulk (scrap and off-has) over the process time, which was calculated from the respective spatial temperature curves. There is a slight offset between the curves, which is due to the dynamics between changing the whole heating curve of the bulk and changing the preheating temperature of scrap in the first cell of the model. This indicates the mathematical correctness of the models regarding the overall energy balance.

4. Conclusions

The described modeling approach can deliver plausible results in short simulation times, usually much less than a minute for reasonable mesh sizes and simulated process times of ~60 min. Of course, the many simplifying assumptions lead to the fact that the shown results are not directly transferable to a real process. These simplifications were developed to clarify the impact of certain model aspects. It was shown that the scrap type for some variation of the different scrap classes will not have a very high impact on preheating temperatures (if the parts can be seen as thermally thin). Furthermore, we showed the importance of radiation modeling for rising temperatures in refining phase 2. It is interesting to note that some of the unknown model parameters, such as c, which have uncertainty and for a real process probably must be tuned against validation data, have much less influence than one could expect. Furthermore, it was shown that the current model is very sensitive (in the convection dominated phase) if low heat transfer coefficient values between 10 and 40 $W/(m^2\ K)$ are used. Nevertheless, these values are low for forced convection, but due to the relatively low velocities in the exemplary process between 0.5 m/s and 3 m/s (according to the assumed conditions and temperature ranges) it cannot be ruled out that these may occur in certain circumstances, as there are no general approximations of heat transfer coefficients in scrap bulks available in the literature now. Furthermore, due to the high temperature dependence of the gas's density, the cooling will lead to a significant slowdown of the gas's flow speed. Therefore, the derivation of heat transfer coefficients laws for different flow velocities through scrap bulks will be necessary to generate sufficiently accurate temperature estimates over the whole bulk.

From the results, it can be concluded that the shown modeling approach is very suitable, especially if radiation and convection phases can be separated. This can be justified by the fact that the convection model is strongly influenced by the number of cells used over the shaft length, while for the bulk radiation model, there is an additional physical implication of the chosen cell size. Separating the two is, therefore, a good way to avoid problems regarding the mesh. For mixed cases where both transmission mechanisms are important at the same time, an additional heat conduction model over the bulk length could help to reduce the numerical error for coarser meshes. Furthermore, the numerical error on a coarser mesh for the convective model could theoretically be mitigated by empirically or numerically determined compensation functions, so that an optimal grid size for the radiation modeling could still be chosen.

Certain cases such as inhomogeneous scrap movement of the bulk, mixed or variously layered scrap bulks may require the model to be extended to 2D or 3D, since those will be hard to account for in a 1D model. The capturing of other relevant aspects, such as varying off-gas properties (due to varying composition), the changing of flow rates (scrap and off-gas) over process time or the variation of heat transfer coefficient according to approximated gas velocities, is currently possible with the current model or rather does not require a lot of effort to implement. Nevertheless, such changes require validation data of existing processes.

5. Summary

To summarize, a fast and versatile scrap preheating model capable of estimating convective and radiation dominated heat transfer inside a scrap bulk, while also considering local heat conduction effects, has been developed. As shown, capturing the effects for the heating over the scrap part thickness is not very relevant for a broad range of different scrap types.

It was shown that in some areas, the model is not subject to strong deviations due to uncertainties in the model's parameters; in many cases, it could therefore be used, for example, to estimate the approximate design of the required length of a scrap preheating shaft. Otherwise, under some circumstances, the model may be very reactive to some parameters. Therefore, it is recommended to start with investigating the model's sensitivity to parameters as mesh size, used heat transfer coefficient and c. Furthermore, as there were no validation measurements performed for this model until now, practical decisions based on these results of our implementation should be carried out with the appropriate caution.

Additionally, there is still further work needed, especially regarding the areas of:

- Simplified modeling of conduction heat transfer between individual scrap parts, for applications where this becomes relevant;
- Validation and parameterization of the blocking modeling;
- Estimation strategies for the c factor and the heat transfer coefficient α_{fs};
- Modeling of more complex scrap bulks with mixed materials;
- Finding suitable optimization factors to fit the model to various experimental cases.

Despite the several simplifications used in the model, the modeling approach itself is very extensible and, therefore, generally applicable to more complex cases, while still being as comprehensive as necessary to allow for real-time control and optimization purposes.

Supplementary Materials: The source code, geometry files and generated results used within this publication are archived under: https://www.mdpi.com/article/10.3390/met11081280/s1. In general, future updates to the code related to within this publication are accessible under: https://github.com/c-schubert/model1d1d.

Author Contributions: Conceptualization, C.S., M.E. and D.B.; software, C.S. and D.B.; writing—original draft preparation, C.S.; writing—review and editing, C.S., D.B., T.E. and H.P.; visualization, C.S.; supervision, M.E., T.E. and H.P.; project administration, M.E.; funding acquisition, T.E. All authors have read and agreed to the published version of the manuscript.

Funding: This research was partly funded by Inteco Melting and Casting technologies GmbH.

Acknowledgments: We would like to thank Michele Manazzone from Inteco for sharing his technical know-how regarding the Ismelt technology.

Conflicts of Interest: The authors declare no conflict of interest.

References

1. Coenen, L. Steel—The Surprising Recycling Champion. Available online: https://www.worldsteel.org/media-centre/blog/2018/blog-steel-surprising-recycling-champion.html (accessed on 20 February 2019).
2. Nardin, G.; Ciotti, G.; Dal Magro, F.; Meneghetti, A.; Simeoni, P. *Waste Heat Recovery in the Steel Industry: Better Internal Use or External Integration?, XXIII Summer School "Francesco Turco"—Industrial Systems Engineering A New Model Proposal for Occupational Health and Safety Management in Small and Medium Enterprises*; Summer School Francesco Turco: Palermo, Italy, 2018.
3. Toulouevski, Y.N.; Zinurov, I.Y. *Electric Arc Furnace with Flat Bath—Achievements and Prospects*; Springer: Berlin/Heidelberg, Germany, 2015; ISBN 978-3-319-15886-0.
4. Yang, Q.; Yang, L.; Shen, J.; Yang, Y.; Wang, M.; Liu, X.; Shen, X.; Li, C.; Xu, J.; Li, F.; et al. Polychlorinated Dibenzo-p-Dioxins and Dibenzofurans (PCDD/Fs) Emissions from Electric Arc Furnaces for Steelmaking. *Emerg. Contam.* **2020**, *6*, 330–336. [CrossRef]
5. Yang, L.; Jiang, T.; Li, G.; Guo, Y.; Chen, F. Present Situation and Prospect of EAF Gas Waste Heat Utilization Technology. *High Temp. Mater. Process.* **2018**, *37*, 357–363. [CrossRef]
6. Lehner, J.; Friedacher, A.; Gould, L.; Fingerhut, W. Low-Cost Solutions for the Removal of Dioxin from EAF Offgas. *Rev. Met. Paris* **2004**, *101*, 49–56. [CrossRef]
7. Dicion, A. *Innovative Energy Conservation through Scrap Pre-Heating in an Electric Arc Furnace*; ACEEE Summer Study on Energy Efficiency in Industry: 2013; p. 5. Available online: https://oaktrust.library.tamu.edu/handle/1969.1/149172?show=full (accessed on 1 July 2021).
8. SMS Group. *Metallurgical Plant and Technology*; Maenken Kommunikation GmbH: Köln, Germany, 2016; pp. 52–57.
9. CISDI Engineering Co. Ltd. (54). Steel Scrap Preheating-Type Electric arc Furnace and Method for Improving Heating Cold Area of Side Wall Charging Electric Arc Furnace. 2018. Available online: https://patents.google.com/patent/US10928136B2/en (accessed on 1 July 2021).
10. Manazzone, M.; Valoppi, A. ISMELT—Inteco Scrap MELting Technology Evolution. Available online: https://www.eases.rwth-aachen.de/dl/AOTK-2021/Manazzone_ISMELT%20%E2%80%93%20Inteco%20Scrap%20MELting%20Technology%20evolution.pdf (accessed on 18 June 2021).
11. Tang, G. Modeling of Steel Heating and Melting Processes in Industrial Steelmaking Furnaces. Ph.D. Thesis, Purdue University, West Lafayette, IN, USA, 2019.
12. Mandai, K. Modelling of Scrap Heating by Burners. Ph.D. Thesis, McMaster University, Hamilton, ON, Canada, 2010.
13. Arink, T.; Hassan, M.I. Metal Scrap Preheating Using Flue Gas Waste Heat. *Energy Procedia* **2017**, *105*, 4788–4795. [CrossRef]
14. Ioana, A.; Preda, A.; Be, L.; Moldovan, P.; Constantin, N. Improving Electric arc Furnace (eaf) Operation through Mathematical Modelling. *UPB Sci. Bull. Ser. B* **2016**, *78*, 185–194.
15. Schubert, C.; Eickhoff, M.; Pfeifer, H. Modelling of Continuous Scrap Preheating. In Proceedings of the 2019 STEELSIM Conference Proceedings, Toronto, ON, Canada, 13–15 August 2019; p. 8.
16. Zhang, L.; Oeters, F. Possibilities of Counter-Current Scrap Pre-Heating with Melting by Use of 100% Fossil Energy. *Steel Res.* **1999**, *70*, 296–308. [CrossRef]
17. Hay, T.; Visuri, V.-V.; Aula, M.; Echterhof, T. A Review of Mathematical Process Models for the Electric Arc Furnace Process. *Steel Res. Int.* **2020**, *92*, 2000395. [CrossRef]
18. Bezanson, J.; Edelman, A.; Karpinski, S.; Shah, V.B. Julia: A Fresh Approach to Numerical Computing. *SIAM Rev.* **2017**, *59*, 65–98. [CrossRef]
19. Bezanson, J.; Bolewski, J.; Chen, J. Fast Flexible Function Dispatch in Julia. *arXiv* **2018**, arXiv:1808.03370v1.
20. StahlDat SX Professional: Register of European Steels. Available online: https://www.stahldaten.de/en/pub/home (accessed on 1 July 2021).
21. GERDAU Midlothian. *Ferrous Raw Materials Manual—Part 3—Ferrous Raw Materialspecifications*; GERDAU Midlothian Grade Specs: 2018. Available online: https://www2.gerdau.com/sites/gln_gerdau/files/downloadable_files/Midlothian%20Ferrous%20Raw%20Material%20Specifications.pdf (accessed on 1 July 2021).
22. Goodwin, D.G.; Speth, R.L.; Moffat, H.K.; Weber, B.W. Cantera: An Object-Oriented Software Toolkit for Chemical Kinetics, Thermodynamics, and Transport Processes. 2021. Available online: https://www.cantera.org (accessed on 30 May 2018).
23. Smith, G.P.; Golden, D.M.; Frenklach, M.; Moriarty, N.W.; Eiteneer, B.; Goldenberg, M.; Bowman, C.T.; Hanson, R.K.; Song, S.; Gardiner, W.C., Jr.; et al. GRI-Mech. Available online: http://www.me.berkeley.edu/gri_mech/ (accessed on 30 June 2020).
24. Rackauckas, C.; Nie, Q. DifferentialEquations.Jl—A Performant and Feature-Rich Ecosystem for Solving Differential Equations in Julia. *J. Open Res. Softw.* **2017**, *5*, 15. [CrossRef]
25. Alberti, M.; Weber, R.; Mancini, M. Gray Gas Emissivities for H_2O-CO_2-CO-N_2 Mixtures. *J. Quant. Spectrosc. Radiat. Transf.* **2018**, *219*, 274–291. [CrossRef]
26. Hottel, H.C.; Sarofim, A.F. *Radiative Transfer*; McGraw-Hill: New York, NY, USA, 1967.
27. Howell, J.R.; Menguc, M.P.; Siegel, R. *Thermal Radiation Heat Transfer*; CRC Press: Boca Raton, FL, USA, 2015; ISBN 1-4987-5774-X.

28. Odenthal, H.-J.; Kemminger, A.; Krause, F.; Sankowski, L.; Uebber, N.; Vogl, N. Review on Modeling and Simulation of the Electric Arc Furnace (EAF). *Steel Res. Int.* **2018**, *89*, 1700098. [CrossRef]
29. Gruber, J.C.; Echterhof, T.; Pfeifer, H. Investigation on the Influence of the Arc Region on Heat and Mass Transport in an EAF Freeboard Using Numerical Modeling. *Steel Res. Int.* **2016**, *87*, 15–28. [CrossRef]
30. Coskun, G.; Yigit, C.; Buyukkaya, E. CFD Modelling of a Complete Electric Arc Furnace Energy Sources. *Innovations* **2016**, *4*, 22–24.

metals

MDPI

Article

Towards Greener Industry: Modelling of Slag Heat Recovery

Reza Safavi Nick [1,2,*], Virpi Leinonen [3,*], Juha Mäyrä [3] and Johan Björkvall [1]

[1] Primary and Secondary Steelmaking Group, Process Metallurgy Department, Swerim AB, SE-974 37 Luleå, Sweden; johan.bjorkvall@swerim.se
[2] Division of Applied Process Metallurgy, Department of Material Science and Engineering, KTH Royal Institute of Technology, SE-100 44 Stockholm, Sweden
[3] SFTec Oy, FI-90570 Oulu, Finland; juha.mayra@sftec.fi
* Correspondence: reza.safavinick@swerim.se (R.S.N.); virpi.leinonen@sftec.fi (V.L.); Tel.: +46-70-927-2939 (R.S.N.); +358-40-712-0276 (V.L.)

Abstract: The steel industry, in accordance with the momentum of greener industry, has welcomed the changes and is actively pursuing that objective. One such activity is the commitment to energy recovery from by-products such as slag since the average energy content of ferrous slags is around 1 to 2 GJ/t_{slag}. The recovered energy could, then, be used in heating or drying process among others. The RecHeat was designed and modelled iteratively to achieve an optimised heat recovery apparatus. The model shows that the temperature of different sections of the heat exchanger part varies from 170 to 380 °C after slag pouring while the average air temperature at the entrance of the heat exchanger is less than 150 °C. Furthermore, the temperature of the fluid medium changes from 125–140 °C to 260–340 °C from one end of the heat exchanger part to the other at the end of the simulation. The outlet temperature at the end of the simulation is calculated to be around 340 °C, which shows an increase by at least 200 °C in the temperature of the air entering the apparatus.

Keywords: mathematical modelling; computational fluid dynamic; slag heat recovery; heat exchanger; drying; slag energy content; heat recovery technology; RecHeat

Citation: Safavi Nick, R.; Leinonen, V.; Mäyrä, J.; Björkvall, J. Towards Greener Industry: Modelling of Slag Heat Recovery. *Metals* **2021**, *11*, 1144. https://doi.org/10.3390/met11071144

Academic Editors: Thomas Echterhof, Ko-Ichiro Ohno and Ville-Valtteri Visuri

Received: 22 June 2021
Accepted: 16 July 2021
Published: 20 July 2021

1. Introduction

Production of iron and steel is a CO_2- and energy-intensive activity. The global steel industry is responsible for one-quarter of industrial CO_2 emissions and 7 to 9% (2020) of global anthropogenic CO_2 emissions. Furthermore, during 2019, 22% of globally used industrial energy was used in steel production. Typically, the cost of energy is around 10–20% of the total production costs of steel [1–4].

Improved energy efficiency is one of the approaches of greener steel production. This is generally seen as one of the major short-term methods of CO_2 emission reduction in the steel industry.

To be able to meet the Sustainable Development Scenario (SDS) set for the industry sector, the energy intensity of crude steel production needs to achieve a 1.2% annual reduction between 2018 and 2030 [5]. Energy or heat recovery, as one of the main fields of a four-stage efficiency methodology of the Step-up program launched by the World Steel association [3], shows the importance of this process.

1.1. Slag Heat Recovery

Energy recovery from hot liquid slag is one of the underused candidates for increasing energy efficiency in steel production. In 2018, around 330 to 375 million tons of blast furnace slag (BFS) and around 250 million tons of steelworks slags (65% basic oxygen furnace (BOF) and ladle furnace (LF) slag, and 35% electric arc furnace (EAF) slag) were produced. It is expected that global steel consumption will increase, meaning that slag volumes will increase in the future [6].

In iron and steel production, slags are the main by-product in terms of mass (90%). These molten slags carry a great amount of unused waste heat and are considered to be value-added products with extra energy output, which raises environmental concerns while offering cost-saving opportunities for industrial applications [2].

Molten slag forms at 1300–1700 °C, and when discharged, a great deal of high-grade heat is carried with it. In the steel industry, slag accounts for 10% of waste energy and 35% of high-temperature waste heat [2]. Therefore, technology development is vital to the recovery of this high-temperature waste heat, which immediately leads to energy savings and emission reduction in the iron and steel industry.

Generally, the current heat recovery technologies can be classified into physical and chemical methods. Lots of research efforts have been directed towards physical methods, like mechanical crushing, air blast and centrifugal granulation process [7,8]. Of the chemical methods, the CH_4 reforming reaction and the coal gasification processes have been widely investigated [3]. For blast furnace (BF) slag, the two European pilot plants with recovery of heat during granulation are the most promising developed technologies [9].

Typically, the energy content of ferrous slags is around 1 to 2 GJ/t_{slag} at the tapping temperature, and part of this is lost due to discharge of the hot slag. The discharged slag is processed either by granulating it directly from the process (typical method for BFS) or tapped to a slag pot and transported to slag handling area. The transported slag, then, is poured on the slag dump area. It is important to emphasise that currently, the energy of the dumped slag is mainly wasted as the slags are cooled down by the atmosphere or by means of water-cooling with no heat recovery.

Over the decades, different types of methods have been developed to recover this unused energy/heat. The heat recovery methods can be divided into two groups, direct contact with slag or indirect contact through air, mist/vapor, or water. Despite the several different types of approaches, there are no widely commercialised use of heat recovery processes available. Wang et al. published an extensive review of the research technologies, indicating that all of them are in the demonstration or experimental phase [2]. The majority of the physical methods concentrate on the development of BFS heat recovery. Less attention has been paid to the development of heat recovery from steelmaking slags; electric arc furnace (EAF), basic oxygen furnace (BOF) and ladle furnace (LF) slag.

1.2. RecHeat Techonology

The RecHeat (Recovery of Heat from molten slag) technology innovation is based on the idea of an "as simple as possible to use", heat exchanger type of technology for molten slag heat recovery. Therefore, the apparatus is constructed of simple metal sheets to enhance the energy exchange through a semi-direct contact of slag with the heat exchange structure surface. Moreover, such a construction benefits from a not-so-expensive recyclable material and localised rebuilding-patching in the case of minor damage. Furthermore, safety, simplicity and practical reason led to the use of air as an energy-transport medium instead of water, even though water has better thermal properties.

In practice, the slag from the steel plant will be carried to the RecHeat using a slag truck. The slag truck, then, tilts the slag pot over the designated area of the RecHeat. As can be imagined, the slag might not be distributed uniformly over the thin steel plate. In such a case, a metallic arm will be used to spread the slag as uniformly as possible so that there will not be any uncovered area. With this arrangement, the slag starts to radiate its energy to the surroundings, where the air will be sucked in into the apparatus through two separate entrances. Of course, since the system should be in an open environment due to security risks and for practical purposes (slag truck accessibility) part of this energy will be lost.

The heated air can, then, be used for drying purposes. This could be a significant gain considering the energy for drying medium covers 50–70% of the operational costs of drying process. Therefore, it can be said that utilization of slag heats is a cost-efficient source of heat for drying. At the same time, coupling of a drying station to the RecHeat draws a path of direct utilization of the recovered heat.

2. Geometry, Modelling and Model Setup

It might not be far from the truth to say that one of the most popular approaches in transferring heat from one medium to another is to use a heat exchanger. This method has proven itself reliable and practical in many industrial and domestic sectors. In addition, with the increasing notion of green industry, recovering heat energy of by-products, e.g., slag in the steel-making process, has become an objective of many industries. Therefore, as mentioned, RecHeat was designed with the notion that it should be usable with minimum effort while recovering tangible energy from the slag.

The objective of the current study, then, is to design an optimum apparatus based on the heat exchanger concept that can retrieve the energy of the slag using air as the medium for the purpose of drying, as mentioned previously. Therefore, the focus of the modelling activity has been to predict the system efficiency by means of computational fluid dynamic (CFD). This approach has the advantage of reducing the cost of construction and the risk involved with multiple design tests. Such a system can be easily modelled using Navier-Stokes and energy equations without any further modification by means of commercial CFD codes. A description of the equations is given below.

2.1. Geometry

Figure 1 shows the overall setup of the RecHeat in practice. As can be seen, the RecHeat is formed of a heat exchanger part which is then connected to a collector. The heat exchanger part is formed of three layers which are stacked on top of each other; each layer consists of four metallic pallets with five channels each. Moreover, as can be seen in the figure, the outlet of the heat exchanger is at the end of the collecting pipe (in the left-bottom corner of the figure) while the air is sucked in through the open ports at the other end (the right side of the figure).

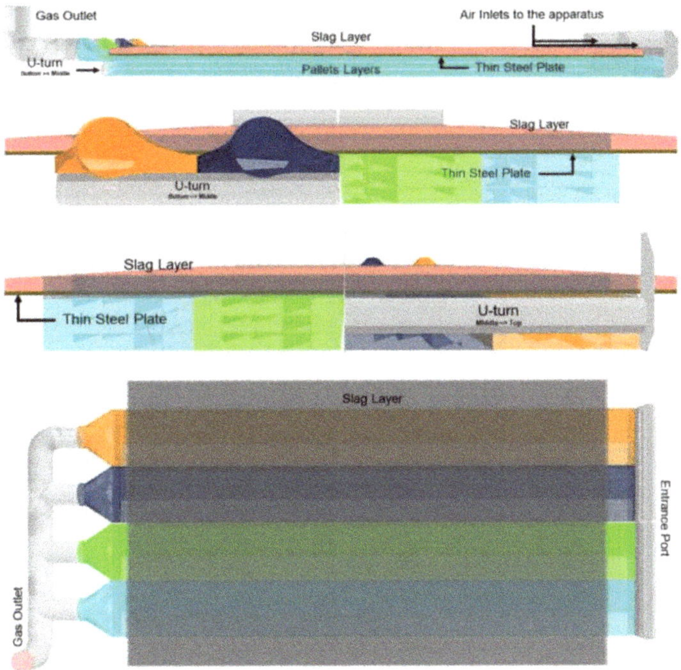

Figure 1. Heat exchanger and the slag layer; side-, front-, back- and top-view of the apparatus from top to bottom.

As mentioned, the heat exchanger section is formed of three stacked layers where the air must travel through the layers (from one side to the other and back) to reach the collecting sections. By looking at the side and the back views, it can be seen that the air enters the heat exchanger section of the apparatus though the bottom channels. Then, the medium will travel towards the other end and with a U-turn, it enters the middle layer. This is visualised in the front-view of the apparatus. Of course, the same happens at the end of the middle layer, where the medium makes a U-turn, entering the top layer (back-view). At the end of the top layer, the air exits the heat exchanger and will be collected and mixed in the piping section.

Moreover, a thin steel plate is placed over the heat exchanger section. This is to prevent the direct contact of the slag with the apparatus which can cause damages to the construction. Of course, the apparatus is also protected by sand on its side to make a pool-shaped area where the liquid slag should be poured in. Then, the liquid slag is transported by the slag truck and poured over, into the designated area. Of course, in real practice, if the slag does not uniformly spread over the surface, it is possible, to a certain degree, to spread it using a mechanical arm.

With such a setup, the structure of the apparatus heats up, extracting energy from the slag on top while the entering air to the system warms up at the two bottom layers until reaching the top layer.

2.2. Modelling Approach

Of course, the next logical question will be "what model setup will, most realistically, represent the current arrangement". To answer this question, one should consider that the apparatus is designed to suck in the air from the surrounding. This medium, in turn, should be affected by the radiative and convective heat due to slag cooling process. Therefore, it is safe to say that the temperature of the air entering the system could naturally differ as the cooling of the slag progresses. Therefore, it can be seen that the generic Navier-Stokes and energy equations could describe the system. These equations are given below.

$$\nabla.U = 0 \tag{1}$$

$$\rho\frac{DU}{Dt} = -\nabla p + \nabla.\tau + \rho g \tag{2}$$

$$\rho c_p\frac{\partial T}{\partial t} + \rho c_p U.\nabla T = \nabla.(k\nabla T) \tag{3}$$

In the above, $\frac{D}{Dt}$ represents the material derivative, U is the velocity, ρ is the density, p is the pressure, τ is the viscous stress tensor, g is the gravitational acceleration, c_p is the specific heat capacity, k is the thermal conductivity and T denotes the temperature. The thermal properties of the slag were reported by Gonzalez-Fernandez et al. [10].

To be able to more realistically set up the model, commercial CFD PHOENICS (v. 2019, Concentration, Heat and Momentum Limited (CHAM), London, UK), which uses the cut-cell method, proved to be a suitable choice. In this application, the numerical domain always consists of a box shown with red lines in Figure 2. In this model setup, the outer faces of the red box are treated as pressure boundary with temperature fixed at the ambient temperature (surrounding environment), while the bottom face of the exterior is treated as a wall mimicking the ground. Hence, the cooling behaviour of the slag and the transient nature of the air entering the RecHeat could be modelled more realistically.

Figure 2. Calculation domain.

2.3. Initial and Boundary Conditions

The initial values and boundary conditions for the model can be summarised in Table 1 as follow:

Table 1. Model initial and boundary conditions.

BC & IV	Object	Type	Magnitude	Unit
Boundary Condition	RecHeat Gas Out	Mass Flow Rate	2.5	kg/s
	Top & Side Faces	Pressure boundary	Ambient P and T	Pa & °C
	Bottom Face	Wall	Adiabatic	-
Initial Value	Slag	Temperature	1300	°C
	Air	Temperature	Ambient T	°C

The side and top faces of the domain (light blue faces of the outer box in Figure 3) are assumed to be pressure boundary faces where the pressure and temperature are equal to the ambient pressure and temperature. It should be noted that since the modelling was to be followed up by a pilot trial, the ambient temperature was set to 20 °C. This is in consideration of the location of the test site and planned season of the trial. However, the pilot trial was delayed due to the pandemic that engulfed all nations, and by the time of the test, the ambient temperature had dropped to a lower magnitude. The bottom face (brown face in Figure 3), then, was treated as an adiabatic wall.

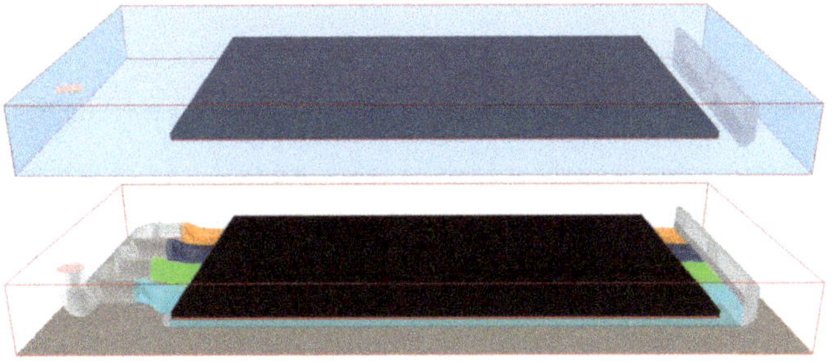

Figure 3. Outer faces of the calculation domain: top and sides faces are treated as pressure boundary and the bottom as adiabatic wall.

Furthermore, the air flow rate was set to 2.5 kg/s, which is equivalent to the expected flow rate that the fan could operate (red circle in Figure 3) and the slag temperature was taken as the average temperature of the slag reaching the testing site. The thickness of the slag was also calculated with respect to the slag pot volume and a spreading area equal to 10 cm. Of course, it should be pointed out that the slag did not spread uniformly over the top surface during the pilot test, but in the model setup, the slag layer was set at a uniform thickness-temperature.

3. Results

The modelling of the RecHeat can be divided into two distinct parts: the structure heating and heat exchanging stages.

3.1. The Structure Heating Stage

In a real process, it can be expected that the apparatus temperature is at equilibrium with the surrounding environment, i.e., ambient temperature. Therefore, the objective of the first stage of the process is to increase the temperature of the structure to the maximum possible magnitude. By pouring the slag over the structure, the body of the apparatus absorbs the energy of the slag through conduction. Simultaneously, the radiative energy of the slag instantaneously heats up the surrounding air.

Figure 4 shows the average temperature of the sucked-in air into the structure during the structure heating stage. It can be visibly seen that the slag radiation energy disperses into the surroundings where the temperature of air is above 250 °C.

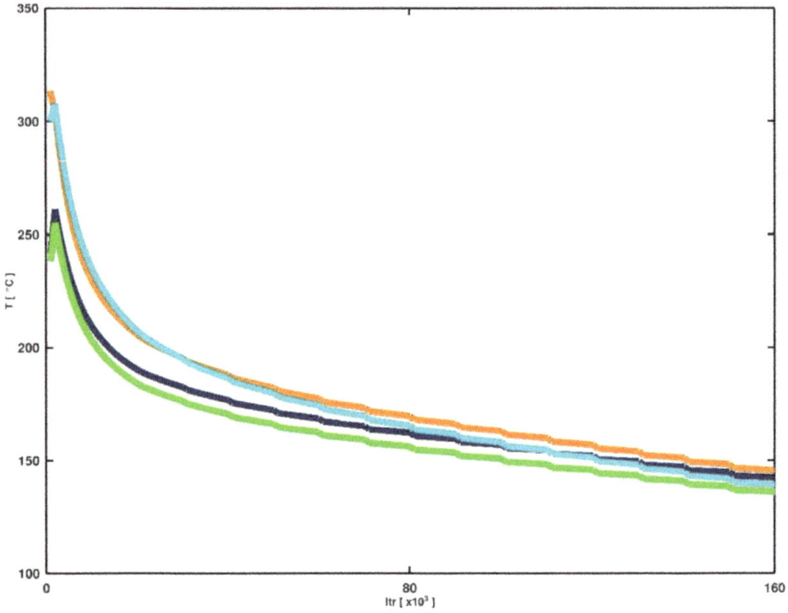

Figure 4. Average air temperature during the structure heating stage; colours correspond to pallet colouring.

Then, the sucked-in heated air begins to exchange its energy with the body of the structure as it flows through the apparatus, since the temperature of the structure is lower than the entering medium. Hence, it can be said that the convection and radiation have a simultaneous effect during the structure heating stage.

Table 2 shows the magnitude of the temperature of each pallet in three layers of the heat exchanger at two instances, the initial stage and the end of the steady state. As expected, the top layer of the heat exchanger registers the greatest temperature magnitude.

Table 2. Temperature magnitude of each pallet in each layer at two instances of the structure heating stage.

Top Layer				Middle Layer				Bottom Layer			
126.00	125.04	125.04	127.27	40.66	39.24	40.00	44.29	43.61	41.22	42.51	47.22
392.14	387.60	384.70	372.51	262.99	264.73	259.63	246.12	170.82	175.02	170.53	162.71

Moreover, it can be seen that the middle and bottom layers register greater temperature magnitude than the ambient, while the middle-layer pallets are at least 2 °C colder than the one in the bottom layer. This behaviour then reverses, and the temperature of the middle layer is more than 80 degrees warmer than the bottom.

To determine the end of the first stage, the change in the magnitude of the temperature of each pallet was monitored. When this variation flatlined near zero, it was concluded that the system was at the end of the structure heating stage. Therefore, the second stage, and heat exchange should be initiated. It should be pointed out that this does not mean the structure heating stage has come to an end; it merely suggests that such a stage is in its converged state.

3.2. The Heat Exchange Stage

3.2.1. Pallet Temperature Profiles

Figure 5 shows the evolution of temperature of each layer for the given stacks, where the colour of the title corresponds to the colour presented in Figure 1. As mentioned, layers are indexed from top to bottom; hence, layer one in each stack registers the greatest temperature magnitude, while layer three (at the bottom) is significantly colder.

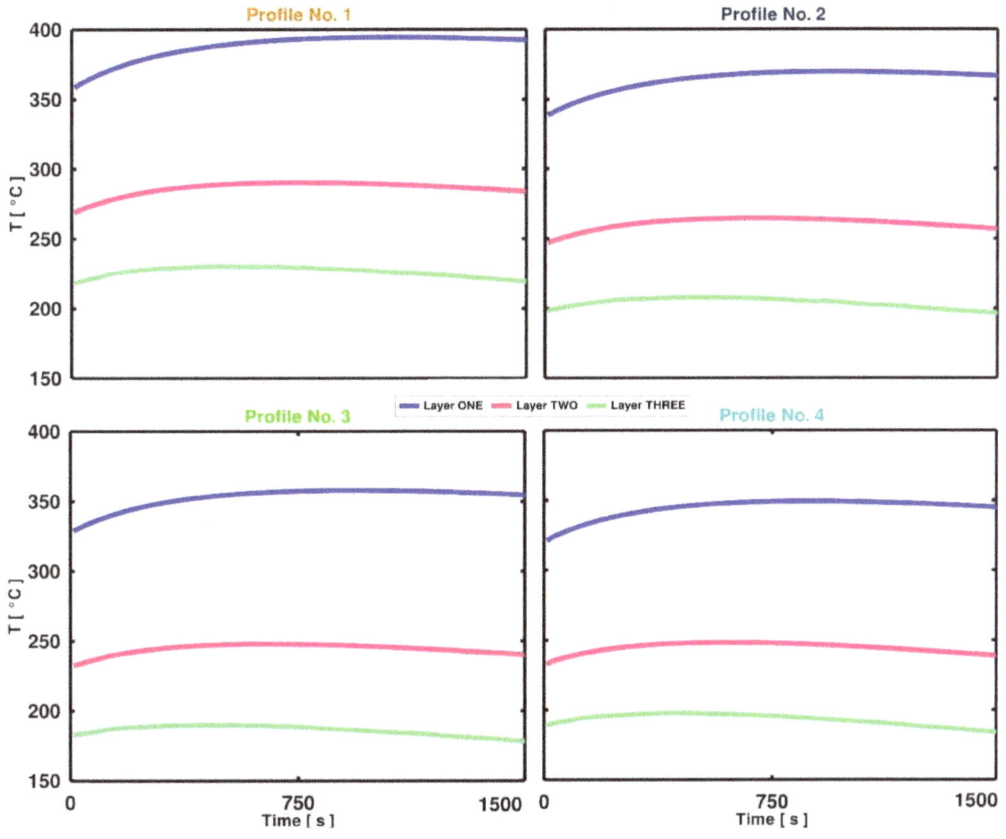

Figure 5. Average temperature of each layer of the stacks; colours of the titles correspond to pallet colouring.

The figure suggests that all the layers of the stack number one (opposite to the outlet) have larger magnitude of the temperature compared to their counterparts in the other three. For stacks number three and four, even though the top pallets have large difference in the magnitude of the temperature, the middle and bottom layers seem to be rather the same.

Moreover, the figure shows that the magnitude of the temperature still increases at the beginning of the heat exchanger stage, but it changes before the midpoint. It can be seen that this change in the slope of the curves appears sooner for the layers in the bottom and the middle in comparison to the top.

3.2.2. Air Flow Temperature Profile

Figure 6 shows the average temperature of the air at the entrance and the exit of the heat exchanger (HEx) section of the apparatus. As can be seen, the temperature of the air entering the heat exchanger still increasing in the magnitude. This corresponds to the energy pick-up from the slag radiation to the environment which continues at the beginning of the transient simulation. However, this behaviour starts to change, and the magnitude of the temperature of the air entering the heat exchanger starts to drop rapidly.

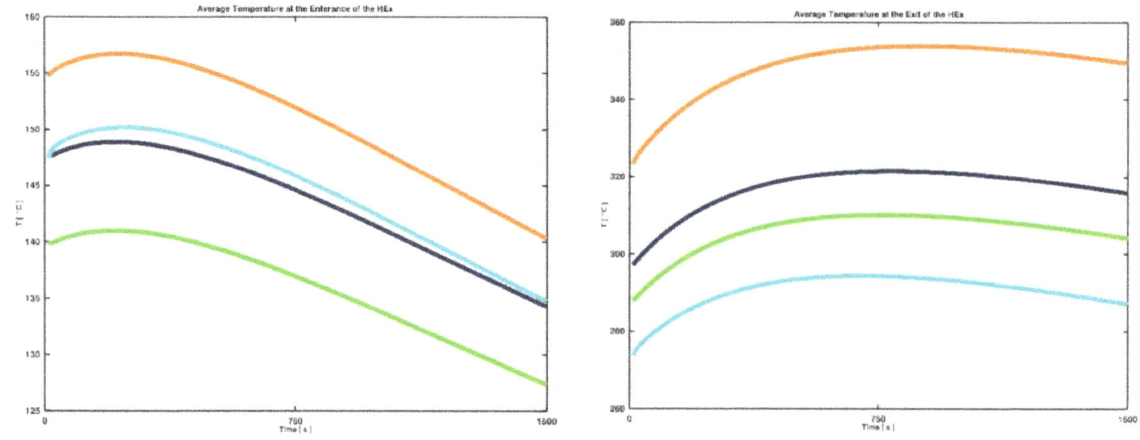

(**a**) (**b**)

Figure 6. (**a**) Average temperature at the entrance of the Hex. (**b**) Average temperature at the exit of Hex.

On the other hand, at the exit ports of the heat exchanger, the magnitude of the temperature increases further nearly till the middle of the simulation. In the case of stack number one (orange pallets) this increase even continues further.

Moreover, Figure 5 shows that even though the temperature of the air entering pallet number 4 is slightly higher than pallet number 2, the exit temperatures of these two pallets differ significantly. This should be due to the differences in the magnitude of the temperature of the stacks (Table 2 and Figure 5).

Table 3 shows the average temperature of the air entering and exiting the heat exchanger at the end of the transient simulation. As can be seen, the air temperature magnitudes increase to more than double at the end of the heat exchanger section of the RecHeat.

Table 3. Average temperature of each stack at the entrance and the exit of the heat exchanger at the final time.

Variable	Location	Stack #1	Stack #2	Stack #3	Stack #4
Temperature	Entrance	140.32	134.29	127.36	134.79
	Exit	349.36	315.76	304.13	287.13
	Diff	209.04	181.47	176.77	152.34

Figure 7 shows the temperature of the air at the outlet of the apparatus. As can be seen, the product temperature at the start of the transient period is equal to 328 °C, which peaks at 350 °C, and by the end of the simulation (at 1500 s) drops to nearly 340 °C. The peak occurs at 650 s into the transient process; at this point, the curve reverses its direction. Therefore, it can be seen that the outlet temperature increases by more than 20 degrees in the first 650 s and it decreases by 12 degrees in 850 s after the peak.

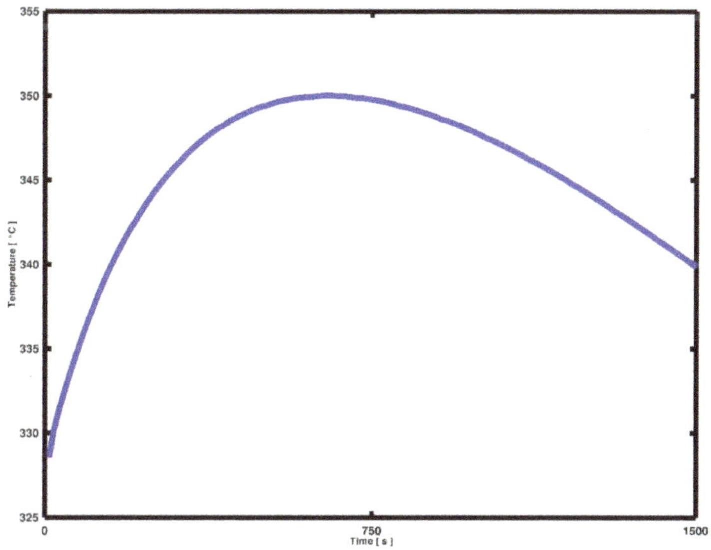

Figure 7. Temperature of the final product (air).

3.3. Heat Energy of the Product

As mentioned, the objective of the RecHeat is to produce hot air to be used in a drying process, and this dried material can be anything such as biomaterial, sludges or others. Therefore, it is logical to calculate the heat energy of the product.

Figure 8 shows the calculated heat energy of the outgoing air. Of course, to calculate the heat of energy one should assess the value using a baseline or reference temperature. In this study, since the air is sucked in from the surrounding environment, it is just logical to choose the ambient temperature as the baseline temperature of the medium. The ambient temperature, considering the location of the test, was chosen as T_{Ref} = 20 °C Using the given value, the heat energy of the air was calculated (Figure 8).

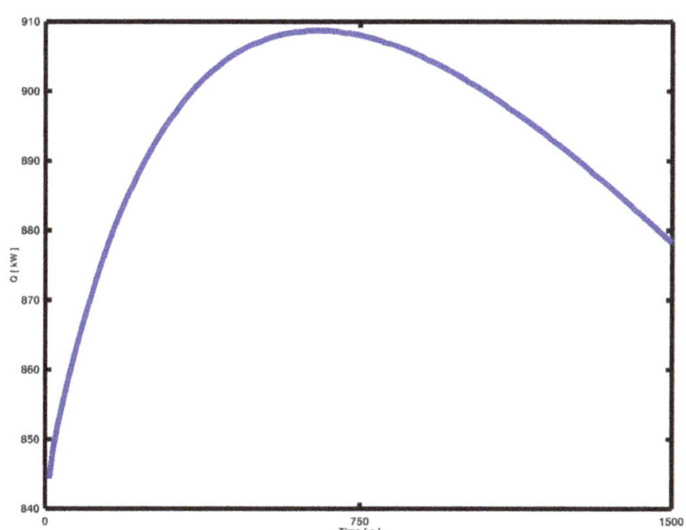

Figure 8. Heat energy of the final product (air).

As can be seen, the heat energy of the product at the beginning of the transient process, at its peak and the end of the simulation are 844, 908 and 878 kW, respectively. As can be seen, the magnitude of the heat energy at the end of the simulation is 34 kW larger than the starting value.

3.4. Streamlines and Air Profile

Figure 9 shows the profile of the air entering the apparatus to the ports at the left and the evolution of the temperature of the medium through the heat exchanger section of RecHeat.

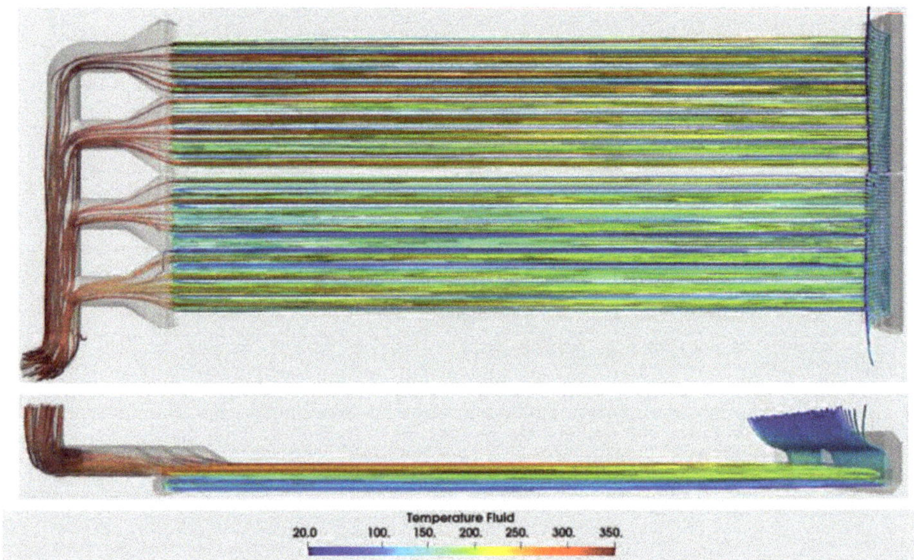

Figure 9. Air streamlines originating from a line across the entrance ports, coloured by temperature.

As can be seen, the source of the streamlines was chosen as the location where the air sucked into the apparatus comes from above the entrance ports. Since the radiation effect decreases, this figure suggests that the medium entering the system is near ambient temperature. Moreover, it can be seen that the temperature of the air at the bottom layers of the heat exchanger section of the apparatus is still not higher than 150 °C. This changes by the end of the middle layer. The figure shows that the air temperature reaches 200 °C and larger at this point; reaching the end of the heat exchanger section, the temperature of the air is close to the exit temperature in three of the four sections. Generally, it can be seen that the temperature of the air in the upper two layers (top view) is greater than the lower two.

4. Discussion

4.1. The Structure Heating Stage

As explained, Table 2 shows that at the beginning of the structure heating stage, the middle layer registers a slightly lower temperature magnitude compared to the bottom layer. Considering that the energy of the slag diffuses in a top-to-bottom direction, it is natural to assume otherwise. This behaviour, therefore, could be due to the heat energy entering the structure through the entrance ports of the RecHeat by means of the sucked-in air.

As Figure 4 shows, the air temperature very rapidly reaches beyond 250 °C at the beginning of the structure heating stage. This is the energy that has radiated from the surface of the slag. This heated air entered the apparatus (and heat exchanger section) at the top layer in the second configuration of the RecHeat, resulting in a different temperature profile.

In the current optimised design of the RecHeat, the heated air enters the heat exchanger section through the bottom layers. Hence, the bottom layers not only receive conductive heat energy of the slag but also absorb part of the radiated heat energy of slag through the flow of the air.

Hence, it is safe to say that this shows the significance of the radiative-convective energy dispersion of the slag during the structure heating process. Moreover, it shows that a change in the arrangement of the air entering the apparatus has a noticeable effect on the temperature profile of the pallets.

Of course, by the end of the structure heating stage, the bottom layer is significantly colder than the middle one which should be due to the stronger effect of the conductive heat through the body absorption from the slag.

4.2. The Heat Exchanging Stage

As mentioned, this stage starts when the energy exchange between the slag and the structure drops to a very low magnitude. At this stage, the temperatures of the pallets still change, but to a lesser extent (Figure 5).

The figure also shows that the change in the average magnitude of the temperature of each pallet is greater for the ones in the bottom and middle layers compared to the top. Therefore, compared to the top layer (blue line), the curves of other layers change their slope. Of course, this change is smaller for the middle layer compared to the bottom. This could be due firstly to the fact that the top layer is still in almost direct contact with the slag; and secondly, the heating process of the air during its transition though the heat exchanger section (absorption of energy of the body by the fluid).

This is more clearly expressed in Figure 9. It can be seen that the ambient air rapidly heats up by entering the inlet port of the RecHeat to 100 °C and increases its magnitude by more than 100 °C when passing through the bottom two layers. Of course, the air temperature increases further by absorbing the energy of the top, while the absorbed energy of the layers by the air should be substituted through conduction of the energy of the slag through its proximity.

Figure 9 also shows that the air exiting the heat exchanger region of profile three and four (closer to the outlet) seems to be colder than the other two. This could be due to the differences in the mass flow rates of the air through different profiles.

Figure 10 shows the mass of air entering each profile. As can be seen, there is a significant difference in the mass flow rate of the air through the two stacks near the outlet. A larger mass flow rate can mean larger energy pick-up leading to lower temperature magnitude.

Meanwhile, even though the air temperature at the end of the heat exchanger varies in the interval of 260–360 °C (Figure 6), the temperature of the air exiting the apparatus is greater than 325 °C and is equal to 340 °C at the end of the transient simulation (Figure 7). Translating this value into heat energy, it can be seen that the RecHeat collects more than 840 kW energy (from the baseline of ambient temperature). Of course, this value seems to be small in comparison to the amount of energy available in the slag, but such a performance proves to be noteworthy considering the temperature range (and the heat energy) required for drying of a wide range of materials.

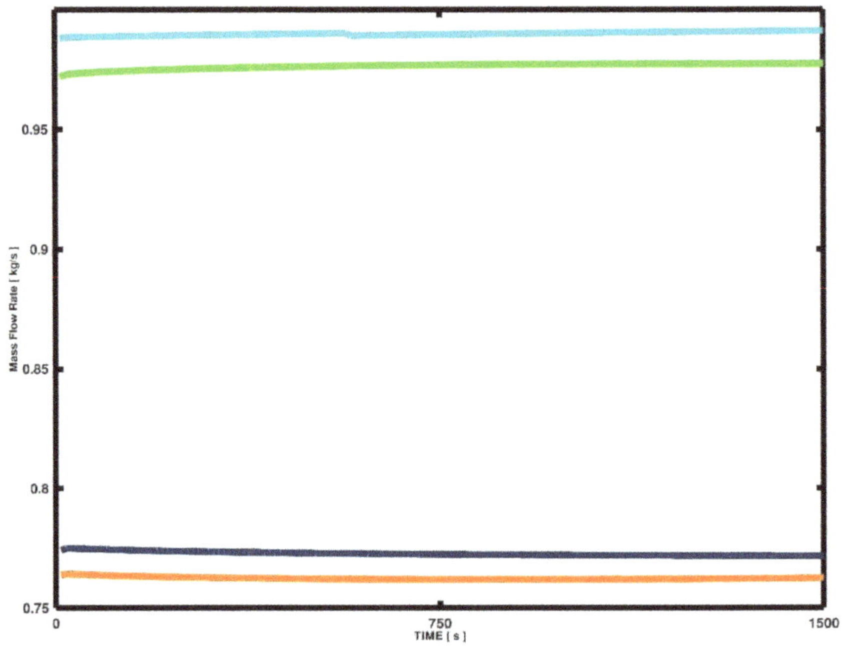

Figure 10. Air mass flow rate through each profile.

5. Conclusions

The recovery of energy from waste materials has become a prime objective in many industries due to social and economic aspects, and one such material and industry is slag and steel-plants, respectively.

To address this objective, and as a part of technological development, a slag heat recovery apparatus, RecHeat, was designed using a one-to-one computer model. Then, the results of the simulation were used to redesign and optimise the structure to maximise the heat recovery process. This was done to prevent the economical exhaustion of the building and testing of such an apparatus. The optimised structure of the RecHeat was then also tested at an industrial pilot scale, as part of the ECOSLAG project (funded by RFCS, Research for steel and coal industry, project no. 800762 [11]).

Two stages were identified during the heat recovery process, i.e., the structure heating and the heat exchanger stages. In the former stage, the body of the apparatus increases its temperature magnitude, while in the latter one, the magnitude of the temperature of the sucked-in air increases when passing through each layer of the heat exchanger section.

The model shows that the temperatures of the layers of the heat exchanger section of the apparatus are around 170, 250 and 380 °C at the end of the structure heating stage, while the average air temperature at the entrance of the heat exchanger section is less than 150 °C.

The temperature magnitudes of the pallets change their slope when the system enters the heat exchanger stage. This could be due to the fact that the air entering the system starts to absorb the energy deposited into the structures. It was shown that the temperature of the fluid medium changes from a 125–140 °C interval to 260–340 °C from one end of the heat exchanger section to the other at the end of the simulation.

The outlet temperature at the end of the simulation is calculated to be around 340 °C, which shows an increase in temperature of at least 200 °C in the air entering the apparatus.

Author Contributions: Initial design, V.L. and J.M.; design optimisation, V.L., J.M., R.S.N. and J.B.; model conceptualisation, R.S.N., J.B. and V.L.; model setup, R.S.N.; writing—original draft preparation, R.SN and V.L.; writing—review and editing, R.S.N., V.L. and J.B. All authors have read and agreed to the published version of the manuscript.

Funding: The research leading to these results has received funding from the European Union's Research Fund for Coal and Steel research programme under grant agreement number: 800762.

Data Availability Statement: Not applicable.

Acknowledgments: The authors would like to acknowledge and thank Anita Wedholm of SSAB Merox and Slag handling staff of SSAB Luleå.

Conflicts of Interest: The authors declare no conflict of interest.

References

1. Birol, F. Energy Technology Perspectives. 2020. Available online: https://iea.blob.core.windows.net/assets/7f8aed40-89af-4348-be19-c8a67df0b9ea/Energy_Technology_Perspectives_2020_PDF.pdf (accessed on 4 June 2021).
2. Wang, R.Q.; Jiang, L.; Wang, Y.D.; Roskilly, A.P. Energy saving technologies and mass-thermal network optimization for decarbonized iron and steel industry: A review. *J. Clean. Prod.* **2020**, *274*, 122997. [CrossRef]
3. Fact Sheet World Steel Association. 2021. Available online: https://www.worldsteel.org/en/dam/jcr:5574d4fd-4dd4-4b2d-9300-2ed7d3de0c7a/step%2520up-vf.pdf (accessed on 28 May 2021).
4. Climate Change and the Production of Iron and Steel. World Steel Association. 2021. Available online: https://www.worldsteel.org/en/dam/jcr:228be1e4-5171-4602-b1e3-63df9ed394f5/worldsteel_climatechange_policy%2520paper.pdf (accessed on 28 May 2021).
5. Iron and Steel International Energy Agency. 2021. Available online: https://www.iea.org/reports/iron-and-steel (accessed on 28 May 2021).
6. Slag Recycling Recovery Worldwide. 2021. Available online: https://www.recovery-worldwide.com/en/artikel/slag-recycling_3528047.html (accessed on 4 June 2021).
7. Feng, Y.H.; Zhang, Z.; Qiu, L.; Zhang, X.X. Heat recovery process modelling of semi-molten blast furnace slag in a moving bed using XDEM. *Energy* **2019**, *186*, 115876. [CrossRef]
8. Zhu, X.; Ding, B.; Wang, H.; He, X.Y.; Tan, Y.; Liao, Q. Numerical study on solidification behaviors of a molten slag droplet in the centrifugal granulation and heat recovery system. *Appl. Therm. Eng.* **2018**, *130*, 1033–1043. [CrossRef]
9. Fleischanderl, A.; Fenzl, T.; Neuhold, R. Dry Slag Granulation—The Future Way to Granulate Blast Furnace Slag. In Proceedings of the 2018 Iron & Steel Technology Conference and Exposition, Philadelphia, PA, USA, 7–10 May 2018.
10. González-Fernández, L.; Ortega-Fernández, I.; Grosu, Y.; Faik, A. Thermophysical Properties of Steel Slag for Heat Recovery and Storage Applicaations. In Proceedings of the 10th European Slag Conference, Thessaloniki, Greece, 8–11 October 2019; pp. 1–3.
11. RFCS, ECOSLAG: Eco-Friendly Steelmaking Slag Solidification with Energy Recovery to Produce a High-Quality Slag Product for a Austainable Recycling. 2018. Available online: https://ec.europa.eu/info/sites/default/files/research_and_innovation/funding/documents/synopsis_of_rfcs_projects_2017-2020.pdf (accessed on 8 June 2021).

metals

MDPI

Review

Continuous Casting Practices for Steel: Past, Present and Future

Roderick I. L. Guthrie * and Mihaiela M. Isac

McGill Metals Processing Centre, McGill University, Montreal, QC H3A 0C5, Canada; mihaiela.isac@mcgill.ca
* Correspondence: roderick.guthrie@mcgill.ca

Abstract: This historical review of casting methods used to produce sheets of steel for automobiles, household products, rocket bodies, etc., all point toward the development of one-step commercial processes, which are capable of casting liquid steel directly into a final sheet product. Progress towards this goal is confirmed by successful advances being made, but there remain major difficulties in reaching it. We concur that the conventional continuous casting method remains the current process of choice for highest-quality steel sheet products, but the ESP TSC (Endless Strip Production—Thin Slab Caster) approach is now highly competitive. Similarly, the original goal of Sir Henry Bessemer to produce a direct strip-making twin-drum caster, in 1856, finally came to lasting commercial fruition at CASTRIP/NUCOR. Nonetheless, a newer approach, promoted by Salzgitter, termed DSP (Direct Strip Production), or promoted by MMPC/MetSim as HSBC (Horizontal Single Belt Casting), has several advantages over CASTRIP in terms of microstructures and productivity. As such, the pros and cons of current methods are reviewed within this brief history of casting.

Keywords: continuous casting; near net shape casting; twin roll (Bessemer) casting; horizontal single belt casting

Citation: Guthrie, R.I.L.; Isac, M.M. Continuous Casting Practices for Steel: Past, Present and Future. *Metals* **2022**, *12*, 862. https://doi.org/10.3390/met12050862

Academic Editors: Thomas Echterhof, Ko-Ichiro Ohno and Ville-Valtteri Visuri

Received: 7 April 2022
Accepted: 12 May 2022
Published: 18 May 2022

1. Introduction

Continuous casting practices for steelmaking have been constantly evolving ever since the early 1930s, when Junghans was first researching ways to pour liquid steel into an open-bottomed, water-cooled mold, to withdraw the partially solidified steel out of it, continuously, in the form of a round or square billet or slab [1,2]. He envisioned that once these continuously cast shapes had become fully frozen, their solidified ends could be cut off for further processing. In this way, they could be transformed into "rebar" to reinforce concrete, or into bars from which nails, bolts, tire cord wire, etc., could be fashioned, etc. Reference 1, edited by Dr. Manfred Rasch, contains detailed articles from many of the key companies, personnel, and timelines of events in his review of the development of continuous casting practices for steel, beginning with Sir Henry Bessemer (1813–1898), and followed by Siegfried Junghans (1887–1954).

Thus, the underlying goal of Junghans' early research was to try to replace the static mold ingot casting systems used in virtually all steel plants up until the 1960s with a continuous process. This would effectively eliminate the ~10–20% losses associated with the "top and tailing" of the thick ingots (up to ~30, or 760 mm) typically being produced. Thus, after the cast ingots had cooled sufficiently (~5 h), to allow them to be safely withdrawn from their ingot molds, they were sent, still partially molten inside, to the soaking pits for temperature homogenization. This could last from approximately 10 h for a hot ingot, or up to 18 h for a cold, ingot. During this treatment time, the micro-segregation of solute elements (C, S, P, Mn, etc.) could be mitigated within the inter-dendritic solidification structure. In the meantime, temperature profiles within the ingot had become uniform, to ~±2 °C, ready for rolling down to semi-finished products. Thus, after proper soaking, the eight or more ingots within the soaking pit would be ready for transfer to the adjacent hot mill for rolling into "flat" or "long" products at temperatures of around 1100 °C.

First, the ingots were sent to the "roughing mill" for ingot "break down". This was accomplished by passing them back and forth under the roughing mill, thereby transforming the original 25–30 inch thick ingot (635–760 mm) into a long, 1–2" (25–50 mm) thick "transfer bar". This bar then had to be "topped" and "tailed"' to eliminate the rolled-in surfaces at the transfer bar's extremities, where surface overlap had occurred during compression and thickness reduction to 30 mm or ~2–3 inches. Similarly, any macroscopic shrinkage holes formed during solidification also needed topping before sending the transfer bar to the stands of the rolling mill. All of this automatically led to a mandatory 10–15% loss in yield.

Therefore, for sheet steel products, following a further removal of iron oxide skin from the transfer bars using high velocity, high pressure water jets just ahead of a six- or seven-stand hot rolling mill, the transfer bars were reduced to much thinner hot rolled steel sheet products, ~1–3 mm thick, using appropriate cooling schedules to create the specific hot grades being targeted (e.g., HSLA, etc.).

Clearly, Junghans' research, which lead to the Conventional Continuous Casting (CCC) method, promised to eliminate much of the material losses associated with the ingot casting approach, and thereby produce a step-up increase in steel productivity.

However, approximately seventy-odd years before Junghans' early research work, Henry Bessemer, inventor of the first tonnage steel process, had proposed a far more elegant process to cast liquid steel into near-net shape products, namely the Twin Roll Casting (TRC) process. He envisioned pouring liquid steel into the cavity produced by a twin set of contra-rotating rolls to freeze the exiting steel directly into ~1–2 mm thick sheet steel products. Following minimal rolling and heat treatment, this would be ready for sale. While many researchers had since tried to fulfill Bessemer's twin-roll casting "dream" and produce a commercial system for thin steel sheet production over the ensuing century before Junghans' research work, none had succeeded in producing high-quality sheets of steel.

We make mention of Clarence Hazelett, founder of the Hazelett Strip Casting Company, U.S.A, who had been researching Bessemer's Twin Roll Caster concepts (TRC) in the early 1930s for non-ferrous systems. He had some success but finally decided to abandon that approach in favor of a Twin Belt Caster (TBC) moving mold system in 1949.

His TBC system was approximately ten times more productive and proved to be a commercial success for copper, lead and zinc alloys, and certain grades of aluminum and magnesium alloys. TBC casters are now in worldwide use within the non-ferrous industry, but not for casting liquid steel. Previous attempts by Hazelett Strip Casting Corp., in collaboration with Sumitomo Metals Corporation, and earlier with U.S Steel/Bethlehem Steel in the 1980s, had failed to develop a satisfactory commercial system for the steel industry.

Therefore, today, the dominant casting process within the steel industry is the Conventional Continuous Caster (CCC), as developed by Rossi and partners with their CONCAST Corporations in Europe and the USA, following Junghans' death and Rossi's sole acquisition of patent rights from Junghans' widow [1,2].

2. The Past

Thus, thanks to the initial research efforts and work carried out by Siegfried Junghans (1887–1954) in Germany, aimed at eliminating the wastage caused by the ingot casting process, a disruptive paradigm was brought to the steel industry. Thus, Figure 1 illustrates the evolutionary sweep in continuous casting developments ever since that seismic change from ingot casting technologies to CCC [3]. It began in the 1950s, following the first designs of continuous casting machines by CONCAST. These casters were developed in collaboration with some early companies and their dedicated researchers, who recognized the important economic implications deriving from this new type of casting machine [1]. For the fixed CCC approach, the formulations for mold powders are important for protecting the surface of molten steel from significant heat losses and by remaining molten at ~900 °C to provide lubrication between the surface of the forming slab and the oscillating walls of the copper mold.

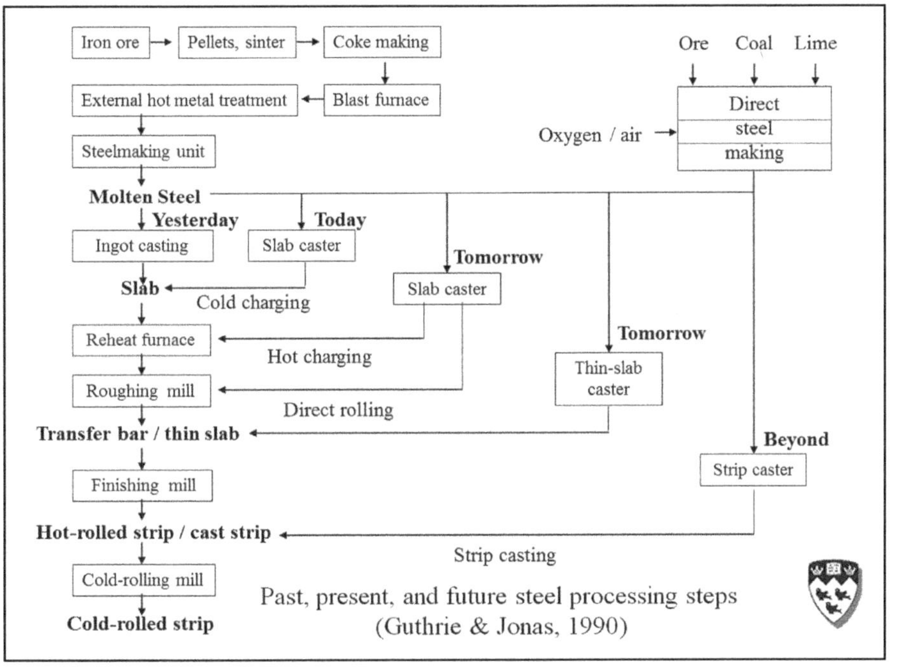

Figure 1. The past, present, and future of steel processing steps, as predicted in 1990 [3].

A key role in the development of the continuous casting method was played by Iain Halliday at Barrow Steel Works, in Scotland. He perfected the Junghans–Rossi oscillation mode (i.e., three-quarters of the cycle down and one quarter up), by imposing a relative velocity during the down-stroke, with the mold going down slightly faster than the strand, which he termed "negative strip". This invention helped to resolve the sticking of the growing frozen steel shell onto the copper mold and allowed him to reach casting speeds of 14.5 m/min in 1958. This casting speed is still a world record for these "fixed or stationary" mold technologies. Conventional continuous casting machines, nearing process perfection in recent years, allow partially solidified slabs, blooms, or billets, to be continuously withdrawn through their open-bottomed molds, at productivity levels equivalent to ingot casting technology, but with a 10–15% yield advantage. Some 1.8 billion tons per year are now cast through them.

Thus, the growing steel industries in Japan and Korea in the 1960s, as well as the well-established companies in the Nordic countries, were the first to adopt this new way of casting. Asian countries went on to become dominant global players in CCC, exporting much of their steel abroad, including to North America. Meantime, many conventional integrated steel plants in North America gradually walked into bankruptcy, some even before being able to adopt this new technology. For example, Dofasco escaped by the skin of its teeth thanks to the warnings and the advice of Dr. Dante Cosma of their Research Department to the President of the Company, Mr. Frank Sherman. He heeded Dante's advice over that of all the other "experts" within his company and canceled the new ingot casting aisle midway through construction. Nonetheless, approximately ten years later, Dofasco could not escape the clutches of ArcelorMittal (AM) in the "race to the bottom" at the start of the second millennium. This take-over seems to have been achieved by selling off Dofasco's valuable iron ore mines in Ontario and Quebec after having first acquired the company, thereby allowing AM to cover its initial expense for Dofasco's acquisition. Naturally, the Board of Directors of Dofasco had already been handsomely rewarded in the original sale of the company.

Nonetheless, and fortunately for North America, the mini-mill concept and its first implementation in the new steelmaking plant built in Oshawa, Ontario, had been confirmed by Dr. Gerald Heffernan and his co-workers in ~1985 [2]. This had been followed a couple or so years later by a succession of NUCOR plants being built in the countryside of the USA; as a new revolution in North America (N.A.), steelmaking practices began to take shape [2]. As such, many newer plants in N.A. have shifted away from the conventional blast furnace-BOF steelmaking model towards a 100% scrap-melting/continuous casting technology. Many of the recent continuous casting technologies now presented, were researched by the conventional integrated steel companies, but were brought into the steel industry through efforts by Minimills to invade the more lucrative niches for steel production. Nevertheless, to the authors' knowledge, all the steel sheet products used for "body in white" automobiles remain within the exclusive domain of advanced CCC steel plants around the world.

3. The Present (2020s)

Referring to Figure 1, the left-hand side of the diagram shows the situation before the advent of continuous casting in the mid-1950s. North America's first commercially successful vertical mold continuous casting machine was commissioned in 1954 to produce slabs of stainless steel at the Welland Plant of Atlas Steels in Ontario, Canada, using a "stick caster". Today, all ingot casting has been eliminated worldwide, apart from some specialty steel applications.

For this new paradigm in casting technology, yield increases of approximately 10–15% have been achieved thanks to the successful development of slag-lubricated, oscillating, open-ended mold machines. These molds, and the behavior of their mold fluxes, are at the very heart of the various continuous casting machines dominating steel casting processes.

Figure 2 shows a conventional continuous casting operation comprising a tundish, a water-cooled, chromium-plated, copper alloy mold, a mold oscillator, a group of cast strand supporting rolls, rolls for bending and straightening the strand, rolls to pinch and withdraw the cast strands, groups of water spray nozzles to extract heat from the strand, a torch cutter for cutting the cast strand, and a dummy bar [4]. This dummy bar fits in the bottom of the open mold at start-up, providing a starting base on which the strand can freeze. The dummy bar then extracts the cast strand, and finally moves away to "storage" once the forming strand has approached the torch cutter.

The overall set-up is generic, in that a "ladle" sits on a rotating "turret" or "swing tower" and empties liquid steel into a "tundish" set below via a protective ladle shroud (or tube). This tundish is used to distribute metal to one, two, four, six, or even eight oscillating molds/strands, set below. In the case of slab casting operations, it is common to have twin slab production. In the diagram, a SEN(Submerged Entry Nozzle) is used to protect the streams entering the mold from re-oxidation. Once the ladle has been emptied of steel, the tundish is allowed to drain while the turret rotates a newly filled ladle into position. In this way, "sequence", or "continuous, continuous" casting can be achieved, with uninterrupted runs in the order of up to 1000 ladle changes being possible, using the flying tundish technique for the rapid replacement of a spent tundish.

Another popular type of fixed-mold conventional continuous caster is the "vertical, curved mold"-type caster, in which, the mold and support rolls are arranged vertically to maximize the chances for inclusions to float out of the deep sump of molten steel before being trapped in the solidifying shell. The strand is then bent and straightened to deliver the strand horizontally, and continuously. In comparison, the first "stick", or vertical mold casters, froze single vertical strands in the same way as today's DC (Direct Chill) casters for the aluminum industry. The fact that curved mold and low-head curved mold machines tend to capture a band of inclusions approximately mid-distance from the top surface of the strand, and its central axial plane, led to the popularity of the vertical-curved bending machine. In technically advanced steel mills, where surface quality is good (freedom from scabs, slivers, and deep oscillation marks), these slabs, blooms, and billets can be hot charged directly to a reheat furnace without surface inspection before further size reduction.

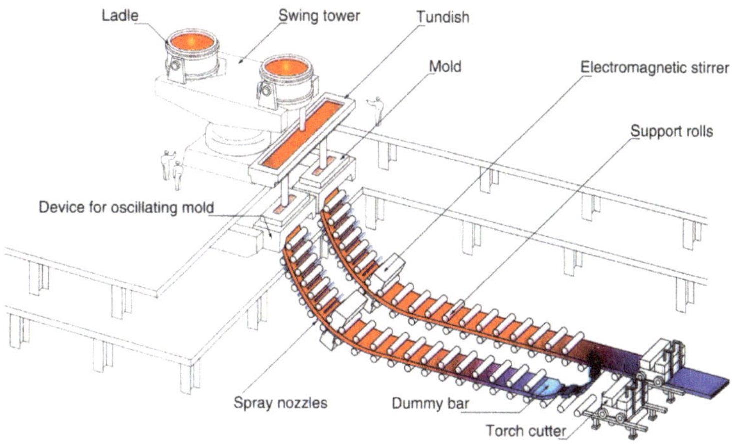

Figure 2. Typical process layout and equipment for various forms of continuous casting; Two-strand continuous slab caster [4].

However, a certain amount of scale (iron oxides) will usually form on the slab's surfaces during their time in the reheating furnace. This can be useful in moderation since surface blemishes can then be removed during passage through the high-powered water jets of the de-scaler. In the case of slabs for sheet products, a roughing mill then rolls the slab down to a "transfer bar" some 25–50 mm (1–2 ins) thick. In most current commercial operations for sheet production in integrated steel plants, the transfer bar is then fed into a six- or seven-stand hot strip mill, where a steel sheet, 1–4 mm thick, is produced at speeds approaching 15 m/s (3000 ft/min). Strip temperatures, alloy compositions, and rates of compression are all carefully controlled to manipulate the solid-state phase transformations and thereby optimize steel microstructures and attendant mechanical properties. Since the early adoption of CCC machines by the large integrated steel mills, there has been considerable pressure on reducing the capital and running costs of these mammoth steel plants. The first development in the commercial continuous casting of steel was the Thin Slab Caster (TSC) in 1985 [1]. These casting machines, illustrated in Figure 3, are aimed at servicing the rapidly growing EAF-based mini-mills in North America, thereby allowing them to enter the lucrative sheet metal market.

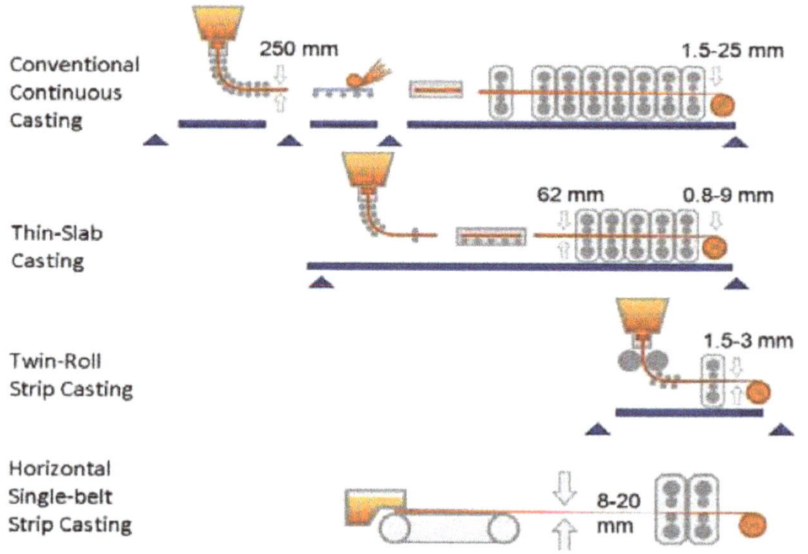

Figure 3. Past and recent commercial developments in strip casting technologies. Typical process layout and equipment for various forms of continuous casting. [5].

4. Thin Slab Casting, TSC

Thus, in the last thirty-odd years, progress towards the goal of process rationalization has been aided through the development of thin slab casting, which is also based on the fixed, oscillating mold. This development has allowed scrap-base-EAF mini-mills such as NUCOR to enter the flat products business [2].

Thin slab casting technology eliminates the need for a conventional rougher/breakdown mill, followed by a seven-stand hot rolling mill, with a muffle furnace for temperature stabilization, followed by more modest in-line rolling requirements. As a result, an integrated line, comprising an advanced EAF-thin slab caster, can produce up to approximately 800,000 tons/year in facilities with lower investment costs, less manpower, and much-reduced energy consumption, compared to an integrated mill. Figure 4 illustrates the layout of such a machine and the funnel-shaped, tapered type of mold that allows for the placement of an SEN (Submerged Entry Nozzle) between its narrow faces. In all TSC cases, casting velocities ~5 m/min are needed to compete with the highly productive conventional casters typically running at 1.2–1.6 m/min.

A more recent approach by Arvedi [6] is a bow-type, low head, endless strip production TSC, developed in 2009. It uses a straight tapered, 80 mm wide, oscillating, fixed mold. This new caster started operations in 2015 in Italy. Its key advantage is that it is an "endless caster", or "Endless Strip Production" (ESP), capable of producing steel strip products endlessly. This means that there is no interruption between pouring liquid steel and producing the final hot strip in one continuous line. This leads to further increases in productivity and environmental benefits.

For instance, the ESP has demonstrated that mill threading can be suppressed, allowing the production of hot-rolled coils in thinner gauges, down to 0.8 mm hot-rolled (HR) strip products. These, in turn, can be used to form cold-rolled (CR) strips down to 200 microns, or 0.2 mm, suitable for electrical grades of steel, for instance. Other advantages claimed vs. standard TSC operations include a halving of operating costs and energy consumption. Mr. Giovanni Arvedi's "Acciaieria Arvedi" steelworks in Cremona has successfully licensed this technology to Rizhao Steel in China, (PRC), where five lines are now operating.

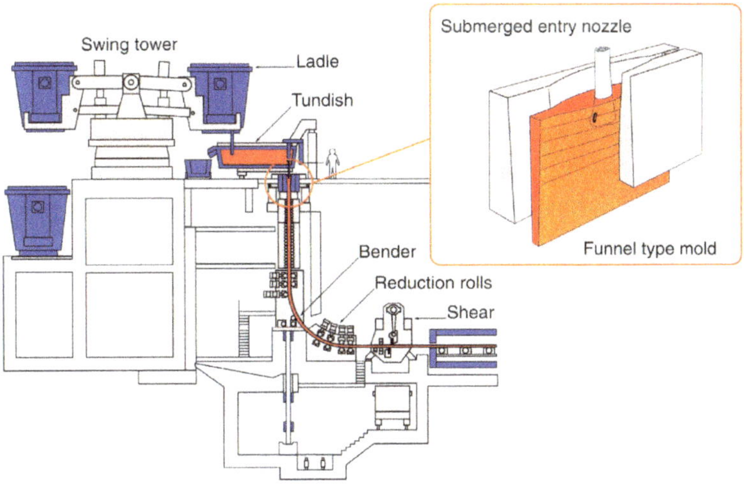

Figure 4. A thin slab caster operation, taken from An Introduction to Iron and steel making [4].

We believe that these innovations represent the apex of current achievements in fixed mold and continuous casting technologies. We see that components of the future steel industry are moving forwards in rationalizing and simplifying the CCC process, leading to a lowering of both capital and operational costs, according to the predictions in Figure 1, by combining casting with rolling in-line.

5. NNSC Moving Mould Technologies

Meanwhile, as noted previously, ever since the 1880s, there have been many efforts to fulfill the dream of Sir Henry Bessemer to cast steel strips directly from a pool of liquid steel using twin-roll casting technologies. Various consortia and companies have researched the characteristics and potential of such Near Net Shape Casting (NNSC) machines. In 2000 AD, Nippon Steel and Mitsubishi Heavy Industries announced the world's first commercial operation of a Twin Roll Caster (TRC) to produce sheets of stainless steel, 2–5 mm thick, 0.76–1.3 m wide, at their Hikari Works in Southern Japan. The roll diameters were approximately 1.2 m, and peripheral roll speeds were in the order of 1 m/s. A pinch roll gathered the strip, which, following minimal hot working, was cooled and coiled directly into the finished product. However, following the second rebuild of their caster, they abandoned their commercial TRC system, quoting "unresolved microstructures" and an uneconomical performance. Similar events occurred in Europe, where the Euro-strip caster, a TRC development, was abandoned without explanation following many years of apparently very promising research work.

Nonetheless, and contrary to those decisions, following initial research work by BHP in Australia to prove the viability of their TRC process in Melbourne, CASTRIP (NUCOR/BHP/IHI) was incorporated, and a commercial twin roll caster was built by IHI to produce low carbon steel strips at NUCOR's Crawfordsville Plant. The building of this commercial strip caster for producing low carbon steel sheet material, 1.7–1.9 mm thick, was announced in 2002 [7]. The NUCOR casters rolls are only 0.5 m in diameter, much the same as in Bessemer's original patent specifications, and run at speeds of 1–2 m/s during casting. Furthermore, the plant continues to operate ten years later. Therefore, some steelmakers have finally realized Bessemer's visionary intuition and achieved the long sought-after rationalization of downstream processing of steel. A second machine was built for NUCOR, and plans have been made to install a CSP (Compact Strip Caster) in the UK, and two in the Peoples Republic of China [8].

However, it is still possible that the economic viability of the TRC process remains in question. Apparently, only two or three consecutive ladles can be cast before a roll change. Apparently, the build-up of frozen steel accretions at the liquid steel/side dam interfaces can open the roll gap settings. This can cause the computer control system to increase the roll separation distance, followed by a "break-out". As such, more than approximately three consecutive ladles seem to be impractical at present (2020), on average. Nonetheless, other reports claim that a ten-ladle sequence is possible. Similarly, while CASTRIP is ideal for mini-mill operations, there is a fivefold mismatch in the maximum productivity of a twin roll caster vs. current integrated steel manufacturing slab casting operations (e.g., ~400,000 tpy vs. ≥2,000,000 tpy).

Such problems regarding TRC operations, together with some basic thermo-mechanical considerations, led to an alternative solution for integrated mills being proposed by Herbertson and Guthrie in 1987, and independently by Professor Schwerdtfeger et al. a little later that year [9]. Both groups independently proposed a Horizontal Single Belt Caster (HSBC) approach, which ideally involves using a high speed (1–3 m/s) horizontal single belt, on which, steel is poured, as shown in the bottom casting system of Figure 3. Since then, the world's first commercial HSBC caster, named Belt Casting Technology (BCT), was designed for producing ~1 MT/year by SMS Demag and was delivered to the Peine plant of Salzgitter. There, casting trials, initially targeting advanced high-strength steels, began in 2012, exploring its potential for casting a variety of novel High-Strength, High-Ductility (HSHD) steels using a 13 m long belt running at ~0.4 m/s [10]. The plant layout is presented in Figure 5.

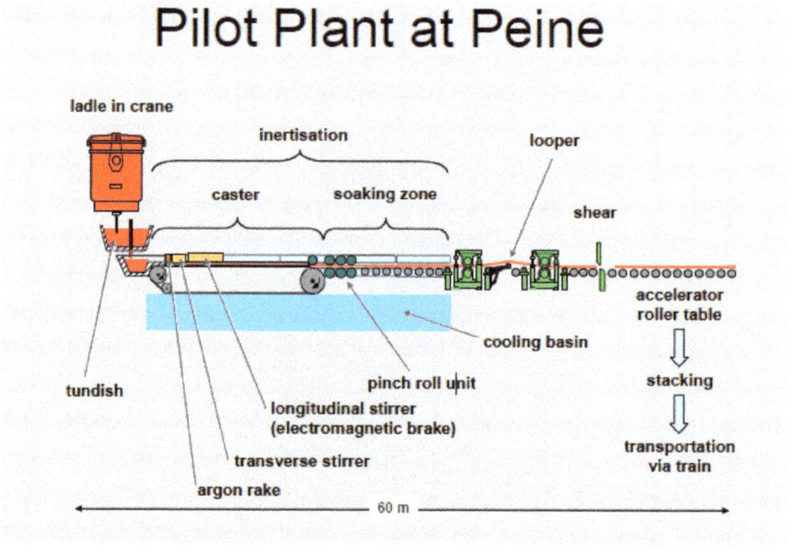

Figure 5. Schematic of the industrial scale pilot plant of Direct Strip Castin (DSC) at Peine, Germany [11].

6. The Future

All continuous casting machines involve the freezing of a shell of steel, forming either against a vibrating "stationary", or against a moving belt, or twin rolls. For the fixed CCC approach, the formulations for mold powders are important for protecting the surface of molten steel from significant heat losses and by remaining molten at ~900 °C to provide lubrication between the surface of the forming slab and copper mold. On that note, one of the key "sticking points" delaying CCC's successful commercial development in the late 50 s and early 60 s, versus the then conventional ingot casting operation, was the development of the oscillating mold. Typically, the molds for slab, billet, and bloom casting are oscillated in a sinusoidal manner (at 1–3 Hz), and mold powder is fed into the meniscus region, where it gradually melts to form a liquid, lubricating slag. This is needed to prevent the newly formed shell from sticking to the copper mold. The oscillatory motion has also been thought by many to contribute to the feeding of molten slag into the mold/strand gap. Based on comprehensive CFD simulations [9], this is true. It certainly helps with strand-sticking problems, which lead to thin shells and costly liquid steel "breakouts". These breakouts of steel can cause the freezing of molten steel onto the caster's guide rolls, leading to extended downtimes. Nonetheless, the well-known OMs (Oscillation Marks) associated with CCC (Conventional Continuous Casters), and shown in Figure 6A, caused by these oscillatory motions, represent surface imperfections or defects, their lengths and frequencies being governed by the length and frequency of the stroke. The first moments of solidification modeled in Figure 6B reveal that the lubricating slag changes the direction of heat transfer from a horizontal direction during the downstroke to a vertical heat flow during the up-stroke when a lower radial rate of heat extraction prevails. This leads to a re-melting of the forming shell. The result is akin to forming a row of linked sausages, as appears in Figure 6B.

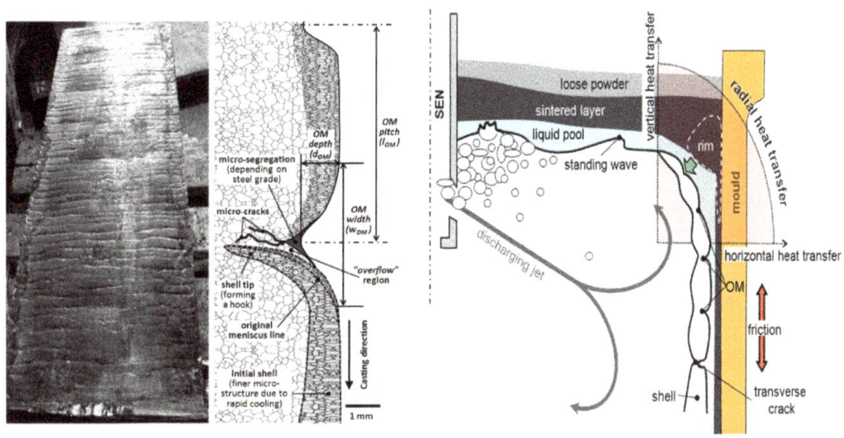

(**A**) (**B**)

Figure 6. (**A**) Typical defect that can lead to oscillation marks on a cast slab in a continuous casting mold, together with a schematic diagram showing the cross-section around an oscillation mark. (**B**) represents the results of a Computational Fluid Dynamic (CFD) predicting how the initial steel shell is formed [12].

The problem with OMs is that they can also be a source of micro-cracks. However, the "un-acknowledged" saving grace for present CCC and most TSC systems is the later oxidation of the outer layers of a cast steel slab, bloom, or billet in the reheat furnaces or muffle furnaces that can effectively "oxidize away" such discontinuities before rolling operations.

A similar situation occurred in the production of "rimming grade steels" via ingot technology, where subsequent oxidation of its blemished surfaces in the ingot soaking pits would also result in blemish-free final surfaces, ready for rolling down to the thin strip auto-body sheet. Thus, ingot casting techniques for "rimming grade" steels for auto companies in North America were practiced up into the 1980s, when equivalent quality, aluminum-killed (AK) steel grades of CCC sheet material finally became available.

For any NNSC process to be able to compete with the blemish-free surfaces of slabs produced using ideally operating CCC and TSC systems, NNSC surfaces must also be completely free of blemishes. Many of us will know that, at very high speeds of rotation, single roll casters can produce perfect bottom surfaces; for example, when producing

~100 micron thick sheets of metallic glasses (e.g., the MetGlas™ process for producing steel sheets for electrical transformers, containing 1% boron and 4% silicon). Therefore, the only question remaining for a moving mold system to usurp the current technological impasse of poor cast surfaces, is what is the critical lower speed for such a moving mold system? Perhaps it is self-evident that the criteria must be (1) how fast does the substrate have to move, to separate the line of initial freezing from the line of initial contact of the liquid metal with the cooling substrate system, and (2) how can we eliminate the high-frequency oscillations of the back meniscus, where the liquid metal first contacts the cooling belt?

We now refer to experimental work carried out at the McGill Metals Processing Centre (MMPC), some fifteen years ago, in which we were able to measure instantaneous heat fluxes using a strip casting simulator, in which we poured liquid aluminum, and later liquid steel, on to a moving, flat copper substrate, 12 mm thick, moving at ~1 m/s. This simulates equivalent initial heat fluxes and the freezing of molten metal onto a 1 mm thick water-cooled steel belt running at the same speed [13]. We were able to characterize the topography of the sand-blasted copper surface and then use an equivalent geometrical construct to predict instantaneous heat fluxes for comparison with experimental results. The modeling result is shown in Figure 7, whose predicted heat fluxes and initial shell formation coincided well with equivalent simulator experiments. Note that the contact is essentially non-wetting, even at the 20–30 μm high peaks of the rectangular pyramids (260 pyramids/mm^2). We must also compensate for sudden gas expansion at the mold-metal interface.

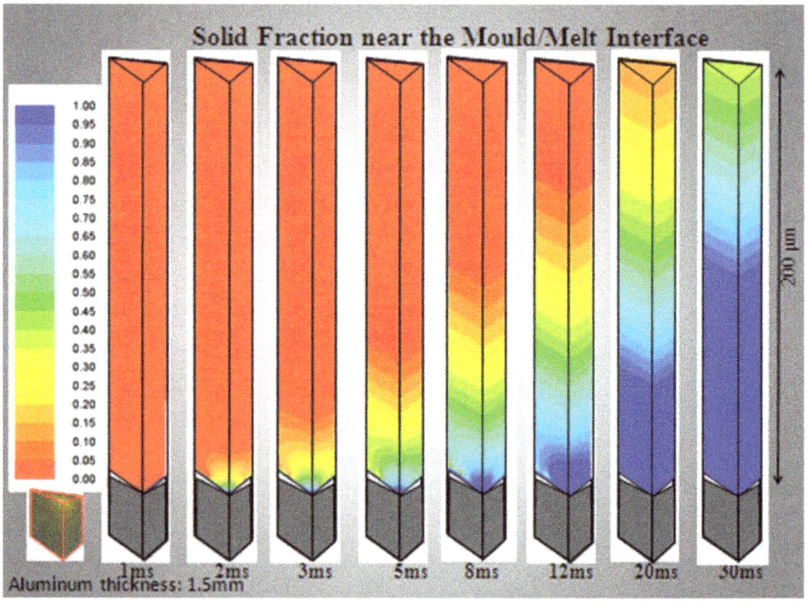

Figure 7. Ab initio computations of the first moments of solidification of an aluminum alloy, AA6111 strip, being cast, iso-kinetically, onto a 1 cm. thick copper substrate (depicted as a grey substrate). The idealized copper surface is modeled as an array of rectangular pyramids, whose peaks contact the melt at discrete points. The interfacial gas conducts the heat from the freezing metal into the copper substrate, as per our experimental results [13].

Armed with this knowledge, one sees that one has perhaps 15 ms to make a perfect surface of the AA6111 sheet before any surface imperfections are stamped into the bottom surface of the metallic sheet. Similar results can, and were, computed for steel alloys, based on their thermal properties and those of the substrate and interfacial gases.

Our experiments have demonstrated that perfect bottom surfaces can be attained, as have equivalent tests at Clausthal University, and Salzgitter Steel. The other question of stabilizing the initial contact of liquid metal onto the cooling mold has also been considered, and we believe that this meniscus line can be held steady, thereby eliminating the so-called chatter marks reported for TRC operations at NUCOR.

Thus, the final question for opting to replace standard CCC processes with far less expensive (capital and operating costs) and far more environmentally friendly NNSC processes in the future, is that of the thickness of the liquid metal cast onto the belt, versus that of the CCC process. Here, it should be clear that NNSC processing overwhelmingly wins, in that casting, for example, 10 mm thick sheet material, or even thinner, eliminates any macro-segregation and large grain sizes associated with the casting of 300 mm steel slabs. NNSC even eliminates the inverse macro-segregation encountered in DC cast aluminum casters. However, the surfaces must remain completely flat for subsequent in-line rolling. Similarly, we have demonstrated that reduced cast thicknesses of ~6 mm or less make the need for side dams and electromagnetic braking to compensate for velocity mismatches between the belt and input metal speeds, redundant for melts of aluminum. For steel and copper melts, we found that side-dams were obligatory, but EM braking was unnecessary at these lowered cast strip thicknesses.

Another question to be addressed is what could be the advantages of HSBC over TRC Bessemer-type casters? Perhaps the biggest advantage is the ability to adjust its design to any plant's steel output. The Peine caster operated by Salzgitter boasts a 13 m (40 ft) long belt capable of producing 10 mm thick sheet material at a rate above 1 MT/an, whereas a TRC would need at least two units and two work crews to produce the same steel sheet output. Alternatively, it would need a roll diameter of approximately 8 m. The second advantage of HSBC is the question of cooling rates during the transformation from liquid to solid. The Bessemer process results in cooling rates for the CASTRIP process in the order of 1000 °C/s, whereas the HSBC results in cooling rates ~100 °C/s. The result is in upper bainite structures for CASTRIP producing low-carbon steels, whereas HSBC can produce steel microstructures much nearer conventional practices and steel properties. The third advantage of HSBC over TRC, among many others, is the fact that HSBC represents unconstrained solidification versus the TBC, TRC, or TBC processes. This can eliminate any problems of macro-porosity in the center of a steel sheet.

Thus, finally, we come to the question of the upper surfaces of HSBC caster systems. For the aluminum alloys tested at the MMPC and at MetSim, in Montreal, we found that AA6111, AA5182, and AA2024, all have perfectly flat upper interfaces, some coated with very thin oxide films of aluminum or magnesium. For steels (TRIP, TWIP, and 3% and 6% silicon steels), and for copper and Cu-Ni melts, we have obtained equivalent results on the quality of upper and lower surfaces on our pilot-scale caster. Porosities are also very low. Similar results have been reported by Professor Karl-Heinz Spitzer [11].

7. Conclusions

One can conclude from this article, that there is an inevitable convergence in the designs of fixed and moving mold casting machines towards the manufacture of steel sheets into one continuous process. The result will be shorter plants, less equipment, fewer personnel, and greater efficiencies. The Arvedi caster for steel [6], the ALCOA micro mill for aluminum alloys [14], and the BCT-Belt Casting Technology (or HSBC) [15] prove that these processes are possible and that they can drastically reduce manufacturing costs while significantly helping to address environmental concerns. As such, they could be poised to revolutionize the tonnage casting of steel and aluminum alloy sheet materials in disruptive technology, removing all of the current bottlenecks to present-day systems and processes. By the same token, we believe that it will be possible to make equivalent advances in the production of long products. It is only a matter of time to accomplish the final logical steps toward the ideal NNSC approaches. Time will tell, but timing is everything!

Author Contributions: M.M.I. and R.I.L.G. both contributed to this paper. All authors have read and agreed to the published version of the manuscript.

Funding: This research was funded by an NSERC Discovery Grant, 224535 plus the MMPC's supporting company's grants.

Institutional Review Board Statement: Not applicable.

Informed Consent Statement: Not applicable.

Data Availability Statement: Not applicable.

Conflicts of Interest: The authors declare no conflict of interest.

References

1. Rasch, M. Anmerkungen zur Fruhgeschichte des Stranggiebens. In *Stranggieben—Continuous Casting; A Technical Revolution in the Steel Industry*; Aschendorff Verlag: Münster, Germany, 2017; pp. 9–40.
2. Guthrie, R.; Isac, M. The development of continuous casting machines for the production of steel in North America. In *Continuous Casting; A Technical Revolution in the Steel Industry*; Aschendorff Verlag: Münster, Germany, 2017; pp. 283–288.
3. Guthrie, R.I.L.; Jonas, J.J. Steel Processing Technology. In *Metals Handbook*, 10th ed.; Properties and Selection; ASM International: Materials Park, OH, USA, 1989; Volume 1, pp. 107–125.
4. Monden, K. *An Introduction to Iron and Steel Processing*; Morita, Z., Emi, T., Enami, T., Hashimoto, O., Eds.; Kawasaki Steel 21st Century Foundation: Tokyo, Japan, 1997.
5. Ge, S.; Isac, M.; Guthrie, R.I.L. Progress in Strip Casting Technologies for Steel; Technical Developments. *ISIJ Int.* **2013**, *53*, 729–742. [CrossRef]
6. Hohenbichler, G.; Mazzolari, F.; Linzer, B.; Jungbauer, A. *Arvedi ESP Technology and Plant Design*; Millenium Steel: Princeton, IN, USA, 2010; pp. 82–88.
7. *ForeCast*; CASTRIP: Charlotte, NC, USA, 2002; Volume II, Issue 2.
8. New Technologies Could Slash the Cost of Steel Production. *The Economist*, 9 March 2017. Business Section—Recasting Steel.
9. Isac, M.; Guthrie, R.I.L. Chapter 4.7: The design of a new casting process; from fundamentals to practice. In *Treatise on Process Metallurgy*; Elsevier Ltd.: Amsterdam, The Netherlands, 2014; Volume 2, pp. 555–583.
10. Spitzer, K.-H.; Schwerdtfeger, K. Production of steel strip with a single-belt process. In Proceedings of the ISS 13th Process Technology Conference, Nashville, TN, USA, 3–7 November 1995; pp. 47–53.
11. Spitzer, K.H. DSC Strip Casting: Process Development and Perspectives for New Materials. In Proceedings of the Roderick Guthrie Honorary Symposium on Process Metallurgy, Montreal, QC, Canada,, 6–9 June 2011; McGill Metals Processing Centre: Montreal, QC, Canada, 2011; pp. 97–106, ISBN 978098696806.
12. Ramirez-Lopez, P.E. Modelling Shell and Oscillation Mark Formation during Continuous Casting via Explicit Incorporation of Slag Infiltration. Ph.D. Thesis, Materials Department, Imperial College, London, UK, February 2010.
13. Guthrie, R.I.L.; Isac, M.; Li, D. Ab Initio Predictions of Interfacial Heat Fluxes in Horizontal Single Belt Casting (HSBC), Incorporating Surface Texture and Air Gap Evolution. *ISIJ Int.* **2010**, *50*, 1805–1813. [CrossRef]
14. Arconic. Faster, Stronger, Smarter. Introducing Arconic's Micromill ™.; YouTube 19 October 2016.
15. Geerkins, C.; Fischer, L.; Wans, J.; Wilmes, R. High Productivity Plants and Innovative Technologies Meet Market Demands and Future Trends. In Proceedings of the Metec & 2nd ESTAD 2015, Dusseldorf, Germany, 15–19 June 2015.

MDPI
St. Alban-Anlage 66
4052 Basel
Switzerland
Tel. +41 61 683 77 34
Fax +41 61 302 89 18
www.mdpi.com

Metals Editorial Office
E-mail: metals@mdpi.com
www.mdpi.com/journal/metals

www.ingramcontent.com/pod-product-compliance
Ingram Content Group UK Ltd.
Pitfield, Milton Keynes, MK11 3LW, UK
UKHW060123010826
14089UKWH00006B/122

9 783036 551531